NATURAL HISTORY.

A MANUAL

OF

ZOÖLOGY

FOR

SCHOOLS, COLLEGES, AND THE GENERAL READER.

BY

SANBORN TENNEY, A. M.,

AUTHOR OF "GEOLOGY, ETC.," AND PROFESSOR OF NATURAL HISTORY IN WILLIAMS COLLEGE.

Illustrated with over Five Hundred Engravings.

"O Lord, how manifold are thy works! in wisdom hast thou made them all: the earth is full of thy riches." — *Ps.* civ. 24.

SEVENTH EDITION.

NEW YORK:
CHARLES SCRIBNER & CO., 654 BROADWAY.
1869.

PRESS WORK BY THE
TROW & SMITH BOOK MANUF'G CO.,
46, 48, 50 GREENE ST., N.Y.

WELCH, BIGELOW & CO.,
ELECTROTYPERS.

TO

THOSE WHO BELIEVE THAT THE LEADING FACTS AND PRINCIPLES
OF NATURAL HISTORY SHOULD BE TAUGHT IN ALL
THE SCHOOLS OF THIS COUNTRY,

THIS VOLUME

IS RESPECTFULLY DEDICATED

BY

THE AUTHOR.

PREFACE.

THIS work is intended to give a general idea of the Animal Kingdom, especially as it is represented in North America, and thereby to make the learner acquainted with the leading facts and principles of the department of Natural History properly called Zoölogy. In its preparation, I have freely used all the materials at my command, but have taken special pains to consult and select from original papers. The general plan of the work is essentially that of Cuvier, with such modifications as have received the sanction of eminent naturalists. For the special way in which the plan is carried out, the writer alone is responsible. In treating of the Mammals, I have followed mainly the writings of Cuvier and Baird; of the Birds, those of Baird, Audubon, and Wilson; of the Reptiles and Batrachians, those of Holbrook, Agassiz, Baird & Girard; of Fishes, those of Cuvier, Storer, and DeKay; of Insects, those of Harris and the authors named in connection with each order; of Crustaceans, those of Dana; of Worms, those of Cuvier, Agassiz, and Weinland; of Mollusks, those of Woodward, Gould, and Lea; of Echinoderms, those of Desor, Agassiz, and Forbes; of Acalephs, those of Agassiz; of

Polyps, those of Dana, Verrill, and Milne-Edwards. Great prominence is purposely given to the Mammals and Birds of this country, that all may have at least an accessible catalogue of these two groups, in which every one is interested; the other groups, however, are not neglected.

The wood-cut illustrations have been designed mainly from original papers, and from nature, and, with few exceptions, have not before appeared in hand-books of natural history. The cuts of the Mammals are mainly from Schinz, Audubon & Bachman, and Richardson; of the Birds, mainly from Audubon and Wilson; of the Reptiles and Batrachians, from Holbrook; of the Fishes, from Storer, Holbrook, DeKay, and from nature; of the Insects, mainly from Harris, Say, Sanborn, and from nature; of the Crustaceans, mainly from nature and Reports; of the Mollusks, from Binney, Woodward, Gould, Lea, Conrad, and from nature; of the Echinoderms, from nature, Agassiz, and Müller; of the Acalephs, from Agassiz; of the Polyps, from Dana, Milne-Edwards, and Verrill; of the Protozoa, mainly from Ehrenberg and Huxley.

The drawings were made with great skill and faithfulness by Mr. Edward S. Morse, well known as a naturalist, Mr. N. Brown, Mr. E. Burrill, and Mrs. J. W. Dickinson. The engraving was done by Messrs. Henry Marsh, N. Brown, and J. F. Richardson. It is sufficient praise for Mr. Morse to say, that the Grizzly Bear, Pronghorn, nearly all of the Mollusks, the Echinoderms, and many others, were drawn by him; for Messrs. Burrill and Marsh, that the Birds were drawn by the former and engraved by the latter; for Mr. Brown, that the Wapiti, Beaver, Otter, Weasel, Wolverine, Lobster, Acalephs, and many Insects, were both drawn and engraved by him, and that he also engraved most of the Fishes; for Mrs. Dickinson, that the Reptiles and Fishes were mainly drawn by her; and for Mr. Richardson, that he engraved many of the Mammals and Insects, and all of the Mollusks and Polyps.

I desire to express my sincere thanks to my personal friends and others who have encouraged and aided me in this undertaking. To Professor Jeffries Wyman I am under obligations for information on several important points, and for the privilege of figuring a beautiful specimen of Astrophyton; to Professor A. E. Verrill, of Yale College, for aid in selecting the cuts of the Mammals, for important suggestions, and for reading and criticising the proofs relating to Polyps; to F. W. Putnam, Esq., Superintendent of the Essex Institute, for reading the proofs relating to the Reptiles and Fishes, and adding most useful notes on nomenclature and other not less important points; to A. S. Packard, Jr., M. D., for aid in the classification of Insects, and for other favors; to Samuel H. Scudder, A. M., Custodian Boston Society of Natural History, for aid in selecting the cuts of the Insects, and for reading and correcting the proofs treating of that group; to Edward S. Morse, Esq., for reading the proofs relating to Mollusks, and for other favors; and to Professor H. James Clark, for reading the proofs relating to Acalephs. I would also thank my beloved wife, and Hon. Joseph White, Professor A. Crosby, Mr. J. Gove, L. R. S. Gove, A. B., T. D. Adams, A. M., and Mr. F. G. Sanborn, for special favors in connection with the work. I am under obligations to Captain Alpheus Hyatt, and the Library of Harvard College, the Library of the Boston Society of Natural History, the State Library, the City Library of Boston, and to Messrs. Piper & Co. of Boston, for the use of rare books. Nor would I omit to express my thanks to the gentlemen of the University Press, whose skill in proof-reading, electrotyping, and printing has done so much to make the book accurate and attractive. And I would here thank my Publishers for their generosity in willingly making the large outlay necessary to issue the work in its present form.

I take this opportunity to say that I shall soon ask my Publishers to issue another volume on Zoölogy. It will contain the same illustrations as the present one, and will be especially adapted to the wants of Grammar Schools, and to the younger classes of readers.

Hoping that the present work may meet, in some degree, a want which has long been felt, and aid the youth and others of our whole country in the delightful study of Natural History, and thereby advance the cause of Learning, I submit it to the kind consideration of the friends of Popular Education.

S. T.

CAMBRIDGE, MASS., *August* 1, 1865.

CONTENTS.

CHAPTER I.

CHAPTER II.

CHAPTER III.

CHAPTER IV.

TENNEY'S NATURAL HISTORY SERIES.

BY SANBORN TENNEY, A. M.,

Author of "Geology, &c.," and Professor of Natural History in Vassar College.

I.

NATURAL HISTORY OF ANIMALS.

By SANBORN TENNEY and MRS. ABBY A. TENNEY.

Containing brief descriptions of the animals figured on Tenney's Natural History Tablets, but complete without the Tablets. 12mo. 261 pages. Illustrated with 500 Wood Engravings. *For Beginners.*

II.

ELEMENTS OF ZOÖLOGY.

(SOON TO BE PUBLISHED.)

Combining the study of the Anatomy and Physiology of Animals with that of Descriptive Zoölogy. Specially adapted for High and Normal Schools, and Academies desiring a larger book than No. I., but not having time to go through with the much more extensive work No. III. 12mo, about 350 pages. Illustrated with more than 500 Engravings.

III.

MANUAL OF ZOÖLOGY.

For Schools, Colleges, and the General Reader. 8vo, 540 pages. With 500 Engravings.

IV.

NATURAL HISTORY TABLETS.

For the School and the Family.

These Tablets are five in number, and contain several hundred Illustrations of Mammals, Birds, Reptiles, Fishes, Insects, Crustaceans, Worms, Shell-Fish, Sea-Urchins, Star-Fishes, Jelly-Fishes, Sea-Anemones, and Corals.

NATURAL HISTORY.

ZOÖLOGY.

CHAPTER I.

STATEMENT OF THE SUBJECT.

NATURAL History is the science which treats of the earth and all natural objects upon its surface and within its crust.

Zoölogy is the department of Natural History which treats of animals. It embraces the study of their forms, structure, development, habits, names, classification, geographical distribution, and the relations which animals lower than himself sustain to Man, the highest representative of the Animal Kingdom.

This science has been established and brought to its present high state of perfection through the labors of such masters as Aristotle, Linnæus, Cuvier, and others scarcely less renowned.

Zoölogy is a science of the highest importance, not only on account of its direct practical relations to the material interests of human society, and its inseparable connection with the great problems of Geology, but especially as an educational branch, securing to its true votaries a spirit of earnest inquiry, habits of accurate observation and careful comparison, vigorous and logical thought, and power of broad generalization; and dealing, as it does, with the highest expressions of matter and of life, its study is eminently adapted to enlarge our ideas of creation and its Great Author. It makes known to us the Plan of Creation, as exhibited in the highest department of nature; and thus we are led to know more of Him who suffers not even a sparrow to fall without his notice.

The Animal Kingdom comprises all organized bodies endowed with sensation and voluntary motion, — that is, all organized bodies except plants. In addition to sensation and voluntary motion, which depend upon special systems of organs peculiar to animals, — the nervous system and the muscular system under its influence, — there are also other characteristics which belong exclusively to members of the Animal Kingdom, and which show still further the differences between them and plants. All, or nearly all, animals possess a more or less well-defined digestive cavity, and most of them other well-defined cavities, which have special functions, or which contain organs which have special functions. In plants, the organs for special purposes are not concentrated and placed in well-defined cavities, but are more or less distributed over the body. Animals feed directly upon plants, or upon other animals that feed upon plants. Vegetation, on the contrary, is nourished by the mineral kingdom. It is the chief province of the vegetable kingdom to convert mineral substances — earth and gases — into food upon which animals can subsist. In animals, the food is received at once into the digestive cavity, whence, after proper elaboration, it traverses and nourishes the whole body. In plants, most of the fluids traverse the whole extent of the body and branches before reaching the foliage, where the process of elaboration is carried on. In respiration, animals consume oxygen, and give off carbonic acid, a gas poisonous, and, when abundant, destructive to animal life; while plants consume carbonic acid, and give off oxygen, so essential to animals. All animals are developed from more or less spherical eggs; plants from seeds, or something analogous to seeds; and the mode of development, and the extent to which growth goes on, are essentially different in the two cases.

It was stated above, that all animals are endowed with sensation. Some, the lowest, have only general sensibility; while others, and all the higher ones, are also endowed with special kinds of sensation, called special senses, of which there are five in number, — the sense of sight, of hearing, of smell, of taste, and of touch, — and dependent upon special organs.

The natural divisions of the Animal Kingdom are Branches or Types, Classes, Orders, Families, Genera, and Species. That is, the animal kingdom is divided into branches, each branch into classes, each class into orders, each order into families, each family into genera, each genus into species, the latter group being composed of individuals essentially alike. These divisions are not the contrivance of man, but exist in nature. According to Agassiz,

Branches are characterized by plan of structure, —

Classes, by the manner in which that plan is executed, as far as ways and means are concerned, —

Orders, by the complication of that structure, —

Families, by form, —

Genera, by details of execution in special parts, —

Species, by the relation of individuals to one another, and to the world in which they live, as well as by the proportion of their parts, their ornamentation, &c.

That is, certain characters determine the Branch, certain others determine the Class, others the Order, others the Family, others the Genus, and others still the Species. These principles of classification, however, are not as yet generally recognized and fully applied; but most writers designate at least the subordinate groups by a combination of characteristics more or less different in their nature.

The number of species of animals is not known, but may safely be estimated as high as a million, or even more, of which the small and microscopic comprise an immense majority.

Cuvier has shown that the animal kingdom comprises four great Branches or Types, — Vertebrata, Articulata, Mollusca, and Radiata. All the animals in any one of these branches are constructed upon the same plan.

The Branch of Vertebrata comprises all animals which have an internal skeleton with a backbone for an axis. Man and all the higher animals belong to this branch.

The Branch of Articulata comprises all animals whose bodies are made up of similar rings or segments, placed transversely to the longitudinal axis; and whose parts which correspond to a skeleton are external. All Insects, Crabs, Lobsters, Shrimps, and Worms belong to this branch.

The Branch of Mollusca comprises soft-bodied animals, such as Cuttle-fishes, Squids, Snails, Mussels, Oysters, and Clams.

The Branch of Radiata comprises animals whose parts are more or less symmetrically arranged around a vertical axis. Sea-urchins, Star-fishes, Crinoids, Jelly-fishes, Coral animals, and Sea-anemones belong to this branch.

Baer has shown that for each of these branches there is a special mode of development in the egg. In Vertebrates the germ divides into two folds, one turning upward and the other downward. In Articulates, the germ lies with its back portion upon the yolk, and absorbs the latter into that part of the body. In Mollusks, the germ lies upon the yolk, and absorbs the latter into the under surface of the body. In Radiates, the germ occupies the whole periphery of the sphere.

Of the four branches of the Animal Kingdom, the Vertebrates are unquestionably the highest in rank, and the Radiates lowest; while of the other two it is not easy to say which, on the whole, is the higher, and which the lower branch. And here a word of explanation is due in regard to the rank of animals.

In one sense, all animals are alike perfect. Each is perfectly adapted to fulfil its own peculiar office in the great economy of nature. In this sense, every animal is perfect. But in regard to organization there is every grade, from those of the most extreme simplicity, and with the most simple functions, to those of the highest possible complication, and with the most numerous, varied, and complicated functions. Now an animal is higher according to its higher complication of structure, and hence more numerous and varied functions.

In the subsequent pages we examine somewhat carefully these four branches of the Animal Kingdom, taking them in the order named above. One chapter is devoted to each Branch, one section to each Class, one sub-section to each Order. Paragraphs marked with small capitals introduce each Family. The Genera are given in italics, the popular name of Species in ordinary type, and the scientific name immediately follows in italics.

CHAPTER II.

THE BRANCH OF VERTEBRATA, OR VERTEBRATES.

The Vertebrata embrace all animals which have a bony or cartilaginous axis, called the spinal column, with an elongated cavity above it, containing the great nervous centre, — the brain and spinal cord, — and another below it, containing the organs of respiration, digestion, circulation, and reproduction. From the brain and spinal cord branch the nerves, in the form of threads, to every part of the body. In all the Vertebrates the skeleton is internal, and constitutes the frame upon which the muscles are placed, the skin, with its appendages, covering the whole. The axis of the skeleton is made up of parts, which are more or less movable upon one another. Each of these parts is called a vertebra, and hence the axis is often called the vertebral axis, or vertebral column, as well as spinal column and backbone.

Fig. 1.

A Vertebra.

All Vertebrates have red blood, which is propelled through the system by a muscular heart. The mouth is furnished with two jaws, usually armed with teeth, which are more or less bony, and often enamelled. Vertebrates exhibit perfect bilateral symmetry, that is, the organs are arranged in pairs on the two sides of the body. The eyes are two, ears two, and the locomotive appendages never exceed four. The animals of this

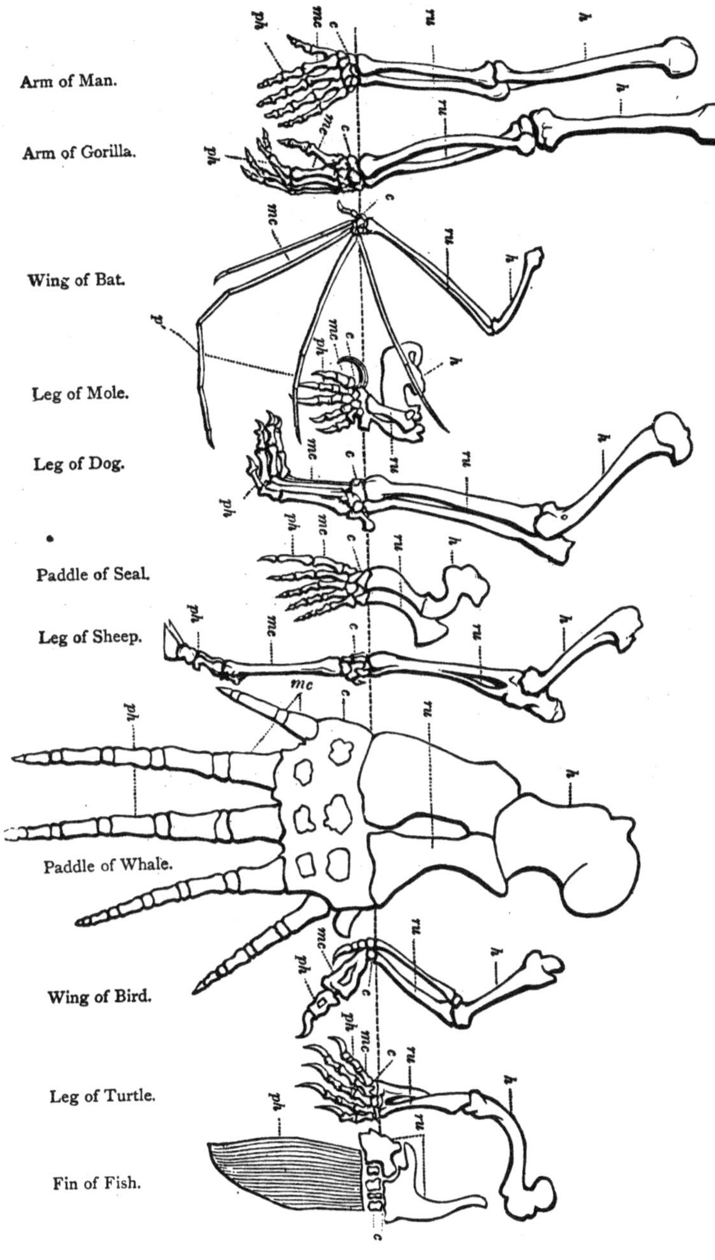

Fig. 2. Arm of Man.

Fig. 3. Arm of Gorilla.

Fig. 4. Wing of Bat.

Fig. 5. Leg of Mole.

Fig. 6. Leg of Dog.

Fig. 7. Paddle of Seal.

Fig. 8. Leg of Sheep.

Fig. 9. Paddle of Whale.

Fig. 10. Wing of Bird.

Fig. 11. Leg of Turtle.

Fig. 12. Fin of Fish.

branch are characterized by higher intelligence than those of any other. Vertebrates comprise five * classes, — Mammalia, Birds, Reptiles, Batrachians, and Fishes. Many writers group the Batrachians and Reptiles together, and give the latter name to the class, thus making the classes only four in number. In all these the fundamental idea of a Vertebrate is plainly manifested, and the principal parts of the skeleton correspond, part to part, head to head, spinal column to spinal column, locomotive members to locomotive members, — the members in each case modified according to the function to be performed, whether it be that of standing, or grasping, or walking, or running, or leaping, or springing, or flying, or creeping, or swimming, but the general plan always the same. Figures 2–12, where corresponding parts are marked by the same letter, show clearly that the anterior locomotive members of different Vertebrates are expressions of one and the same fundamental idea.

SECTION I.

THE CLASS OF MAMMALIA, OR MAMMALS.

THE Class of Mammalia comprises all Vertebrates which bring forth their young alive, and nourish them with milk from their own bodies. They are all furnished with a solid skeleton, which is divided into well-defined regions, as the head, trunk, and extremities; the upper jaw is fixed to the cranium, the lower formed of only two pieces; the teeth are enamelled, and the neck, with few exceptions, has only seven vertebræ. The brain is com-

* Agassiz, in his "Essay on Classification," recognizes eight classes in the Branch of Vertebrates, — Mammals, Birds, Reptiles, Amphibians, Selachians, Ganoids, Fishes proper, and Myzontes.

posed of two hemispheres, and a muscular diaphragm separates the cavity of the chest from that of the abdomen. They all breathe air by means of lungs; the blood is warm, and their circulatory system consists of a heart with four cavities, arteries which carry the blood from the heart to all parts of the body, veins which conduct the blood back to the heart, and capillary vessels which connect the termination of the arteries with the beginning of the veins. The nose of mammals forms part of the face; the eyes are protected by two lids, which are generally furnished with eyelashes; the ears are composed of three parts; and they are all endowed with vocal organs. Mammals are the highest in rank of all the animals of the globe. The number of species is about two thousand, distributed among the following orders: Bimana, or Man; Quadrumana, or Monkeys; Cheiroptera, or Bats; Insectivora, or Insect-eaters; Carnivora, or Flesh-eaters; Marsupialia, or Marsupials;* Rodentia, or Gnawers; Edentata, or Edentates; Pachydermata,† or Pachyderms; Ruminantia, or Ruminants; and Cetacea, or Cetaceans.

SUB-SECTION I.

THE ORDER OF BIMANA, OR MAN.

THE structure of Man is essentially the same in kind as that of other mammals, differing only or mainly in degree; yet the degrees of difference separate him widely from all other animals, and place him in an order by himself, and far above all other organized beings. He is the only animal to which the erect position is natural;

* Marsupials are now considered as a Sub-Class parallel with other mammalian quadrupeds. In fact, Cuvier so regarded them, but treated them as an Order.

† Many modern systematists unite the Pachyderms and Ruminants in one Order, called Herbivores, or Plant-eaters.

Fig. 13.

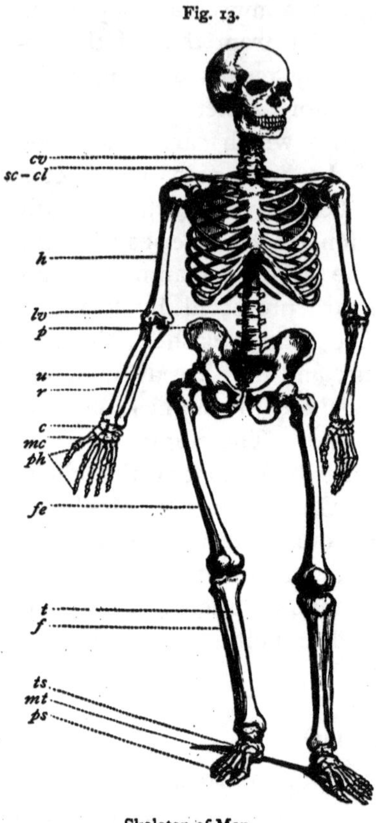

Skeleton of Man.

cv, cervical vertebræ; *sc-cl*, scapula and clavicle; *h*, humerus; *lv*, lumbar vertebræ; *p*, pelvis; *u*, ulna; *r*, radius; *c*, carpus; *mc*, metacarpus; *ph*, phalanges; *fe*, femur; *t*, tibia; *f*, fibula; *ts*, tarsus; *mt*, metatarsus; *ps*, phalanges.

his whole organization is adapted to that attitude. His brain is the largest in the Animal Kingdom, excepting only that of the elephant and of the whale, and in its organization is far superior to that of any other animal. His face is a model of beauty, and endowed with a wonderful power of expression. The hand of man is superior in its structure and in its functions to the corresponding member of any other animal. Man alone truly speaks a language. Even physically considered, he is the highest possible expression of a vertebrate. But Man is the highest representative of the Animal Kingdom, not only on account of his superior form and higher physical organization, but, above all, on account of those high mental and spiritual endowments which belong to him alone, and which enable him to understand and appreciate the wonderful and sublime harmonies of the material and moral world, and his own relations to the Author of Nature and of Revelation.

Fig. 14.

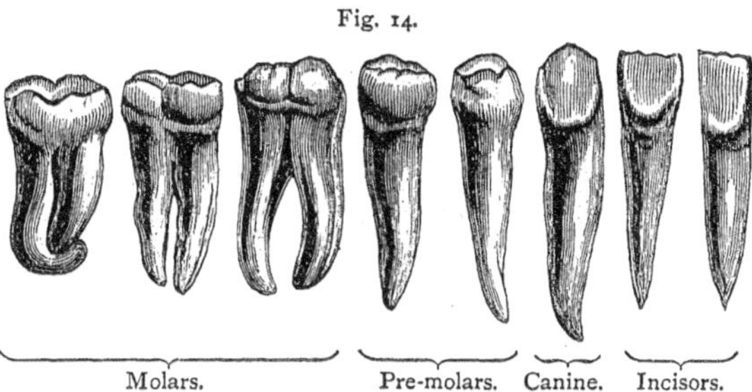

Human Teeth.

While each species of the lower animals is limited to a given region or country, and in many cases cannot survive a removal, Man's home is the whole earth,—he alone is truly cosmopolite. But while Man is found in all zones and climates, he differs greatly both in his physical and mental nature in different regions. In the examination of vegetable life, and life as revealed in animals lower than Man, we find that both reach their highest expression in the hot regions. Not so with Man. It is not in the moist, warm air of the trop-

Fig. 15.

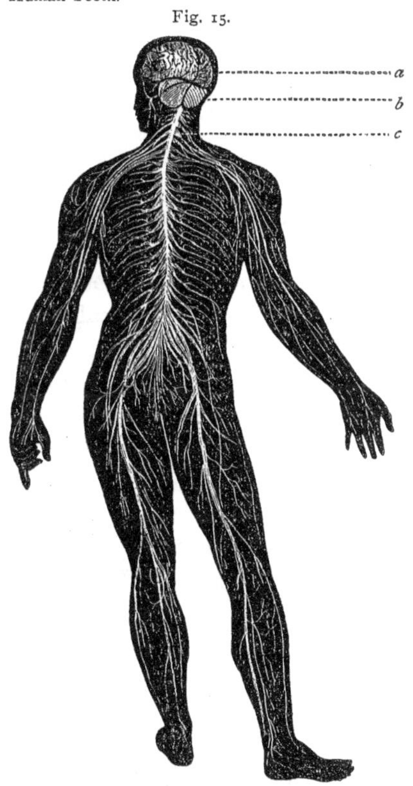

Nervous System of Man.

a, cerebrum, or principal brain, called the hemispheres; *b*, cerebellum, or smaller brain; *c*, spinal cord giving off its branches of nerves.

ics, where all his wants are supplied by the spontaneous productions of the soil, nor in the frigid regions, where he wages a constant warfare with nature to secure food and clothing, that Man appears in his highest stage of development; but it is in the temperate regions that he reaches his highest expression physically, mentally, and morally. Ethnologists recognize three principal varieties, or types, of Man, each having certain characteristics not possessed by the others, — the Caucasian, the Mongolian, and the Ethiopian. Besides these, there are the Malayan and the American, which are regarded as subordinate types, and derived from the Mongolian. With this explanation, we may regard the human family as composed of five types, or races, — the Caucasian, Mongolian, Ethiopian, Malayan, and American.

The Caucasian race occupies Europe, except Lapland and Finnland, Southwestern Asia, Northern Africa, Iceland, and the United States. The Caucasian type is characterized by a round or oval head, smooth skull, vertical and symmetrical features, fair complexion, ample beard, and medium stature. The facial angle varies from 75° to about 90°, and the capacity of the skull from 75 to 109 cubic inches. The Caucasians are possessed of a considerable degree of muscular strength, energy, and endurance, and in many cases these exist in a marked degree of superiority. They are more beautiful in person, and exhibit higher intelligence and refinement, than any other race. They are represented in Southwestern Asia by the Hindoos, Persians, and Syro-Arabians; in Europe by the Teutonians, who inhabit Iceland, Scandinavia, Germany, the eastern and southeastern portions of Scotland, the eastern part of England, and the northeastern part of Ireland; and by the Celts, who inhabit Western and Southern Europe, including most of Scotland and Ireland, Wales, France, Spain, Portugal, Italy, and adjacent islands; and by the Slavonians, who inhabit Turkey, part of Prussia, Poland, and Russia.

The Mongolian race embraces the inhabitants of Central, Eastern, and Northern Asia, the Laplanders and Finnlanders of Europe, and the Esquimaux of Greenland and North America. The Mongolian type is characterized by a broad head, angular face, oblique eyes, tawny skin, rough, straight hair, scanty beard, rather low stature, long body, and short extremities. The facial angle is from 70° to 80°, and the capacity of the skull from 69 to 73 cubic inches. The Mongolians have less physical strength and energy, and less mental power, than the Caucasians. They are shrewd, crafty, insincere, obstinate, cruel to vanquished foes, and contented with a stationary civilization.

The Ethiopian race embraces the inhabitants of Africa south of the Tropic of Cancer, together with their descendants in the United States. The Ethiopian type is characterized by an elongated, narrow cranium, crisp, curly hair, projecting jaws, thick lips, and black or dusky skin. The true Negroes of Western and Central Africa, and the Caffres and Hottentots of Southern Africa, are two prominent divisions of the Ethiopian race. The Negroes, in addition to the characteristics just mentioned, have the skull thick and heavy, the facial angle from 65° to 70°, the mouth wide, face narrow and lower part greatly projecting, the chin retracted, eyes prominent, iris black, and the vessels of the eye suffused with a bilious tinge. Their beard is scanty, and chiefly confined to the point of the chin. The body is muscular, strong, and symmetrical. The bones of the forearm are somewhat elongated, the shin-bones slightly bent forward, the calves placed high up, the feet broad and heavy, the soles flat, and the heel bone considerably projecting. They can endure hard and protracted labor under a broiling sun, and in marshy districts, where other races would sink under disease. They are patient, honest, and passionately fond of simple melodies.

The Caffres have the cranium higher and more rounded, the jaws less prominent, and the nose less depressed, than the true Negroes, and the skin varies from dark brown to clear yellowish brown. The Hottentots are low in stature, the head is flatter and the body less athletic than in the Caffre, and the color is brownish-yellow, besides various other marked peculiarities.

The Malayan race extends from Madagascar on the west to Easter Island in the Pacific, and from the island of Formosa to New Zealand. Within these limits there are two or three well-marked divisions of the inhabitants, — the Malays proper, the Papuans, and the Australians.

The Malayan type proper is characterized by a rather small head, with a capacity of 64 to 89 cubic inches, the dome of the skull high and rounded, the forehead low, face broad and flat, cheek-bones high, nose short and expanded, but not flat, mouth wide, upper jaws projecting, angle of lower jaw very prominent, the auditory opening placed high up, and the orbital ridges prominent and overhanging. The Malayans are short and robust, the skin varies from clear brown to dark olive and bright yellow, the hair is black, straight, shining, generally rough, and the beard scanty. The Malay proper is treacherous, ferocious, and implacable. The more civilized indulge in narcotics to great excess, and the more savage in cannibalism and piracy.

The Papuans constitute the sole inhabitants of New Guinea and the smaller islands immediately adjacent. They so much resemble the Negroes and Caffres that they are popularly known as the Papuan Negroes. The native Australians are characterized by a spare form, a marked lankness of the limbs, large head, projecting brows, broad nose, wide mouth, straight, dark hair, and a skin varying from chocolate-brown to black. The Australians and Papuans may truly be considered among the lowest specimens of the human family.

The American race includes all the aboriginal inhabitants of America except the Esquimaux. The native American head resembles that of the Mongolian, but is more rounded than the latter; the forehead is low and narrow, cheek-bones high, the hair straight and black, the body of good size and well-proportioned, and the complexion copper-colored. The members of this race are active, but not capable of long-continued hard labor. Though some tribes exhibit a good deal of ingenuity, as a race they have never made much progress in the arts of civilized life.

SUB-SECTION II.

THE ORDER OF QUADRUMANA, OR MONKEYS.

The Order of Quadrumana comprises all animals popularly known as apes, baboons, and monkeys. As the name indicates, they have their four extremities hand-like, the fingers being long and flexible, and the thumb opposable to the fingers. Some of the animals of this order bear a general resemblance to the members of the human family; but there are the widest differences between the highest of the Quadrumana and Man, even when physically considered. The erect posture is natural to Man; not so with any of the Quadrumana. Although some of them may stand, and even walk, somewhat erect, it is an unnatural and insecure position, — the foot then resting on its outer edge only, and their narrow pelvis being unfavorable to an equilibrium. Their so-called hands, it is true, resemble human hands; but although admirably adapted to grasping and climbing, they are

Fig. 16.

Chimpanzee, *T. niger*, Geoff.

vastly inferior to the perfect hand of Man in delicate structure and functions. In many species the face presents something human-like in appearance; but the elongated muzzle in most cases reminds us rather of the quadrupeds. The Quadrumana are generally selfish, crafty, malicious, and thievish. Many species are docile, and can be trained to perform remarkable feats; but none have ever been trained to render useful service for man. They inhabit the warm regions of both hemispheres, and are most numerous on the wooded tablelands. They may be divided into three families, — the Simiadæ, or Simiæ catarrhinæ, comprising all the true monkeys of the Eastern Hemisphere; the Cebidæ, or Simiæ platyrrhinæ, comprising all the monkeys of America; and the Lemuridæ, comprising monkey-like animals, which are known under the name of Makis, and which are found most numerous in Madagascar.

SIMIADÆ, OR OLD-WORLD MONKEYS. — The Old-World

monkeys proper, that is, all except the Lemuridæ, are characterized by oblique nostrils, which are near together, a human-like system of teeth, thirty-two in number, by the presence of cheek-pouches in many species, and by the absence in all cases of a prehensile tail. They comprise the highest of the Quadrumana, or those which bear the closest resemblance to Man, and may be divided into two groups, according to the absence or presence of a tail.

1. Tailless or Anthropoid Apes. The members of this group have no cheek-pouches, no callosities, and their fore legs or arms are much longer than the hind ones. The Genera are *Troglodytes*, including the Chimpanzee and Gorilla, and *Simia*, which includes the Orang-Outang, and *Hylobates*, including the Gibbons.

The Chimpanzee, *T. niger*, Geoffroy, of tropical Western Africa, is four to five feet high when erect, covered with dark hair, lives in troops, constructs huts of leaves and branches, arms itself with clubs and stones, and thus repulses the attacks of man and other enemies. When domesticated, the Chimpanzee learns to walk, sit, and eat like a human being.

Fig. 17.

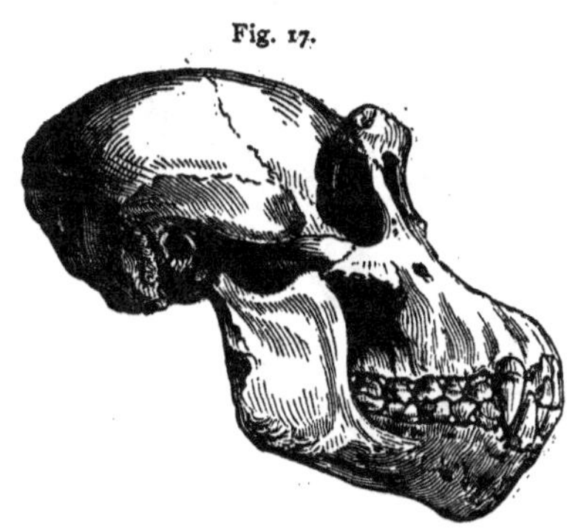

Skull of Chimpanzee, *T. niger*, Geoff.

The Gorilla, *T. gorilla*, Savage, of tropical Western Africa, is five to six feet high, and its heavy frame, large and powerful jaws, wary and ferocious disposition, and gigantic muscular strength, render it one of the most formidable of all the apes.

The Orang-Outang, *S. satyrus*, Linnæus, of Borneo and vicinity, is about five feet high when erect, and is covered with coarse, red hair; the face is bluish, and the hind thumbs are very short compared with the toes. Its home is among the trees, and it has the habit of weaving branches together in order to form a resting-place.

Fig. 18.

Orang-Outang, *S. satyrus*, Linn.

The leading characteristic of Genus *Hylobates*—Long-armed Apes, or Gibbons — is very long arms, reaching even to the ground, when the animal is standing erect. The Gibbons rarely exceed four feet in height, and inhabit the secluded parts of India and the Eastern Archipelago.

2. Tailed Monkeys of the Old World. The Genus *Semnopithecus* — Solemn Apes — is characterized by a long, slender body, and long, slender, straight tail, and by the absence of cheek-pouches. The solemn apes are remarkable for their mildness of disposition and great

intelligence. They belong to Asia and the Asiatic Archipelago.

The Kahau, *S. nasica*, Schr., is celebrated for its extremely long nose.

Fig. 19.

Kahau, *S. nasica*, Schr.

The Genus *Presbytis* is represented by the Tailed Gibbon, which inhabits Sumatra.

The Genus *Cercopithecus* — Guenons — is characterized by a moderately prominent muzzle, long tail, and the last of the inferior molars with tubercles like the rest. The species are numerous, and of great variety of size and color. They live in troops, and commit great havoc in gardens and cultivated fields. They are easily tamed.

The Genus *Colobus* — Thumbless Apes — comprises a few African species closely related to the solemn apes.

The Genus *Macacus* — Macacos — is characterized by a fifth tubercle on their last molars, and their limbs are shorter and thicker and muzzle more projecting than in the Guenons.

The Genus *Inuus* — comprising the Barbary Ape — differs from the Macacos mainly in having a small tubercle instead of a tail. This ape inhabits the precipitous sides of the Rock of Gibraltar.

Fig. 20.

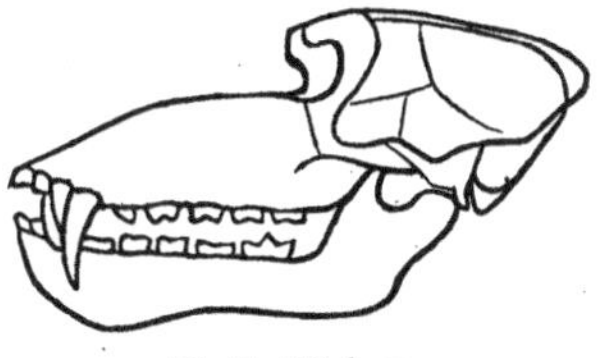

Skull of Baboon.
Cynocephalus.

The Genus *Cynocephalus* — Baboons — is characterized by a dog-like muzzle. The Baboons are mainly inhabitants of Africa and the Philippine Islands, and are large, ferocious, and dangerous animals. They are essentially constructed to live upon the ground, and are inferior to the preceding apes and monkeys, approaching more nearly ordinary quadrupeds. They are known as Dog-headed Monkeys and Mandrills.

CEBIDÆ, OR NEW-WORLD MONKEYS.—This group comprises the monkeys of the New World, in all ninety-one species. They are characterized by a more or less rounded head, by nostrils opening on the sides of the nose and wide apart, by thirty-six teeth, and in many cases by a long prehensile tail, and by the absence of cheek-pouches and callosities; they are in general smaller and less ferocious than those of the Eastern Hemisphere, and as a whole seem to be inferior to them. They inhabit almost the whole territory from Central America to 35° or 38° south latitude. Only one species, however, is found west of the Andes.

The prehensile tail of these monkeys is capable of being twisted firmly around branches of trees, and some species are thus able to sustain the entire weight of the body. The tail is also sensitive, and thus becomes both an organ of feeling and prehension, enabling the possessor to obtain small objects which are in situations where the hand cannot be inserted.

1. Sapajous, or those with a prehensile tail.

The Genus *Mycetes*—Howlers—is prominent among the New-World monkeys, and found throughout the whole length of the territory occupied by the American monkey tribes. Howlers have the head pyramidal, and are provided with a vocal apparatus by which they produce the loudest and most frightful yells or howls, often making night hideous. These monkeys are mostly of large size, three feet long, with a tail about as long as the body, and they are more ferocious than any other American species.

The Genus *Ateles*—Spider Monkeys—comprises those which inhabit chiefly Brazil and Guiana, and which are mild, timid, and slow in their movements. They take their name from their long and sprawling legs, and correspond to the Semnopitheci of Asia and the Colubi of Africa.

Fig. 21.

Spider Monkey, *A. belzebuth*, Briss.

The Genus *Lagothrix* comprises the Gluttonous Monkeys of the interior of South America.

The Genus *Cebus*—Weepers—comprises monkeys which derive their name from their plaintive cry. They are mild in disposition, quick in their movements, and easily tamed. This genus is the richest of all in species, and is most fully represented in Guiana and Brazil.

2. The Sagouins, or Sakis, or those with non-prehensile tails.

The Genus *Pithecia* comprises the Fox-tailed Monkeys, so called from their long and bushy tails.

The Genus *Callithrix* comprises the Squirrel Monkeys, which are of small size and prettily colored. They are found chiefly on the banks of the Orinoco.

The Genus *Nyctipithecus* comprises the Night-Monkeys, characterized by their large nocturnal eyes.

Fig. 22.

Marmoset, *H. chrysomelas*, Pr. Max.

The Genera *Hapale* and *Midas* comprise diminutive monkeys of an agreeable form, and known respectively as Ouistitis—also as Jacchus—and Tamarins. The term Marmoset is also applied to the members of both genera. They have the tail longer than the body, the fur long and bushy, soft, and of beautiful and brilliant colors. They live chiefly in Brazil, but extend also to the northern countries of South America.

LEMURIDÆ, OR MAKIS. — This group comprises Quadrumana which differ from the true monkey tribes both in their more general resemblance to ordinary quadrupeds, and in their teeth; and their first hind finger is armed with a pointed raised nail. The Lemuridæ comprise all the Quadrumana of Madagascar and adjacent islands. They are also found in Central and Western Africa, Southern Asia, and in the Indian Archipelago.

The Genus *Lemur* — Lemurs or Makis proper — comprises those which have six incisors in the lower jaw, compressed and slanting forward; four in the upper that are straight, the canines trenchant, six molars on each side above, and six on each side below. These belong wholly to Madagascar and vicinity, and feed upon fruit.

Fig. 23.

Lemur, *L. catta*, Linn.

The Genus *Indris* comprises those which have only four teeth in the lower jaw. The only species inhabits Madagascar. It is tamed, and used like a hound.

The Genus *Loris* has sharper points to the grinders, the body slender, the tail wanting, and the eyes near together. The Lorises live in the East Indies, are nocturnal, and feed upon insects.

The Genera *Galago* — Galagos — of Africa, and *Tarsius* — Tarsiers — of the Moluccas, are distinguished from other Lemuridæ by elongated tarsi, tufted tail, large eyes and ears. They are nocturnal, and feed upon insects.

The Genus *Galeopithecus*—Flying Lemurs, so called—is placed here by De Blainville and others. The Galeopitheci are closely related to the Bats, but differ from the latter by the fingers of the forward extremities being furnished with trenchant nails, which are no longer than those of the feet, so that the membrane which occupies the spaces between them, and which is continued as far as the tail, can perform no other functions than those of a parachute. Hence they cannot properly fly, but are able, by the sustaining membrane, to make leaps of several hundred feet in extent. They live on trees in the Indian Archipelago, and feed upon insects, and probably fruits to some extent.

Fig. 24.

Galeopithecus.

SUB-SECTION III.

THE ORDER OF CHEIROPTERA, OR BATS.

The Order of Cheiroptera comprises mammals whose distinguishing characteristic consists in a fold of skin which, commencing at the sides of the neck, and extending between the four members and fingers of the anterior extremities, supports the animal in the air, and enables it to fly. In most cases they have the arms, forearms, and fingers excessively long, forming, with the membrane mentioned above, true wings of great extent of surface, so that they are able to fly long and rapidly, and execute movements as varied and complicated as those of birds. Their eyes are excessively small, ears generally large, thumbs short and armed with a sharp hooked claw, hind feet weak and divided into five toes armed with trenchant and pointed nails. They are nocturnal in their habits, and during the day remain in caves, hollow trees, or other dark places, suspended by their hooks, or by the nails of their hind feet. In cold and temperate regions they pass the winter in a state of lethargy. Although their eyes are so small, their large ears and broad wings possess such a delicate sensibility that bats are enabled to fly unharmed through the most winding and complicated passages, and that, too, after their eyes have been destroyed. Bats may be divided into two great groups.

1. Frugivorous Bats, or Rousettes, have trenchant incisors in each jaw, grinders with flat crowns, and feed chiefly upon fruit, but also capture birds and small quadrupeds. About forty species are known, inhabiting mainly the East Indies and tropical Africa. They are the largest of the bats, and the flesh of some is used for food. The Genus *Pteropus* is the principal one.

The Black Rousette, *P. edulis*, Geoff., of the Straits of Sunda and the Moluccas, is of a blackish-brown color, and measures nearly four feet between the extremities of the wings. Its loud cry resembles that of the goose.

2. Insectivorous Bats have three grinders on each side in each jaw, bristled with conical points, that are preceded by a variable number of false molars. About two hundred species are known.

The Genus *Phyllostoma*—Vampires—is characterized by a membrane in the form of a leaf, which is reflected crosswise on the end of the nose. They belong wholly to tropical America, and have the reputation of inflicting severe wounds upon men and animals, which they bite in order to suck their blood.

The Vampire Bat, *P. spectrum*, Linn., is of the size of a magpie, reddish brown, and has the leaf in the form of a funnel.

The Genus *Vespertilio*—Common Bats—has the muzzle without leaf or other appendages, the incisors two to four above, six beneath, and the tail involved in the membrane. This is the most numerous genus of all, its species being found in all parts of the world. A half-dozen or more North American bats are usually referred to this genus.

The Red Bat, *V. noveboracensis*, Linn., is three to four inches long, with a spread of wings of ten to twelve inches; color reddish tawny.

The Hoary Bat, *V. pruinosus*, Say, is over four and a half inches long, the spread over fifteen inches; color grayish above, the throat with a fawn-colored band.

Fig. 25.

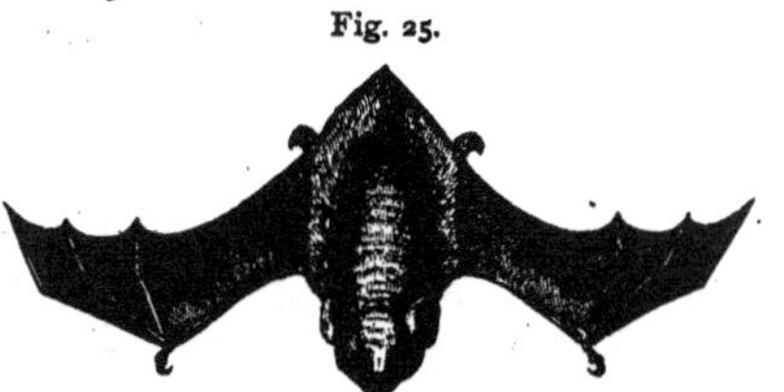

Hoary Bat, *V. pruinosus*, Say.

The Little Brown Bat, *V. subulatus*, Say, is over three inches long, the spread nine inches; color olive-brown above, grayish beneath.

The Silver-Haired Bat, *V. noctivagans*, Le Conte, is about three and a half inches long, spread ten to eleven inches; the color black, with a whitish collar across the shoulders.

The Carolina Bat, *V. carolinensis*, Geoff., is nearly four inches long, the spread twelve inches, and the color chestnut.

V. monticola, Bach., of a fulvous color, and *V. virginianus*, Bach., sooty brown, are additional species, from Virginia.

The Genus *Molossus*, comprising *M. cynocephalus* and *M. fuliginosus*, Cooper, of the Southern States, is characterized by a large head and muzzle, canines two to four in the upper jaw, none in the lower, and bifid upper incisors.

The Genus *Plecotus*, comprising *P. Lecontii* of the Southern States and *P. Townsendii* of Oregon, is characterized by greatly dilated ears, and by two fleshy crests between the eyes and nostrils.

The rapid flittings, turnings, and curious gyrations of bats, observable in early evening, or on cloudy days, or in the deep shade of woods, are for the purpose of capturing insects, of which they devour immense numbers.

SUB-SECTION IV.

THE ORDER OF INSECTIVORA, OR INSECT-EATERS.

THE Order of Insectivora comprises mammals which feed wholly or mainly upon worms and insects. They are mostly nocturnal in their habits, and in cool climates many of them remain in a torpid state during the winter. Their incisors and canines vary in proportion and relative position; their molars are studded with acute points, feet short, plantigrade, and clavicles perfect. Wagner has recognized five families: — Dermoptera, characterized by the body being margined with a hairy membrane. Such are the Galeopithici, already noticed with the Lemuridæ. — Scandentia, characterized by a squirrel-like appearance, except that the muzzle is attenuated and elongated. They climb trees with agility, and in this respect differ from all the other Insectivora. Such are the Banxrings and their allies, confined wholly to the Indian Archipelago. These two families demand no further attention in the present work; but the three remaining ones — Soricidæ or Shrew Family, Talpidæ or Mole Family, and Aculeata or Hedgehog Family — which are universally recognized as distinct families, require somewhat special attention.

Fig. 26.

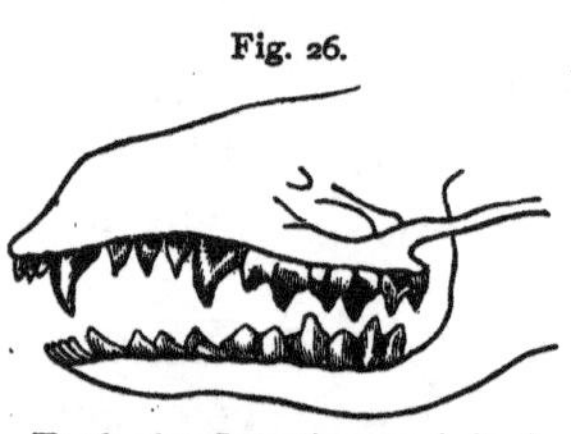

Teeth of an Insectivorous Animal.

Soricidæ, or Shrew Family.—This Family is characterized by a general mouse or rat-like appearance, elongated and tapering muzzle, ears with distinct concha, and fore feet smaller than the hind ones. On either flank, or at the base of the tail, are situated peculiar glands which emit an unpleasant odor. In true Shrews the teeth vary from twenty-eight to thirty-two in number. The dental formula is generally regarded as, incisors $\frac{1-1}{1-1}$, premolars $\frac{3-3}{2-2}$ to $\frac{5-5}{2-2}$, molars $\frac{4-4}{3-3}$. The two large incisors in each jaw are directed nearly horizontally forwards, the upper pair curved into a hook, and the lower are straighter, and with the trenchant upper edge more or less lobed. Shrews are found all over the world, and more than twenty species belonging to North America, and distributed among three genera, have been described. They live under rubbish, and in burrows which they dig in the earth.

The Genus *Neosorex* is characterized by rather short, valvular ears, partly furred on both surfaces.

The Water Shrew, *N. navigator*, Cooper, found at Fort Vancouver, is two and one twelfth inches long to the tail, which is three inches; the color dark sooty-brown above, grayish-white beneath.

Fig. 27.

Water Shrew.
N. navigator, Cooper.

The Genus *Sorex* is characterized by large valvular ears, concha directed backwards, tail about as long as the body without the head, feet of moderate size and not fringed, the upper forward incisor with a second basal hook, and a small angular process on the inner side near the point.

Trowbridge's Shrew, *S. Trowbridgii*, Baird, of Western North America, is two and a half inches long to the tail, which is two inches; the color sooty-brown above, paler beneath.

The Wandering Shrew, *S. vagrans*, Cooper, of Western North America, is two inches long to the tail, which is one inch and two thirds; the color above olive-brown varied with hoary, dusky yellowish-white below.

Suckley's Shrew, *S. Suckleyi*, Baird, of Western North America, is two and a quarter inches long to the tail, which is about one inch and a half; the color light chestnut-brown above, grayish white beneath.

The Thick-tailed Shrew, *S. pachyurus*, Baird, of Minnesota to Fort Ripley, is two and a quarter inches long to the tail, which is one inch and three quarters; the color light olive-brown above, ashy white beneath.

Forster's Shrew, *S. Forsteri*, Rich., of Eastern North America, is two and three quarters inches long to the tail, which is one inch and two thirds; the color smoky-brown above, pale grayish-ash beneath.

Richardson's Shrew, *S. Richardsonii*, Bachm., of Northern North America, is two and three quarters inches long to the tail, which is about one inch and a half; the color above rusty iron-gray, paler beneath.

The Eared Shrew, *S. platyrhinus*, Wagner, of Northeastern United States, is two inches long to the tail, which is one inch and a half; the color chestnut above, pale cinereous beneath; ears large.

Cooper's Shrew, *S. Cooperi*, Bach., of Labrador to Nebraska, is about two inches long to the tail, which is less than two inches; the color light chestnut-brown above, lighter beneath.

Hayden's Shrew, *S. Haydeni*, Baird, of Western North America, is one inch and three quarters long to the tail, which is less than one inch and a half; the color grayish chestnut-brown above, whitish beneath.

The Masked Shrew, *S. personatus*, Geoff., of the Southern Atlantic States, is one inch and three quarters long to the tail, which is over one inch; the color light chestnut-brown above, dull white beneath.

Hoy's Shrew, *S. Hoyi*, Baird, of Wisconsin, is one inch and three quarters long to the tail, which is one inch and a quarter; the color olive chestnut-brown above, dull rusty white beneath.

Fig. 28.

Thompson's Shrew, *S. Thompsonii*, Baird.

Thompson's Shrew, *S. Thompsonii*, Baird, of Nova Scotia to Ohio, is two inches long to the tail, which is one inch and a quarter; the color dark olive-brown above, ashy white beneath.

S. palustris, Rich., of Hudson's Bay, is three and a half inches long to the tail, which is over two and a half inches; the color hoary black above, ash-gray below.

S. fimbripes, Bach., has been found only in Pennsylvania. It is two and one eighth inches long to the tail, which is one inch and three quarters; the color brown above, buff below.

The Genus *Blarina* is characterized by a stout body, tail shorter than the head, or nearly equal to it, and coated with short bristly hairs, and with a small bunch at the tip. The hands are large in proportion to the feet, the palms as broad or broader than the soles, and the latter usually hairy at the heels. The fore claws are longer than the hind ones, external ear and auditory opening invisible, the skull short and broad, anterior upper incisors with the points simple, teeth rarely in contact, and the lower anterior process of the lower jaw short and stout.

The Mole Shrew, *B. talpoides*, Gray, of Nova Scotia to Georgia, is three and a half inches long to the tail, which is one inch; the color dark ashy-gray.

The Short-tailed Shrew, *B. brevicauda*, Gray, of Illinois to Nebraska, is four inches to the tail, which is over one inch; the color dark brownish-plumbeous; feet and edge of lips whitish.

The Carolina Shrew, *B. carolinensis*, Bach., of South

Carolina to Missouri, is about two and a half inches long to the tail, which is three quarters of an inch; the color dark leaden-gray.

B. angusticeps, Baird, of Vermont, is over two and a half inches long to the tail, which is one inch; the color plumbeous. The skull is very narrow.

The Ash-colored Shrew, *B. cinerea*, Bach., of Pennsylvania to Florida, is two and a half inches long to the tail, which is three quarters of an inch; the color above iron-gray, light gray beneath.

The Short-footed Shrew, *B. exilipes*, Baird, of Texas to Mississippi, is less than two inches long to the tail, which is over half an inch; the color above hoary olive-brown with a chestnut tinge, grayish white beneath.

The Genus *Mygale* comprises the Desmans, which differ from the Shrews in having two very small teeth between the two great incisors of the lower jaw, and in their two upper triangular and flattened incisors. The muzzle extends into a long and flexible proboscis; their feet are webbed, and they are aquatic in their habits. They inhabit Southern Russia, and one species is found about the streams of the Pyrenees.

TALPIDÆ, OR MOLE FAMILY. — This family is characterized by a stout thick body, with no visible neck or external ears, very short limbs, greatly expanded fore-feet, and strong fossorial claws. The tail is usually short, sometimes nearly as long as the body; the fur is soft, compact, and velvet-like. Moles are found all over the world except in the inter-tropical regions; each country, however, has its characteristic genus or genera.

The Genus *Scalops* is distinguished by a long, depressed muzzle, nostrils at the extremity, and either superior or lateral, hidden eyes, short tail, toes more or less webbed to the claws, teeth thirty-six to forty-four, and the two anterior upper ones very large.

Fig. 29.

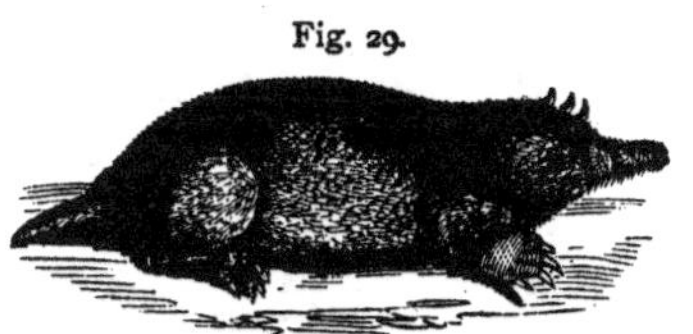

Common Mole, *S. aquaticus*, Cuv.

The Common Shrew-Mole, *S. aquaticus*, Cuv., of the United States east of the Mississippi, is four to five inches long to the tail, dark plumbeous, the feet and tail white. The teeth are thirty-six in number, nostrils superior, palms broader than long, tail nearly naked, and feet fully webbed. Its eyes are so small and completely hidden by the fur, that the casual observer readily supposes it to be blind. In fact, the aperture for the eye is only about the diameter of a human hair, and the eye-balls are smaller than a grain of mustard-seed. It inhabits both dry and wet lands, burrowing in every direction at a little depth beneath the surface, and throwing up at intervals the little hills of loose earth so familiar to every observer. It burrows with great rapidity, moves swiftly through its winding and complicated galleries, and its strength is wonderful. It is a remarkable fact, certified by good authority, that these animals come to the surface daily at twelve o'clock.

The Silver or Prairie Mole, *S. argentatus*, Aud. & Bach., of the Western States and southward, is somewhat larger than the last, palms scarcely broader than long, tail nearly naked, and the color silvery plumbeous.

The Oregon Mole, *S. Townsendii*, Bach., of the Pacific coast of North America, is six inches long, nearly black, nostrils superior on the tip of the snout, teeth forty-four, and the tail scantily haired.

The Hairy-tailed Mole, *S. Breweri*, Bach., of New England to Ohio, is about five inches long from the nose to the root of the tail, dark plumbeous, glossed with ashy brown, the ear opening rather large, nostrils lateral, palms rather narrow, teeth forty-four, and tail densely hairy.

The Genus *Condylura* is distinguished by the fringe of elongated caruncles encircling the end of the nose.

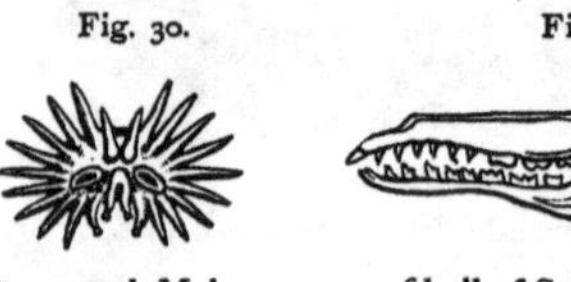

Fig. 30. Star-nosed Mole, — end of Muzzle.

Fig. 31. Skull of Star-nosed Mole. *C. cristata*, Illig.

The Star-nosed Mole, *C. cristata*, Illig., is about four inches long to the tail, which is about as long as the body without the head; the color dark brownish-black. This mole lives near the banks of streams, and in moist meadows, where it digs its numerous and apparently interminable burrows, but, unlike the Shrew-mole, raises few hills of loose earth.

The Genus *Talpa*, comprising the Common European Mole, *T. europæa*, Linn., is confined to Europe and Asia; and the Genus *Chrysochloris* — Golden-green Moles — to Africa. The latter presents the only example of mammals with splendid metallic tints like those which adorn so many birds, fishes, and insects.

The Genus *Urotrichus* has the muzzle prolonged into a cylindrical tube, terminating in a simple naked bulb, and the nostrils cylindrical and lateral. Found in Northwestern America and in Japan.

ACULEATA, OR HEDGEHOG FAMILY. — This Family is characterized by the back being covered with spines or bristles, and the tail short or wanting. The species all belong to the Old World.

Fig. 32. Hedgehog, *E. europæus*, Linn.

The Genus *Erinaceus* —

Hedgehogs — has bristles instead of hair, and the skin of the back is furnished with such muscles that the animal, by bending his head and paws towards his abdomen, can shut himself up as in a bag, and present bristles on all sides to the enemy.

The European Hedgehog, *E. europæus*, Buff., inhabits woods and hedges, passes the winter in its burrow, and feeds upon insects and fruit.

The Genus *Centetes* — Tenrecs — has the body covered with spines or bristles like the hedgehog; but the animals of this genus have not the power of rolling themselves so completely into a ball. Three species inhabit Madagascar. They pass three months of the year in a state of lethargy, although inhabiting the torrid zone.

Fig. 33.

Tenrec, *C. semi-spinosus*, Cuv.

SUB-SECTION V.

THE ORDER OF CARNIVORA, OR CARNIVOROUS ANIMALS.

The Order of Carnivora comprises all the Mammalia which feed wholly or mainly upon flesh, and with few exceptions they capture the animals upon which they prey. They are distinguished from all other animals, not only by their general appearance, but especially by their sharp teeth and claws, and by their internal digestive apparatus. They have six incisors in each jaw, the lateral ones the largest, a long, stout canine in each side of both jaws just behind

Fig. 34.

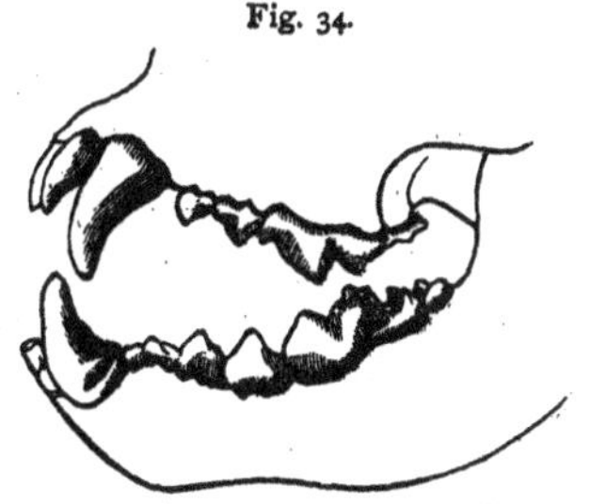

Teeth of carnivorous animal.

the incisors, and a variable number of molars, which are wholly trenchant, or in part with tuberculous crowns. Thus the molars of Carnivora are not properly grinding teeth, but they cut the prey like a pair of shears. These animals are so much the more exclusively carnivorous as their teeth are more completely trenchant; and the relative proportions of their vegetable or animal food may be calculated from the extent of the tuberculous surfaces of the teeth as compared with the portion which is trenchant. The clavicles of Carnivora are imperfect or wholly wanting; the stomach is simple, and intestines short, — perfectly adapted to their easily-digestible food. The Carnivora are found all over the globe, and two hundred to three hundred species have been described. They are divided into seven families, — the Felidæ or Cat Family, Hyenadæ or Hyena Family, Canidæ or Dog Family, Viverridæ or Civet Family, Mustelidæ or Weasel Family, Ursidæ or Bear Family, and the Phocidæ or Seal Family.

Felidæ, or Cat Family. — This Family comprises the Carnivora which are the most dexterous and rapid in their movements, and endowed with the keenest senses, and which are the most rapacious and formidable. The head is short and broad, the teeth and claws excessively sharp, and the latter are concealed in a sheath while the animal is walking or at rest, but are instantly thrust forth when occasion requires their use. The tongue is covered with sharp prickles pointing backwards; the number of mature teeth is twenty-eight or thirty, and the dental formula is, incisors $\frac{3-3}{3-3}$, canines $\frac{1-1}{1-1}$, premolars and molars $\frac{4-4}{3-3}$ or $\frac{3-3}{3-3}$. The feet are digitigrade, with five toes before and four behind. The soles are densely hairy, with naked pads on the ball of the foot and under each toe. The cushion-like nature of the bottoms of their feet enables these animals noiselessly to approach their prey, which they seize by a sudden spring.

The Genus *Felis* — true Cats — is characterized by a long and tapering, sometimes tufted tail, and by the molars, which are $\frac{4-4}{3-3}$.

The Lion, *F. leo*, Linn., inhabiting Africa except the Nile countries, Arabia, Persia, and a large part of India, is the strongest and most courageous of all the cats. It is readily distinguished by its uniform tawny color, by the tuft of hair at the end of the tail, and by the long, flowing mane of the male. The largest individuals are eight or nine feet long, exclusive of the tail. The lioness is smaller than the lion, and has no mane. The appearance of this animal is noble, his gait stately, his strength very great; his roar is tremendous and often terrific.

The Royal Tiger, *F. tigris*, Buff., of India and northward, is as large as the lion, with a longer body and rounder head. The color is lively fawn above, irregularly striped with black; below, pure white. There is no animal that he will not attack, even engaging in conflicts with the lion, which sometimes prove fatal to both.

The American Tiger, or Jaguar, *F. onca*, Linn., of Texas to Patagonia nearly, is somewhat smaller than the last, clear brownish-yellow above; beneath white, spotted. On the sides of the body there is a series of polygonal figures with borders composed of black blotches enclosing a light area, with a few blotches in the centre.

The Panther, *F. pardus*, Linn., of Africa and India, is fawn-colored above and white beneath, with six or seven rows of black spots or blotches on each flank.

The Leopard, *T. leopardus*, Linn., of Africa and India, is similar to the last, but has ten rows of smaller spots. This and the Panther are smaller than the Jaguar.

The Puma, or American Panther, *F. concolor*, Linn., of Canada to Southern Patagonia, is the largest of the American Cats except the Jaguar, being larger than the largest dogs, and weighing in some cases a hundred and fifty pounds. The color above is uniform pale brownish-yellow, finely mottled by dark tips to all the hairs; beneath, dingy white. It preys upon deer, sheep, and smaller quadrupeds, and has been known to attack and

Fig. 35.

Puma, *F. concolor*, Linn.

kill a human being. It is sometimes called Catamount, and Cougar.

The Ocelot, *F. pardalis*, Linn., of Texas to 30° south latitude, is about the size of the American wild-cat, of a gray color, with large fawn-colored spots, bordered with black, forming oblique bands on the flank.

The Tiger Cat, *F. eyra*, Desm., of Texas to Guiana, is about the size of the common domestic cat, but with a longer neck, and the color uniform brownish red.

The Yaguarundi, *F. yaguarundi*, Desm., from the Rio Grande to Paraguay, is larger than the common cat, with a much longer body, and the prevailing color a grizzled brownish gray.

The Domestic Cat, *F. catus*, Linn., is originally from

the European forests. In its wild state it is grayish-brown, with darker transverse undulations.

The Genus *Lynx* — Lynxes — is distinguished from Felis by the molars, which are always $\frac{3-3}{3-3}$, and by the tail, which is scarcely longer than the head, and abruptly truncate at the tip.

The American Wild-Cat, *Lynx rufus*, Raf., is about thirty inches long to the tail, which is about five inches, the fur full and soft, the color above and on the sides pale rufous overlaid with grayish; beneath white, spotted. The tail has a small black patch above at the end, and the inner surface of the ear is black, with a white patch.

The Texas Wild-Cat, *L. rufus*, var. *maculatus*, Aud. & Bach., has the fur short and rather coarse, and quite distinct dark spots upon the back and sides.

The Red Cat, *L. fasciatus*, Raf., of Washington Territory, has the fur soft and full, the color above rich chestnut-brown, ears black inside, pencilled, and the terminal third of the tail black above.

Fig. 36.

Canada Lynx, *L. canadensis*, Raf.

The Canada Lynx, *L. canadensis*, Raf., is about forty inches in total length, the general color grayish hoary, waved with black, ears grayish with a narrow black margin on the convexity, and tipped with a black pencil, and the end of the tail black; feet very large, densely furred beneath in winter.

HYENADÆ, OR HYENA FAMILY. — This Family comprises digitigrade Carnivora, which have the fore legs longer than the hind ones, the claws non-retractile, feet four-toed, tongue rough, the dental formula, incisors $\frac{3-3}{3-3}$, canines $\frac{1-1}{1-1}$, premolars $\frac{4-4}{3-3}$, molars $\frac{1-1}{1-1}$. The premolars are very large and blunt, and these animals are able to crush the bones of the largest prey, and swallow the fragments without the slightest mastication. So powerful are the muscles of the neck and jaws, that it is almost impossible to wrest anything from between their teeth; and among the Arabs their name is the symbol of obstinacy. Hyenas live in caves, are nocturnal, voracious almost beyond the power of description, and feed chiefly upon prey which they find dead. Many superstitious traditions are connected with them. They belong to Africa and Asia, and the largest have a total length of five feet or more.

The Striped Hyena, *H. vulgaris*, Buff., of Africa and India, is gray with dark stripes, and a mane which is erect when the animal is angry.

The Brown Hyena, *H. brunnea*, Thumb., and the Spotted Hyena, *H. crocuta*, Schr., inhabit Southern Africa.

In the caves of England and on the continent of Europe are found in abundance the bones of an extinct species of Hyena, *H. spelæa*, together with the bones of many other extinct animals, which bear unmistakable marks of its teeth.

CANIDÆ, OR DOG FAMILY. — This Family comprises digitigrade Carnivora without retractile claws, and with all the feet apparently four-toed; the forward ones, however, with a rudimentary thumb high up.

The Genus *Canis* — Wolves — is distinguished by the

post-orbital process of the frontal bone being very convex, and curving downwards, and by the circular pupil of the eye. Wolves are crafty, ferocious, and greedy; feeding upon whatever they can kill, and also gorging themselves upon the bodies of dead animals, which they scent at great distances. They hunt in packs, and are thus able to overpower animals which singly they could not master. In newly settled districts wolves often make great havoc among sheep, calves, and other domestic animals.

Fig. 37.

Wolf, *C. occidentalis.*

The White and Gray Wolf, *C. occidentalis*, var. *griseo-albus*, Rich., found throughout North America, is pure white to grizzly gray.

The Dusky Wolf, *C. occidentalis*, var. *nubilis*, Say, found from the Missouri to the Pacific, is light sooty or plumbeous brown.

The Mexican Wolf, *C. occidentalis*, var. *mexicanus*, Gm., is gray and black, with neck more maned than usual, and a dark band encircling the muzzle.

The Black Wolf, *C. occidentalis*, var. *ater*, Rich., of Florida and other Southern States, is wholly black.

The Red Wolf, *C. occidentalis*, var. *rufus*, Aud. & Bach., of Texas, is mixed red and black above, lighter beneath.

The Prairie Wolf, or Coyote, *C. latrans*, Say, of the

countries west of the Mississippi, is considerably larger than the common fox, dull yellowish-gray on the back and sides, with a clouding of black; under parts dingy white.

The Jackals, several species, of Africa and Asia, are closely related to the wolves, belonging in the same genus.

The Domestic Dog, *C. familiaris*, Linn., is distinguished from the other species of Canis by the recurved tail. Some naturalists consider the dog a wolf; others, that he is a domesticated jackal; and yet those dogs which have become wild again on desert islands resemble neither the one nor the other of these. Of all animals, he is the only one that has followed man to every part of the globe.

The Genus *Vulpes* — Foxes — is characterized by a slender head, elliptical pupil of the eye, scarcely lobed upper incisors, and the post-orbital process of the frontal bone is bent but little downwards, and the anterior edge turned up.

The Common American Red Fox, *V. fulvus*, Desm., is notorious for his nocturnal depredations upon farm-yards, whence he carries away chickens, geese, and turkeys to the dense thickets, where he spends most of the daytime. The general color is reddish-yellow, the back behind grizzled with grayish, throat and a line on the belly and tip of tail white, feet and ears black.

The Cross Fox, *V. fulvus*, var. *decussatus*, Desm., takes its name from the black cross formed by a dark band between the shoulders crossed by another over the shoulders. It is found in the northeastern parts of the United States and Canada.

The Silver Fox or Black Fox, *V. fulvus*, var. *argentatus*, Shaw, is entirely black except the hind part of the back, which is more or less grizzly; and the tip of the tail is white.

The Prairie Fox, *V. macrourus*, Baird, of the central parts of North America, closely resembles the Red Fox, but is larger, has longer fur and a longer tail, and is regarded as the most interesting species known.

The Swift Fox, *V. velox*, Aud. & Bach., of Oregon, is smaller than the Red Fox, the general color yellowish gray above, back conspicuously grizzled, sides and portions of the legs pale reddish-yellow; the under parts whitish, tail tipped with black.

The Arctic Fox, *V. lagopus*, Rich., of the Arctic regions, is smaller than the Red Fox, of a white color, the tail very full and bushy, and the soles of the feet densely furred.

The Gray Fox, *V. virginianus*, Rich., of the United States, is mixed hoary and black.

The Coast Fox, *V. littoralis*, Baird, of the island of San Miguel, coast of California, is the smallest North American fox, being scarcely larger than the common house cat; the color similar to that of the gray fox.

VIVERRIDÆ, OR CIVET FAMILY. — This Family comprises small animals of the average size of the domestic cat, but more elongated, and with a more pointed muzzle, and with a long tail. In most cases the feet are digitigrade, with hairy soles and retractile claws. The dental formula is, incisors $\frac{3-3}{3-3}$, canines $\frac{1-1}{1-1}$, premolars $\frac{4-4}{4-4}$, molars $\frac{2-2}{2-2}$. The dentition differs from that of the dog family in having one tubercular true molar less on each side of the lower jaw. They secrete in a sort of pouch or gland a substance formerly much used in perfumery, and which was long an important article of commerce. With one exception, they belong to the Old World.

The Genus *Bassaris* is represented in North America by the Civet Cat, *B. astuta*, Licht., of Texas to California. This animal is about the size of the domestic cat, but more slender; the color above brownish yellow mixed

Fig. 38.

Civet Cat, *B. astuta*, Licht.

with gray beneath, and the tail white, the latter with six to eight black rings. They are arboreal, easily tamed, and favorite pets with the miners.

The Genus *Viverra*, according to its old limits, comprises the four following species.

The Civet, *V. civetta*, Linn., of Africa, is ash-colored, irregularly barred and spotted with black. There is a mane along the whole back and tail which the animal raises at will. This species furnishes the musky substance called civet.

The Common Genet, *V. genetta*, Linn., of Southern Europe to Cape of Good Hope, is gray, spotted with brown or black; the tail annulated with black and white. Its skin is an important article of trade.

The Mangouste or Ichneumon, *V. ichneumon*, Linn., of Egypt, is about the size of the domestic cat, very slender, of a gray color, tail long and terminated with a black tuft. It hunts chiefly the eggs of the crocodile, but also preys upon all sorts of small animals. It is kept in houses, like the common cat. The Europeans at Cairo call it Pharaoh's Rat.

The Mangouste, *V. mungos*, Linn., of India, smaller than the last, is celebrated for its combats with the most dangerous serpents, and for having led us to the knowledge of the *Ophiorhiza mungos* as an antidote for their poison.

MUSTELIDÆ, OR WEASEL FAMILY. — This Family comprises elongated and slender-bodied animals, with five-toed plantigrade or digitigrade feet, and with a single tubercular molar tooth only on either side of each jaw. The Mustelidæ comprise all the animals known as Fishers, Martens, Sables, Weasels, Minks, Otters, Badgers, and

Skunks. Nearly all of this family have glands which secrete a fetid liquid, and in some cases of a most disagreeable odor.

The Genus *Mustela*—Martens—is characterized by a slender body, long tail, and thirty-eight teeth, the formula being, incisors $\frac{3-3}{3-3}$, canines $\frac{1-1}{1-1}$, premolars $\frac{4-4}{4-4}$, molars $\frac{1-1}{2-2}$. They are arboreal in their habits, and some of them yield furs of great value.

The Fisher, *M. Pennantii*, Erxl., of the United States, is the largest known species of this genus, being two feet long to the tail, which is more than a foot in length; the legs, belly, tail, and hind part of the back are black; the back towards the head has an increasing proportion of grayish.

The American Sable or Pine Marten, *M. americana*, Turton, of Northern Maine and of the Adirondac Mountains, N. Y., thence northward and westward, is seventeen inches long to the tail, which is about ten inches; the general color reddish-yellow clouded with black, legs and tail blackish, a broad yellowish patch upon the throat widening below so as to touch the legs, the central line below sometimes yellowish, and the feet are densely furred. The fur is very full and soft, with many long hairs interspersed. The highly prized fur known as Hudson Bay Sable is furnished by this species.

Fig. 39.

American Sable, *M. americana*, Turton.

The Pine Marten, *M. martes*, Linn., of Europe, is brown, with a yellow spot under the throat.

The Common Marten, *M. foina*, Linn., of Europe, is brown, with the whole under part of the throat and neck white.

The Sable, *M. zibellina*, Pall., of Siberia, so celebrated for its rich fur, known as the Russian Sable, is brown, spotted with gray about the head, and its feet furred. It inhabits the coldest regions, and the hunting to obtain it, in the midst of winter and tremendous snows, is attended with the greatest privations and dangers.

The Genus *Putorius* — Weasels — is characterized by a very slender body, long tail, and thirty-four teeth, the dental formula being, incisors $\frac{3-3}{3-3}$, canines $\frac{1-1}{1-1}$, premolars $\frac{3-3}{3-3}$, molars $\frac{1-1}{2-2}$. The lower sectorial tooth is without an inner tubercle.

Fig. 40.

Weasel.

The Least Weasel, *P. pusillus*, Aud. & Bach., from New York to Puget's Sound, is six inches long to the tail, which is less than two inches; the color brown above, white beneath.

The Small Brown Weasel, *P. Cicognonii*, or *fuscus*, Aud. & Bach., of North America, is about eight inches long to the tail, which is three inches; the color in summer, brown above, whitish beneath; in winter, white; the tail with the tip black.

The Little Ermine, *P. Richardsonii*, Bonap., of North America, is eight inches long to the tail, which is over five inches; the color in summer dark chestnut-brown above, whitish beneath; in winter white, tail with a black tip.

The Common White Weasel, *P. noveboracensis*, De Kay, of the United States, is about ten inches long to the tail, which is about six inches, the color in summer chestnut brown above, whitish beneath; in winter, white; tail tipped with black one third of its length.

The Long-tailed Weasel, *P. longicauda*, Rich., of Western North America, is about eleven inches long to the tail, which is nearly seven inches; the color in summer, light olive-brown above, brownish-yellow beneath; in winter, white; the tail with a black tip about one fourth its length.

Kane's Ermine, *P. Kaneii*, Baird, of Behring's Straits and Siberia, is eight and a half inches long to the tail, which is about four inches; the color in summer, brown above, in winter white; the tail tipped with black one half its length.

The Bridled Weasel, *P. frenatus*, Aud. & Bach., of Texas and Mexico, is eleven inches long to the tail, which is about seven inches; the general color above chestnut-brown, yellowish beneath, and the tail tipped with black. The head above is dark brown, with three white marks.

The Yellow-cheeked Weasel, *P. xanthogenys*, Gray, of California, is closely related to the last.

The Common Mink, *P. vison*, Rich., of the United States, is about seventeen inches long to the tail, which is about half the length of the body; the general color dark brownish-chestnut, tail nearly black, and the end of the chin white.

Fig. 41.

Mink, *P. vison*, Rich.

The Little Black Mink, *P. nigrescens*, Aud. & Bach., of the United States, is smaller than the last, the color chestnut-brown glossed with black; the tail almost entirely black, and the end of the chin white. This species furnishes the most valuable of the mink furs. The furs sold under the name of American Sable, are mink.

The European Ermine, *P. erminea*, Linn., celebrated for its valuable and well-known fur, is about ten inches long to the tail, which is about five inches; the color olivaceous-brown in summer, white in winter, and the tail has a long black tip.

The Ferret, *P. furor*, Linn., of Spain and Barbary, is celebrated from its being employed to ferret out rabbits from their holes. The Common Polecats of the Old World all belong to this genus.

The Genus *Gulo* — Gluttons — is characterized by a stout body, bushy tail, densely hairy soles with six naked pads, and thirty-eight teeth; the formula for which is the same as in the Mustelidæ.

The Wolverine, *G. luscus*, Sabine, of Northern New York, thence northward and westward, is about three feet long to the root of the tail, which is over a foot in length;

Fig. 42.

Wolverine, *G. luscus*, Sabine.

the color dark brown, the tail, legs, and under parts black. There is a lighter broad band on the flanks passing over the base of the tail and rump, and a grizzled light patch upon the temples. The most extravagant stories have been told of this interesting animal. It is safe to say that, for its size, it is very powerful, ferocious, and exceedingly voracious. It is very troublesome to the sable-hunters, by breaking up their wooden traps and destroying the bait, or game; it also destroys *caches* of provisions. The Glutton of Russia is a closely related species.

The Genus *Lutra* — Otters — is characterized by a flat head, elongated body, short, palmated feet, distinct digits, the central longer than the exterior ones, and tail depressed and rounded at the sides. Otters are aquatic, and are found in all parts of the world. They feed upon fish, which they pursue with such dexterity that few are swift enough to elude them. Otters have a singular and amusing habit of sporting. Selecting a bank of snow in winter, or a clayey bank in summer, they scramble to the top, and then slide head foremost to the bottom. If their sliding-place leads into the river, as is generally the case, they go plump into the water, whence they quickly come forth again to repeat an operation which evidently gives them great satisfaction.

Fig. 43.

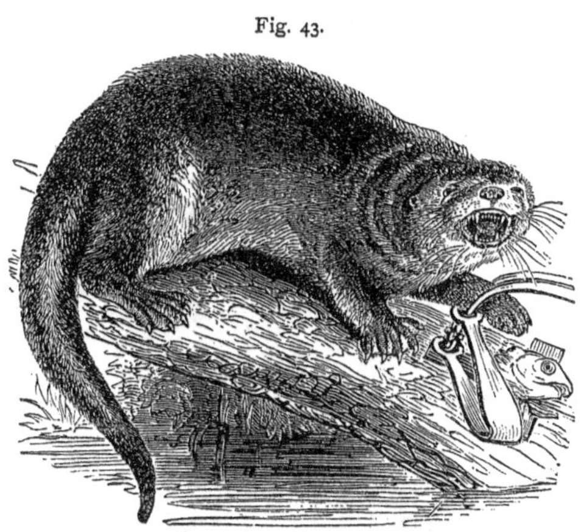

American Otter, *L. canadensis*, Sabine.

The American Otter, *L. canadensis*, Sab., of North America east of the Rocky Mountains, is about four and a half feet long, including the tail, which is eighteen inches in length. The color is liver-brown above, slightly lighter beneath, sides of head and neck dirty-whitish. The fur is of two kinds; one long, somewhat coarse, and scattered, the other shorter, fine, and dense.

The California Otter, *L. californica*, Gray, of the Pacific coast, closely resembles *L. canadensis*.

The Genus *Enhydra* comprises the Sea Otter, *E. marina*, Flem., of the Pacific coast of both hemispheres, which reminds us of the seals. It is about twice the size of the common otter, the color blackish, fur very long and soft, and very valuable.

The Genus *Mephitis* — Skunks — is characterized by an elongated body, pointed nose, fossorial feet, long and bushy tail, upper hind molar very large and nearly square, and black color with white markings. The skunks are, moreover, characterized among all other animals by their well-known and almost intolerable odor, which they emit when disturbed. The dental formula is about the same as that of Putorius. Skunks are nocturnal, and feed upon beetles and other small animals, and upon eggs. They walk on most of the sole of the foot, with the back much curved and tail erect. Five species are found in North America.

The White-backed Skunk, *M. mesoleuca*, Licht., of Texas and Mexico.

The Long-tailed Skunk, *M. varians*, Gray, of Texas and Mexico.

The California Skunk, *M. occidentalis*, Baird, of the western coast of the United States.

Fig. 44.

Skunk, *M. chinga*, Tied.

The Common Skunk, *M. chinga*, Tied., of the United States north of Texas and east of the Missouri plains.

The Little Striped Skunk, *M. bicolor*, Gray, of Southern Texas and California.

The Genus *Taxidea* — Badgers — is characterized by a stout, robust, depressed body, very short tail, much enlarged fore claws adapted to digging, and by a wedge-shaped skull.

The American Badger, *T. americana*, Waterh., of Arkansas, thence northward and westward, is rather less than two feet long to the tail, which is about six inches; the general color grayish above, light beneath. The hair is long, especially on the hind part of the body, whence it extends so far towards the extremity of the tail as almost to conceal the latter; and the ears appear as if they had been clipped. The Mexican Badger, *T. Berlandieri*, Baird, is a closely related species. Badgers live in burrows, and dig with astonishing rapidity.

Fig. 45.

American Badger, *T. americana*, Waterh.

URSIDÆ, OR BEAR FAMILY. — This Family comprises the true plantigrade carnivora, — those which walk on the whole sole of the foot. They are five-toed, and the toes are distinctly separate. Their teeth are the same in number as those of the Dog Family, but the sectorial teeth and the molars behind them are tuberculated. They have no cœcum. Though carnivorous, they feed more or less upon vegetable food. Many of the species are ready climbers. Those which inhabit cold climates pass the winter in a torpid state. The Ursidæ comprise the Raccoons, Pandas, and Bears.

The Genus *Procyon* — Raccoons — is characterized by a stout body, pointed muzzle, and moderately long tail.

The Common Raccoon, *P. lotor*, Storr., of the United States, is less than two feet long to the tail, which is about a foot; the general color light gray, tinged with pale rusty

Fig. 46.

Common Raccoon, *P. lotor*, Storr.

across the shoulders, and much overlaid with black-tipped hairs. The under parts are of a similar gray, but without the black tips; and over the whole body the dull-sooty under-fur shows through. The tail has five distinct black rings, and a tip of the same color, the interspaces being grayish-white. The end of the muzzle is whitish, and there is a black patch upon the cheek and another behind the ear. The Raccoon is nocturnal in its habits, and feeds upon roots, birds, and other small animals. It is easily tamed, and is said to dip its food in water before eating it.

The California Raccoon, *P. Hernandezii*, Wagler, of Western North America, is larger than *P. lotor.*

The Genus *Ailurus* comprises the Shining Panda, *A. refulgens*, Fred. Cuv., of Northern India, of the size of the common cat, the fur soft and thickly set, the color above of the most brilliant cinnamon-red, behind more of a fawn color, the head whitish, and the tail marked with brown rings; beneath black. Cuvier calls this the most beautiful of known quadrupeds.

The Genus *Ursus*—Bears—is characterized by a large, thick, clumsy body, broad head, short tail, wholly plantigrade feet, with naked soles and long nails. The dental formula is, incisors $\frac{3-3}{3-3}$, canines $\frac{1-1}{1-1}$, premolars $\frac{4-4}{4-4}$, molars $\frac{2-2}{3-3}$. Four North American species have been described.

The Grizzly Bear, *U. horribilis*, Ord., of the plains of the Upper Missouri to California, is six to eight feet in length, hair coarse, an erect mane between the shoulders, and the color grizzly. The feet are very large, and the fore claws are twice as long as the hind ones, and on the largest individuals are six inches in length. This animal is one of the most powerful and most ferocious of all the Bear tribe. When excited by hunger or anger, it attacks man or any animal it can overtake. Even the Bison sometimes falls a victim to its ferocity, and is dragged

Fig. 47.

Grizzly Bear, *U. horribilis*, Ord.

away whole to be eaten at leisure. The Grizzly is not easily brought down, unless shot through the head or heart, and when wounded is exceedingly dangerous to the hunter.

The Brown Bear, *U. arctos*, Linn., of Europe, closely resembles the last, but is regarded as distinct.

The Black Bear, *U. americanus*, Pallas, of North America generally, weighs from two hundred to four hundred pounds, is of a uniform black or deep brown color, and the hair is comparatively soft and glossy. Under ordinary circumstances this bear is not very ferocious, seldom attacking man unless wounded or much excited by hunger. The Black Bear subsists upon roots, berries, and living animals. The Cinnamon Bear of the Rocky Mountains and Oregon is probably a variety of this species.

The White or Polar Bear, *U. maritimus*, Linn., of the Arctic regions of both hemispheres, is eight feet long, and attains the weight of one thousand to fifteen hundred pounds. It is snow-white, wholly carnivorous, and feeds upon seals and other animals.

PHOCIDÆ, OR SEAL FAMILY. — This Family comprises amphibious mammals, — the Seals and the Walrus. Though they spend more or less time upon the land, their home is in the sea. They have an elongated, muscular, flexible body, clothed with short thickly-set hair. Their locomotive appendages are the same in number as those of other Carnivora, but are so enveloped in the general covering of the body that the only service they can render upon land is to enable the animal to crawl; but, as the intervals of the toes are filled with a stout membrane, they become excellent paddles for swimming. The hind locomotive members project backwards nearly on a line with the body, and appear to the casual observer somewhat like the tail of a fish placed horizontally. Their toes are all terminated with pointed nails.

The Genus *Phoca* is the one in which Linnæus included all the seals, but later the various species have been distributed among several genera.

Seals have six or four incisors above, four or two below, canines and grinders to the number of twenty-two or twenty-four, all trenchant or conical, and without any tuberculous part whatever. The head resembles that of the dog, whose intelligence and expressive look it also possesses. Seals are easily tamed, and soon become attached to those who feed them. They live upon fish and other aquatic animals, eat in the water, and close their nostrils by a sort of valve when they dive. A dozen or more species have been described.

Fig. 48.

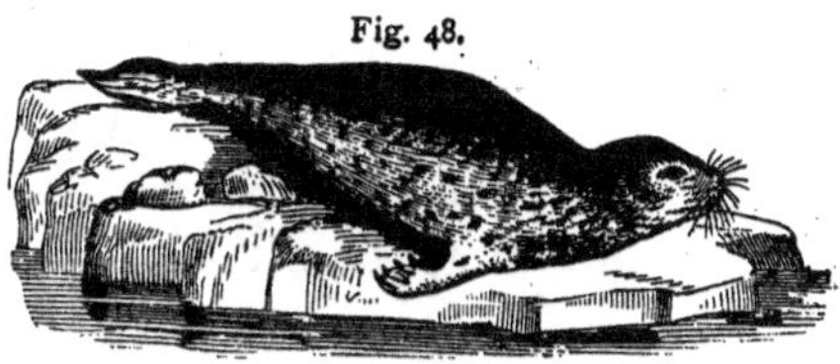

Common Seal, *Ph. vitulina*, Linn.

The Common Seal, *Ph. vitulina*, Linn., abundant in the cool and frigid regions, is three to five feet long, yellowish-gray, and more or less shaded and spotted with brown according to age. The eyes are large, dark, and lustrous.

The Harp Seal, *Ph. groenlandica*, Hooded Seal, *Ph. cristata*, Gm., and Bearded Seal, *Ph. barbata*, Fabr., are other species from the Northern regions.

The Monk Seal, *Ph. monachus*, Gm., of the Mediterranean, is ten to twelve feet long, and is the one best known to the ancients.

The Elephant Seal, *Ph. leonina*, Linn., of the Southern Ocean, is twenty to twenty-five feet long, and the muzzle of the male is terminated with a wrinkled snout, which becomes inflated when he is angry. This seal is hunted for its oil, of which it yields a large quantity.

The Sea Lion, *Ph. jubata*, Gm., of the Pacific, is fifteen to twenty feet in length.

The Sea Bear, *Ph. ursina*, Gm., of the Arctic regions, is eight feet long.

The Genus *Trichecus* comprises the Walrus or Morse, *T. rosmarus*, Linn., of the Arctic regions, which resembles the large seals, but is especially distinguished from them by having no canine or incisor teeth in the lower jaw, and by having the upper canines enormously developed, forming tusks which in many cases are two feet long. The Walrus is of the size of the largest ox, and attains the length of twenty feet, and is covered with short brown hair. It uses its strong tusks to lift itself from the water upon the rocks or ice-banks, where large numbers bask together in the sunshine. The tusks also serve as means of defence, and for obtaining sea-weed, which with fish constitutes its food. When attacked it is fierce, and becomes a formidable antagonist, especially if attended by young. It readily smashes a boat with its tusks, or, rising in the water, hooks them over the side, and upsets it.

SUB-SECTION VI.

THE ORDER OF MARSUPIALIA, OR MARSUPIALS.

The Order of Marsupialia comprises animals whose special characteristic is that their young are brought forth in an exceedingly premature state of development, and, in most instances, are received into a peculiar pouch on the abdomen of the mother, where they are nourished till they have acquired a degree of development corresponding to that in which other mammals are born. The young, after they are able to walk, also resort to the pouch of the mother for safety in times of danger. With the exception of one family found in America, the Marsupials are all confined to Australia and islands immediately adjacent; and it is a singular fact, that all the mammals of Australia, over a hundred species of which are known, belong to this order.

DIDELPHIDÆ, OR OPOSSUM FAMILY. — This Family comprises all the Marsupials of North and South America, and is peculiar to this continent. Opossums are mostly small animals, the largest scarcely exceeding the common cat, and the smallest but little larger than a mouse. They are plantigrade and five-toed. The dental formula is, incisors $\frac{5-5}{4-4}$, canines $\frac{1-1}{1-1}$, premolars $\frac{3-3}{3-3}$, molars $\frac{4-4}{4-4}$. Their food consists of birds, birds' eggs, insects, and other small animals. The tail is long, prehensile, and nearly naked. More than twenty species are known. In some, the pouch is rudimentary.

Fig. 49.

Opossum, *D. virginiana*, Shaw.

The Genus *Didelphys* comprises the Common Opossum, *D. virginiana*, Shaw, of the United States west of the Hudson. It is twenty inches long to the tail, which is about fifteen inches. The hair is whitish with brown tips, imparting a dusky shade. It lives upon trees, and feeds upon fruits, eggs, and small animals. Its movements are not rapid, and it often lies motionless for hours in the warm sunshine. When captured, or slightly wounded, it has the habit of feigning itself dead, and by this artifice often escapes from the inexperienced hunter. The young, which at birth weigh only three or four grains, are placed in the pouch, where they remain growing very rapidly till four or five weeks old, when they begin to venture forth, but for a long time keep close to the mother, generally clinging to her by their tails.

The Texas Opossum, *D. californica*, Bennet, is found from Texas westward.

The Genus *Cheironectes* is characterized by palmated feet. It is represented by a small species in Brazil.

The Genus *Thylacinus* is distinguished from the true Opossums by two incisors less in each jaw, a non-prehensile tail, and the absence of a thumb on the hind feet. A species about the size of a wolf, but with shorter legs, is found in Australia. An extinct species has been found imbedded in the plaster quarries of Paris, in France.

Dasyuridæ, or Dasyurus Family. — This Family comprises those which have two incisors and four grinders less in each jaw than Opossums, a non-prehensile tail, and the thumb of the hind feet rudimentary or wanting. They vary in size from that of a mouse to that of a wolf.

PARAMELEIDÆ comprise burrowing marsupials.

PHALANGISTIDÆ, OR PHALANGER FAMILY.—This Family comprises those which have the two toes next the thumb united by a membrane as far as the last phalanx. Such are the true Phalangers of the Moluccas, which live upon trees, and, according to Cuvier, at the sight of man, suspend themselves by the tail, and, if gazed at steadily, at length fall to the ground; and the Flying Phalangers of Australia, which have the skin of the flanks extended between the legs, which enables them to suspend themselves in the air and make greater leaps. The species of this family vary in size from that of a mouse to that of a cat, or larger.

MACROPODIDÆ, OR KANGAROO FAMILY.—This Family comprises Marsupials which are specially characterized by the remarkable development of their hinder parts. The hind legs and the tail are long and powerful; the fore legs very short and weak, and little used in progression, which is accomplished mainly by leaping, for which their whole structure is most admirably fitted. They sit mainly upright upon their haunches, supported in part by the tail. They feed upon fruits and plants, are perfectly harmless, and easily tamed. Forty

Fig. 50.

Kangaroo.

species are known, varying from the size of a hare to that of a deer, or larger.

The Genus *Macropus* includes the Greater Kangaroo, *M. major*, Shaw, which is the largest animal of Australia. It is six feet high as it sits upright, of a grayish color, and makes leaps of enormous extent. The young, at birth, are only an inch long; they resort to the pouch even after they are old enough to graze, which they actually accomplish in that position while the mother herself is feeding.

PHASCOLOMYIDÆ, OR WOMBAT FAMILY.—This Family comprises animals having large, flat heads, short legs, and a body that appears as if crushed, and without a tail. They have two incisors in each jaw, similar to those of Rodents, and each of their grinders has two transverse ridges. They are sluggish, feed upon grass, and burrow in the ground. The Wombat is of the size of a badger, and both this and *Lipurus*, a closely related genus, live in Australia.

Fig. 51.

Wombat, *Ph. ursinus*, Cuv.

Fig. 52.

Skull of Wombat.

SUB-SECTION VII.

THE ORDER OF RODENTIA, OR GNAWERS.

The Order of Rodentia comprises all the gnawing Mammalia. They are readily distinguished by their teeth. In each jaw they have two chisel-shaped incisors, between which and the molars there is a wide space without teeth. The incisors are covered with enamel only in front, so that their posterior edges wear away faster than the anterior edges, thus always keeping these teeth sharp, however much they are used; and they

Fig. 53.

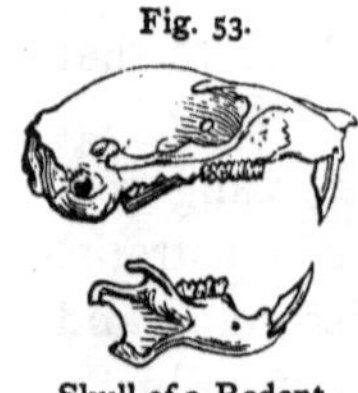

Skull of a Rodent.

grow at the base as fast as they wear away at the summit. The lower jaw is articulated with the skull in such a manner that the jaws have no horizontal motion, except backwards and forwards, as is requisite in the act of gnawing. The enamelled ridges of the molars are transverse, thus in opposition to the horizontal forward and backward motion of the jaw, and exactly adapted to the process of trituration. The form of Rodentia is generally such that the hind parts considerably exceed those of the front; they are thus adapted to leaping instead of walking. Their fore-arms have little or no power of rotation, and the bones of the fore-arm are in many cases united. Rodents have simple or but little divided stomachs, long intestines, brain without convolutions, and eyes directed laterally. The number of species is great, six hundred or more having been described. Most of them are small, the beaver, with one or two exceptions, being the largest. Rodents are found in all parts of the world, and are especially numerous in America. They comprise at least five families, — the Sciuridæ or Squirrel Family, Saccomyidæ or Gopher Family, Muridæ or Rat Family, Hystricidæ or Porcupine Family, and Leporidæ or Hare Family.

Sciuridæ, or Squirrel Family. — This Family comprises the Squirrels and their allies, which have the tibia and fibula distinct, and the molars $\frac{5-5}{4-4}$ or $\frac{4-4}{4-4}$. It includes three sub-families.

1. Sciurinæ, characterized by a distinct post-orbital process, and by molars, rooted, $\frac{5-5}{4-4}$.

The Genus *Sciurus* — True Squirrels — is characterized by compressed incisors, long ears, divided snout and upper lip, long tail, with the hairs arranged mainly on the sides, absence of cheek pouches, and inner lines of the upper molars parallel. Squirrels are lightly built, agile, live upon trees, and feed on fruit and nuts. There are about fifty American species, of which twelve or more belong to the United States.

The Southern Fox Squirrel, *S. vulpinus*, Gm., of the Southern States, from North Carolina to Texas, is twelve inches long to the root of the tail, which is fifteen inches; the color varies from gray above and white beneath, through all shades of rusty to pure uniform lustrous black; the ears and nose white. This is the largest North American Squirrel that has been described.

The Fox or Cat Squirrel, *S. cinereus*, Linn., of New Jersey to Virginia, and west to the Alleghanies, is about twelve inches long to the tail, which is fourteen inches; the body heavy, color varying from light gray above and white beneath, through all shades of pale rusty, to a grizzly above and black below; the ears and nose never white.

The Western Fox Squirrel, *S. ludovicianus*, Custis, of the Mississippi Valley, is about twelve inches long to the tail, which is about the same length; color grizzly rusty-gray above, and bright ferruginous beneath; the nose and ears never white.

The Gray and the Black Squirrel, *S. carolinensis*, Gm., of the United States east of the Missouri, is nine to eleven inches long to the tail, which is about an inch longer than the head and body; the color in the gray variety, grizzled light yellowish-gray above, pure white beneath. The Southern gray squirrel and the Northern gray squirrel are generally regarded as distinct species; but Baird considers them as varieties of one species, for which he retains the name given above. The Southern variety is smaller than the Northern, and, according to Audubon, has different habits. The gray squirrel occurs of every shade from gray to jet-black; and the black and dusky varieties have also been regarded by some as a species distinct from the gray. Gray squirrels are remarkable for their occasional extensive migrations. Assembling in immense numbers, they make their way across the country, swimming streams, and turning aside for no obstacle.

Fig. 54.

Gray Squirrel, Southern var., *S. carolinensis*, Gm.

The Texas Fox Squirrel, *S. limitis*, Baird, is somewhat smaller than *S. carolinensis.*

The California Gray Squirrel, *S. fossor*, Peale, is about the size of *S. vulpinus*, but more slender.

The Chestnut-Backed Squirrel, *S. castanonotus*, Baird, of the Rocky Mountains, is about the size of *S. cinereus.*

The Tuft-Eared Squirrel, *S. Aberti*, Woodh., of San Francisco Mountains, is about eleven inches long to the tail. This is considered the handsomest squirrel in America.

The Red Squirrel, or Chickaree, *S. hudsonius*, Pallas, of the United States east of the Missouri, and north to Hudson's Bay, is seven to eight inches long to the tail, which is about six inches; the color above and on the sides mixed black and grayish-rusty, with a broad band of bright ferruginous along the back and upper surface of the tail; beneath, dingy white. These squirrels are seen at all seasons of the year, and in all kinds of weather. In the Northern forests the deepest snows of winter are soon covered with their tracks, and penetrated by holes bored to find the cones of spruce, pine, and the nuts scattered beneath, or which they had hidden the previous autumn. They often sit for

Fig. 55.

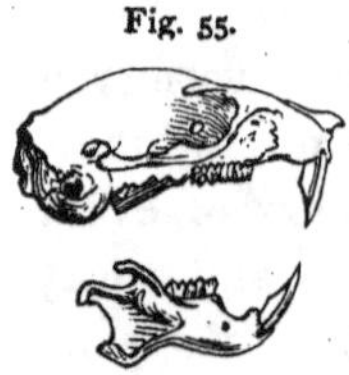

Skull of Red Squirrel.

hours upon a stump or limb of a tree close to the trunk, and, holding a cone or a nut in their fore paws, gnaw it briskly till they get all the food it contains. Disturbed while upon the ground, the Chickaree ascends the nearest tree, and, making for the outer branches, leaps from these to another tree, and, passing thus from tree to tree, is soon out of sight. Sometimes, however, when suddenly startled, it ascends a tree a short distance, and commences chattering with great fury, and leaping about as if in defiance of its intruder.

The Mountain Gray Squirrel, *S. Fremontii*, Aud. & Bach., of the Rocky Mountains, resembles the Red.

Richardson's Squirrel, *S. Richardsonii*, of Western North America, resembles the Red Squirrel, but is larger.

The Oregon Red Squirrel, *S. Douglassii*, Bach., also resembles *S. hudsonius*, but is larger.

The Genus *Pteromys* — Flying Squirrels — is characterized by a densely furred membrane extending laterally from the sides between the fore and hind feet, by means of which the animal is enabled to glide from one tree to another, supported as by a parachute. There are long, bony appendages to the feet, which support a part of this lateral membrane. Four North American species have been described.

The Common Flying Squirrel, *P. volucella*, Desm., of the United States east of the Missouri, is about five inches long to the tail, which is a little less in length than the head and body; the fur very soft and silky, the color light yellowish-brown above, and creamy white beneath.

Fig. 56.

Flying Squirrel, *P. volucella*, Desm.

The Northern Flying Squirrel, *P. hudsonius*, Fischer, of the Northern States, is larger than *P. volucella*.

The Rocky Mountain Flying Squirrel, *P. alpinus*, Rich., has the lateral membrane with the border straight, and the tail longer than the body exclusive of the head.

The Columbia River Flying Squirrel, *P. oregonensis*, Bach., is seven inches long to the tail, which is over six inches, the color yellowish brown above, and dull white beneath.

The Taguan, *P. petaurista*, Linn., of the Indian Archipelago, is nearly as large as a cat, the male a lively marroon above and red beneath; the female brown above.

Fig. 57.

Aye-Aye, *C. madagascarensis*, Cuv.

The Genus *Cheiromys* — Aye-Ayes — may be mentioned here, but its true place is doubtful. Some naturalists regard this curious genus as belonging to the Quadrumana, to which it seems related in the structure of its head, and in the opposable thumb of the hind feet. The teeth in general position are essentially those of a rodent, but the lower incisors are much compressed and extended from before backwards. Only one species of this monkey-like rodent is known. It inhabits Madagascar, is of the size of a hare, of a brown color, and burrows in the ground.

The Genus *Tamias* — Striped Squirrels — is characterized by ample cheek pouches, tail shorter than the body and not bushy, three to five dark dorsal stripes, and four permanent upper molars. This genus comprises only a few species, two of which are found in Europe and Asia, and the remaining four or five in North America.

The Chipping, Striped, or Ground Squirrel, or Chipmunk, *T. striatus*, Linn., from Montreal to Virginia and westward to the Missouri, is five to six inches long to the tail, which is four to four and a half inches; the general color above, finely grizzled yellowish-gray and brown, the back and sides with five longitudinal black stripes. The dark stripes are bordered by chestnut-brown, and the rump is pale chestnut. A variety is wholly black. The writer has one specimen of this color from New Hampshire. The Striped Squirrel makes its hole near the roots of a stump or tree, into which it carries its stores for winter, and where it stays, without once coming out, so long as the cold weather lasts. In autumn these squirrels may be constantly seen hurrying towards their holes, their cheek-pouches distended to the utmost capacity with nuts and grain.

Fig. 58.

Stiped Squirrel, *T. striatus*, Linn.

The Missouri Striped Squirrel, *T. quadrivittatus*, Rich., of Missouri and westward and southward, is four to five inches long to the tail, which is about as long as the body; the back with five dark stripes, their intervals forming four grayish-white lines; the sides of the body deep ferruginous, the under parts dingy grayish-white. There is a light stripe along the top of the head, with branches above and below the eyes.

The Gila Striped Squirrel, *T. dorsalis*, Baird, of New Mexico, is distinguished by its single distinct dorsal dark stripe.

Townsend's Striped Squirrel, *T. Townsendii*, Bach., of the Pacific coast, is the largest of this genus yet seen in North America.

The Genus *Spermophilus* — including Ground Squirrels, Spermophiles, and Gophers — is characterized by a squirrel-like body, variable ears, well-developed cheek-pouches, and absence of the thumb claw. They are all burrowing animals. This genus is represented in North America by at least fourteen species.

The California Ground Squirrel, *S. Beechyii*, Rich., is about the size of the cat squirrel, *S. cinereus*, the color above mixed black and light yellowish-brown; beneath, pale brownish-yellow. It is notorious for its depredations upon the farm products, and for its extensive excavations.

The Columbia Ground Squirrel, *S. Douglassii*, Rich., of Columbia River, is very similar to the preceding.

The Line-tailed Squirrel, *S. grammurus*, Bach., is found from the sources of the Arkansas to Sonora.

The Black Ground Squirrel, *S. Couchii*, Baird, is found in Northern Mexico.

Say's Striped Squirrel, *S. lateralis*, Rich., is found from the Rocky Mountains to the Cascade Range.

Harris's Spermophile, *S. Harrisii*, Aud. & Bach., is found in the Mohave Desert.

The Gray Gopher, *S. Franklini*, Rich., of Northern Illinois and Wisconsin and to the Saskatchewan, is about nine inches long to the tail, which is five and a half inches; the color above light yellowish-brown varied with black, the top and sides of head and neck hoary gray, and under parts whitish.

The Round-tailed Spermophile, *S. tereticaudus*, Baird, found in California, is five to six inches long to the tail, which is half an inch to an inch less.

The Striped Gopher, or Leopard Spermophile, *S. tridecem-lineatus*, Aud. & Bach., of Michigan to the plains of Missouri and southward, is about the size of the Red Squirrel, *S. hudsonius*, with the tail something more than

Fig. 59.
Leopard Spermophile,
S. tridecem-lineatus, Aud. & Bach.

half the length of the body, the color dark brown above, with light stripes and lines of light spots alternating with each other, there being six of the former and five of the latter. This is one of the most beautiful animals of the genus. Its burrows are quite deep and branching, and into these it at once disappears with a chirp whenever it is alarmed.

The Mexican Ground Squirrel, *S. mexicanus*, Wagner, of Texas and Mexico, resembles the Striped Gopher.

The Sonora Ground Squirrel, *S. spilosoma*, Bennett, is found from New Mexico to the Gulf of California.

Parry's Marmot, *S. Parryi*, Rich., found upon the shores and islands of the Arctic Seas, is about the size of the Fox Squirrel.

The Yellow Gopher, *S. Richardsonii*, F. Cuv., of Michigan to the Rocky Mountains, is rather larger than the Red Squirrel.

Townsend's Spermophile, *S. Townsendii*, Bach., of the Rocky Mountains to the north, is about the size of the Red Squirrel.

The Genus *Cynomys* — Prairie Dogs — is characterized by rudimentary cheek-pouches, short ears and tail, five distinct claws to all the feet, and very large molars.

The Prairie Dog, *C. ludovicianus*, Baird, of the Missouri region, and westward and southward, is about the size of the Fox Squirrel, but heavier, appearing much like a miniature woodchuck. Its color is reddish-brown above, and brownish-yellow below. These animals utter a sharp chirp, which is called barking; hence their name. They live in burrows, and large numbers are often found in the same locality, forming communities

Fig. 60.

Prairie Dog,
C. ludovicianus, Baird.

which the hunters call "dog-towns." Before each hole there is a little mound of earth upon which a Prairie Dog is almost always sitting, on the lookout for intruders, and, on the slightest alarm, dives into its hole, but soon cautiously appears again. It is well known that their holes are the home of the burrowing owls and rattlesnakes, with whom the dogs seem to live in perfect harmony; but it is more probable that they are intruders who are tolerated from necessity.

The Short-tailed Prairie Dog, *C. Gunnisonii*, Baird, is similar to *C. ludovicianus*.

The Genus *Arctomys* — Woodchucks — is characterized by a large, thick, depressed body, rudimentary cheek-pouches, rudimentary thumb armed with a small flat nail, and naked soles. They pass the winter in a torpid state.

The Woodchuck or Ground Hog, *A. monax*, Gm., of the Northern States and southward, is fifteen to eighteen inches long to the tail, which is about half as long as the head and body; color varying from black all over to grizzled above and bright chestnut-red beneath; the feet always black or dark brown, and the tail generally black. They often commit great havoc in fields of clover, upon which they like to feed.

The Yellow-footed Marmot, *A. flaviventer*, Bach., is found in Nebraska.

The Hoary Marmot, or Whistler, *A. pruinosus*, Gm., of Northwestern North America, is about the size of the common woodchuck.

The Alpine Marmot, *A. alpinus*, Linn., of the high mountains of Europe, immediately below the region of perpetual snow, is about the size of a hare; the color yellowish-gray. It is often tamed, and is very gentle and playful.

2. Myoxinæ have no post-orbital process, molars rooted, $\frac{4-4}{4-4}$, and no cœcum.

The Genus *Myoxus* — Dormice — is characterized by laterally compressed incisors, four grinders on each side of both jaws, the crowns of which are divided by enamelled lines. Dormice are pretty little squirrel-like animals, of the size of rats and mice, with soft fur, hairy and tufted tail, and lively eye. In the winter they become torpid. So far as known, they belong to Europe and other parts of the Old World.

3. Castorinæ have no post-orbital process, molars rootless, $\frac{4-4}{4-4}$, or in Aplodontia $\frac{5-5}{4-4}$.

The Genus *Castor* — Beavers — is characterized among all the Rodentia by the broad, horizontally flattened, and scaly tail. Beavers have five toes to each foot, the hind feet webbed, and the second hind toe has a double claw. With one exception, they are the largest of living Rodents, and are wholly aquatic in their habits; their feet and tail are admirably adapted for swimming, and their chief food is bark and aquatic plants. Their incisors are very sharp and powerful, enabling them to gnaw down trees of the hardest wood. Beavers prefer running water, in order that the wood which they cut may be carried to the spot where it is to be used. They keep the water at a given height by dams, which they build of trees and branches mixed with stones and mud, and build houses for winter with the same materials. Each house consists of two stories, and serves for two or three families. The upper story is above water and dry, for the shelter of the animals themselves; the lower is beneath the water, and contains their stores of bark and roots. The only opening to the hut is beneath the water. They have burrows

Fig. 61.

American Beaver, *C. canadensis*, Kuhl.

in the banks, whither they retire when their houses are attacked. The general color of the beaver is a uniform reddish-brown, and the fur is of the best quality, and was formerly very valuable. Only two species are known.

The American Beaver, *C. canadensis*, Kuhl., found all over North America; and the European Beaver, *C. fiber*, Linn., of the Old World. These are much alike.

The Genus *Castoroides* comprises the Fossil Beaver, *C. ohioensis*, Foster, of New York and westward, known only from its skull, which shows that this beaver-like Rodent was six times the bulk of our living species.

The Genus *Myopotamus* comprises the Couia, *M. coipus*, Cuv., of South America, which resembles the beaver in size and habits, but has the tail round and long. The fur is yellowish-gray, and is known among the hatters under the name of *nutria*.

The Genus *Aplodontia* comprises the Sewellel, *A. leporina*, Rich., of Puget's Sound, which is about the size and general appearance of the muskrat, but with the tail very short and much depressed.

SACCOMYIDÆ, OR POUCHED GOPHER FAMILY. — This Family comprises Rodents which have large and distinct external cheek-pouches, pelage composed of stiff hairs with no under fur, molars $\frac{4-4}{4-4}$, and the upper lip not cleft.

1. Geomyinæ comprise those which have the body thick-set and clumsy, skull massive, incisors very large and thick, limbs very short, fore claws five in number and enormously developed. They are burrowing and nocturnal in their habits. The Genera *Geomys* and *Thomomys* comprise the Pouched Gophers of North America, of which there are more than a dozen species.

Fig. 62.

Pouched Gopher, *G. bursarius*, Rich.

The Pouched Gopher, Pocket Gopher, or Pouched Rat, *G. bursarius*, Rich., of the Northwestern States, is eight to ten inches long to the tail, which is one to two inches; the color reddish brown above, paler beneath, with a plumbeous tinge along the vertebral region. Its cheek-pouches are very large, extending as far back as the shoulders, and lined with short hair; and, as in other members of this family, are used mainly or wholly to convey food into the burrows, to be stored up or eaten at leisure.

The Salamander, *G. pinetis*, Raf., of Florida to Alabama, is a gopher readily distinguished by the single deep groove of the upper incisors, dividing the surface into two unequal portions.

The Pecos Gopher, *G. Clarkii*, Baird, is found in Texas.

The Chestnut-faced Gopher, *G. castanops*, Lec., is found in the Upper Arkansas region.

The California Gopher, *Thomomys bulbivorus*, Baird, is about the size of *G. bursarius*, cheek-pouches completely furred inside and white to their very margin, which is dark brown. The color above is reddish-chestnut-brown, finely lined by dusky tips to the hairs; beneath paler; the chin dusky, with the extremity white.

The Broad-headed Gopher, *T. laticeps*, Baird, and the Oregon Gopher, *T. Douglassii*, Giebel, are other species respectively from California and Oregon.

2. Saccomyinæ comprise those which have the body graceful and slender and motions agile, the skull delicate, muzzle long and tapering, tail very long, hind feet long, and the fore claws moderate, but exceeding the hind ones.

The Genus *Dipodomys* — Kangaroo Rats — is characterized by a broad, depressed head, large, rounded ears, acute snout, ample cheek-pouches opening externally, very long hind legs, and a long tail with a brush-like tip.

The Kangaroo Rat, *D. Ordii*, Woodh., of the eastern slope of the Rocky Mountains, is about five inches long to the tail, which is about as long as the body, with an erect crest of long hairs towards the end. The color above is yellowish-brown; below, white.

The Jumping Rat, *D. Phillipii*, Gray, and the *D. agilis*, Gambel, belong to Western North America.

The Genus *Perognathus* — comprising the Tuft-tailed Mouse, *P. pencillatus*, Woodh., and five or six closely allied species, all of Western North America — differs from *Dipodomys* in having the ears small, tail of moderate length, under surface of the soles naked or sparsely hairy, the molars rooted and the transverse ridges tuberculated. The Tuft-tailed Mouse is three to four inches long to the tail, which is four to five inches, with a pencilled crest at its extremity. The color above is yellowish-brown; the under parts, hind feet, and fore legs, white.

Muridæ, or Rat Family. — This Family comprises Rats, Mice, and their immediate allies, in all more than three hundred species, some of which are found in every country on the globe. None are of large size, the musk-rat being the largest, and some are the smallest quadrupeds known, except the shrews. The dental formula is, incisors $\frac{2}{2}$, molars usually $\frac{3-3}{3-3}$, rooted or rootless. Of this large family, the most extensive of the whole order of Rodents, there are three, at least, well-defined sub-families.

1. Dipodinæ are characterized by unequal, generally rooted molars, and greatly elongated hind legs.

The Genus *Dipus* — Jerboas — is characterized by a large head, long, densely hairy, and tufted tail, hind legs which are exceedingly long in comparison with the forward ones, and by the metatarsus of the three middle toes which is formed of a single bone, and by the three upper molars on each side. The Jerboas move about on their hind feet, making great leaps. The ancients called them Biped Rats. They belong to Africa and Asia.

The Genus *Jaculus* has the hind legs and tail very long, the latter thinly haired, the hind feet five-toed, and the upper molars four on each side.

The Jumping Mouse, or American Jerboa, *J. hudsonius*, Baird, of Labrador and southward and westward to the Pacific, is about three inches long to the tail, which is four to six inches; the color above light yellowish-brown lined finely with black, beneath white, and the sides yellowish-rusty, sharply defined against the colors of the back and belly. When startled it progresses by very long and rapid leaps, and there is probably no other mammal of its size that can make its way over the ground with so great rapidity, or so quickly escape from its pursuers.

Fig. 63.

Jumping Mouse, *J. hudsonius*, Baird.

The Genus *Gerbillus*, — Gerbils, — of the warm parts of the Old World, belongs near this group, if not in it.

2. Murinæ have compressed incisors, molars $\frac{3-3}{3-3}$ or $\frac{2-2}{2-2}$, and rooted, the largest anterior, and the smallest posterior. They comprise Mures and Sigmodontes.

a. Mures, or the Old-World Rats, are characterized by very large and broad molars, and those in the upper jaw have three tubercles in each transverse series.

The Genus *Mus* — Rats — is characterized by upper divided lips, acute snout, whiskers in five series, large and nearly naked ears, and long tail, the scaly whorls of which are very distinct. Over fifty species of this genus are known, four of which have taken up their abode in the United States.

The Norway or Brown Rat, *M. decumanus*, Pallas, is eight to ten inches long to the tail, which is somewhat shorter than the head and body; the color above grayish-brown mixed with rusty, beneath ashy white. This rat is known all over the world, and is very destructive in its habits. It belonged originally to Central Asia; crossing the Volga in large troops in 1737, it stocked Russia, and subsequently overrun all Europe. In 1775 it found its way to North America. It is often called Wharf Rat.

The Black Rat, *M. rattus*, Linn., is readily distinguished from the Brown Rat; its color being sooty-black above, passing into dark plumbeous or paler beneath. Its original locality is unknown. It has been the house-rat of Europe from earliest times, and was introduced into America in 1544. This species is rare, or wholly wanting, in localities where it was formerly very abundant; for it always disappears before its more formidable rival, the Brown Rat. Both these species devour everything edible that they can secure, often capturing living prey.

The Roof Rat, or White-bellied Rat, *M. tectorum*, Savi, of the Southern States, is smaller than the Brown Rat. It is originally from Egypt, where it frequented the thatched roofs of the houses; hence its name.

The House Mouse, *M. musculus*, Linn., originally from Europe and Asia, but now found all over the world, is grayish-brown, finely lined with darker, passing into ashy plumbeous, with a reddish tinge on the belly; the feet are ashy brown.

The Genus *Cricetus*, comprising the Hamsters of Northern Europe and Asia, differs from rats in having cheek-pouches and a hairy tail.

b. Sigmodontes, or New-World Rats and Mice, are characterized by narrower molars than in Mures, and those of the upper jaw have two tubercles in each transverse series.

The Genus *Reithrodon* — Harvest Mice — is characterized by short, hairy ears and tail, and upper incisors with a longitudinal channel along the anterior face. There are four species in the United States.

The Harvest Mouse, *R. humilis*, Baird, of South Carolina and westward, is less than two inches and a half long to the tail, which is a little shorter than the head and body; the color above grayish-brown, grayish-white beneath. The region about the mouth, and the chin, and feet are white.

The Long-tailed Harvest Mouse, *R. longicauda*, Baird, of California, is very similar to *R. humilis*.

The Genus *Hesperomys* — White-footed Mice — is characterized by a murine appearance, variable and scantily-haired tail, molars diminishing from first to last, and elongated, the sides indented and the crowns with a single longitudinal furrow. Most of the species have white feet, and the tail whitish with a darker stripe above. Fifteen or more species inhabit the United States.

The White-footed, or Deer Mouse, *H. leucopus*, Wagner, of Labrador to Virginia and westward to the Mississippi, is three to four inches long to the tail, which is nearly as long as the head and body; the color yellowish-brown, with generally a dorsal wash of darker, and the under surface of the tail pure white.

Fig. 64.

White-footed Mouse, *H. leucopus*, Wagner.

The Red Mouse, *H. Nuttalli*, Baird, of Pennsylvania to Georgia and westward, is of the size of *H. leucopus;* the general color bright yellowish-cinnamon, the feet and under portion of the tail white.

The Cotton Mouse, *H. gossypinus*, Lec., is somewhat larger than *H. leucopus*, rusty yellowish-brown above, ashy-white beneath.

The Gray-bellied Mouse, *H. cognatus*, Lec., is closely related to *H. leucopus*, but smaller, the color yellowish-brown, the under parts and feet dingy white.

The Long-tailed Mouse, *H. Boylii*, Baird, of Western North America, is larger than *H. leucopus*, body stout, the color above mixed glossy-brown and pale yellowish-brown; the lower parts white.

The Hamster Mouse, *H. myoides*, Baird, of Canada and New York, is of the size of *H. leucopus* or larger, with moderate cheek-pouches, the color above cinnamon-brown lined with dusky, the under parts and feet pure white.

The Prairie Mouse, *H. michiganensis*, Wagner, of Michigan and westward, is about three inches long to the tail, which is half an inch or an inch shorter than the head and body; the color above blackish-brown, the under parts snowy white.

The Great-eared Mouse, *H. californicus*, Baird, is one of the largest of the genus, sooty-brown above, white below.

The Desert Mouse, *H. eremicus*, Baird, of California, is readily distinguished by its naked soles.

The Missouri Mouse, *H. leucogaster*, Baird, of the Upper Missouri, is grayish-brown above, white below.

The Rice-field Mouse, *H. palustris*, Wagner, of South Carolina and Georgia, is of a mixed black and pale brownish-ash color above, ashy white beneath.

The Genus *Neotoma* — Wood Rats — is characterized by a rat-like appearance, large and nearly naked ears, long and more or less densely hairy tail, and hairy heels.

The Wood Rat, *N. floridana*, Say & Ord., of Florida and northward and westward, is of the size of the Black Rat, grayish-brown mixed with rusty above, and the under parts and feet white.

The Bush Rat, *N. mexicana*, Baird, of Texas to California; the Black Wood Rat, *N. micropus*, Baird, of Texas and Mexico; the Brown-footed Rat, *N. fuscipes*, Cooper, of California; the Hairy-tailed Rat, *N. occidentalis*, Cooper, of Oregon; and the Rocky Mountain Rat, *N. cinerea*, Baird, are additional species of this genus. *N. magister*, Baird, is found fossil in the caves of Pennsylvania.

The Genus *Sigmodon*—Cotton Rats—is characterized by the shape of the enamel on the two last molars in the lower jaw, which is in the form of the Greek letter sigma.

The Cotton Rat, *S. hispidus*, Say & Ord., of the Southern States, is about half as large as the Norway Rat; the color above reddish-brown lined with dark brown.

The Texas Cotton Rat, *S. Berlandieri*, Baird, is of a lighter color, and with a larger tail.

3. Arvicolinæ are characterized by incisors as broad as deep, molars $\frac{3-3}{3-3}$, rootless, ears short and hidden, muzzle broad and rounded, tail very short and mostly clothed thickly with hair, and the whiskers as in Murinæ.

The Genus *Arvicola*—Field Mice—is characterized by small size, soles naked anteriorly, tail rather short, cylindrical, and hairy. The posterior upper molar is composed of five or six prisms, and the posterior lower one of three. This genus is represented in the United States by more than twenty species, about half of which belong to the western portions.

The Red-backed Mouse, *A. Gapperi*, Vigors, of the Northern States, is about the size of the common house mouse, the back with a broad stripe of bright rufous brown, sides yellowish-gray mixed with brown, and the under parts yellowish-white.

The Meadow Mouse, *A. riparia*, Ord., of the Northern and Middle States, is four and a half inches long to the tail, which is about two inches, the feet large, the color dark brown above, ashy-plumbeous below.

The Gray Mouse, *A. Breweri*, Baird, of the Eastern United States, is about four and a half inches long to the tail, feet very broad and stout, fur coarse, and the color grayish yellow-brown above, ashy-white beneath.

The California Arvicola, *A. californica*, Peale, is about the size of *A. riparia*, the fur very long and soft, color lustrous light yellowish-brown above, grayish-white beneath.

The Prairie Meadow Mouse, *A. austera*, Lec., of the Mississippi valley, is about the size of *A. riparia*, pale cinnamon-rufous, variegated with black, below brighter.

The Upland Mouse, *A. pinetorum*, Lec., of the Atlantic States and westward, is three and a half inches long to the tail, dark chestnut-brown above, hoary plumbeous below.

The Genus *Myodes* — Lemmings — comprises little mouse-like animals, the largest hardly as large as a rat, with a broad skull, large fore feet, long claws fitted for digging, and very short tail. Lemmings inhabit the northern regions of both continents, and are celebrated for their occasional extensive migrations. Norway, Sweden, and Lapland are sometimes overrun with these animals. Coming, in countless numbers, no one knows whence, and going no one knows whither, they sweep onward in a straight line, swimming rivers and lakes, nor turn aside for scarcely any obstacle; and they destroy everything edible in their course.

The Genus *Fiber* comprises the Muskrat, *F. zibethicus*, Cuv., abundant throughout North America, and twelve to fourteen inches long to the tail, which is ten to eleven inches; the body thick and clumsy, tail much compressed, and in its natural position sickle-shaped, the convex portion being above, and the hind feet partly webbed. The general color is dark brown above, rusty brown below. Muskrats feed upon mussels and roots of grasses and aquatic plants, and build winter houses of mud, sticks, and grass, and having an entrance under the water which leads to a dry apartment above. In summer they dig burrows of great extent along the banks, in which they bring forth their young. They are good swimmers, moving with ease and considerable rapidity. At early evening, or on a moonlight night, they may be seen swimming from bank to bank, or log to log, and often sporting together in the most playful manner.

4. Spalacinæ, comprising the Blind Rat-Moles of the Old World, may perhaps be considered a fourth subfamily of Muridæ.

HYSTRICIDÆ, OR PORCUPINE FAMILY. — This Family comprises a large number of Rodents, which at first view seem very different from one another, but which are united by important characters. The molars are $\frac{4-4}{4-4}$, and the terminal portion of the muzzle is clothed with small hairs. They are mainly American, and chiefly confined to the southern portion of the continent.

The Genus *Erethizon* — Porcupines — is characterized by a flat cranium, short muzzle, medium-sized tail, and spines which are short and half hidden in the hair.

The White-haired or Canada Porcupine, *E. dorsatus*, F. Cuv., of Northern United States and Canada, is about two feet long to the tail, which is seven inches. The tail and upper parts are covered with a mass of white spines with dusky and bearded tips. The general color of the fur is dark brown, among which are long hairs with white tips. This animal is extremely sluggish, making but little effort to escape from man or beast; but its formidable armor is an effectual defence. It readily climbs trees, and feeds upon bark, leaves, and tender ears of Indian corn. It lives in hollow trees and in holes, among the rocks.

The Yellow-haired Porcupine, *E. epixanthus*, Brandt, of the Upper Missouri region and the whole Pacific coast of North America, is nearly the size of the Beaver; general color dark brown, and the long hairs of the body tipped with greenish-yellow.

Fig. 65.

Porcupine, *E. dorsatus*, F. Cuv.

The Genus *Hystrix* belongs to the Old World.

The Crested Porcupine, *H. cristata*, Linn., of Southern Europe and Barbary, and Southwestern Asia, is of a grizzly dusky black, and the upper part of the head and neck with a crest of long, lighter-colored hairs. Its body is armed with striated spines, the longest of which are a foot in length, and in the middle about the size of a large goose-quill. These are banded with black and white, and terminated by very sharp points. The tail is short, and furnished with hollow, truncated tubes attached to slender pedicles, which make a noise when shaken. When the animal is at rest, the quills lie flat upon the body, the points directed backwards; when attacked or excited, they are raised, and thus constitute formidable weapons of defence.

The popular notion that Porcupines have the power to throw their quills at an enemy is entirely erroneous.

The Genus *Dasyprocta* — Agoutis — has four toes before and three behind. The Agoutis belong to South

America and adjacent islands, where they seem to take the place of hares and rabbits, which in general appearance they much resemble. *Cœlogenys*, comprising the Pacas, is a closely related genus, also of South America.

The Genus *Dolichotis* comprises the Patagonian Cavy, weighing about twenty pounds, and resembling a hare.

The Genus *Chinchilla* — Chinchillas — and closely allied genera inhabit the mountain regions of Chili and Peru. Chinchillas are scarcely larger than rats, with a short tail, and are covered with ashen-colored fur of the finest and softest quality, which is extensively used.

The Genus *Cavia* — Cobayes or Guinea Pigs — comprises animals which in general appearance are miniatures of the next genus, except in their separate toes.

The Common Guinea Pig, *C. cobaia*, Pallas, is indigenous to South America, but is now found also in a domestic state in all parts of the world. It is in no way related to the pig. The head and nose resemble those of a hare, and the eyes are large and round. It is said that rats will not stay in houses where these animals are.

The Genus *Hydrochœrus* is characterized by large size, four toes before and three behind, and all armed with large nails and united by membranes.

The Capybara, *H. capybara*, Cuv., of South America, is the largest known Rodent, being three feet long, and exceedingly bulky. Its muzzle is thick, limbs short, hair coarse, and tail almost wholly wanting, and the general color yellowish-brown. It is aquatic in its habits, is hunted as game, and its flesh is quite good for food.

Leporidæ, or Hare Family. — This Family is distinguished from all other Rodents, not only by many external characters, but especially by the upper incisors, which are double, each principal incisor having a smaller one behind it. The dental formula is, incisors $\frac{4}{2}$, molars $\frac{6-6}{5-5}$ or $\frac{5-5}{5-5}$. All the incisors are less deeply implanted in the jaws than in other Rodents, and are always white; and the molars are always rootless. The fore feet are five-toed, and the hind ones four-toed, and all well developed; and at the lower part of the shank the tibia and fibula are always united. The members of this family have the feet clothed with hair beneath, and the inner surface of the cheeks lined with hairs. The tail is short and bushy, and is carried erect; or it exists only in a rudimentary condition. They feed upon bark, tender twigs, and leaves. Some live in burrows, but most have merely a *form*, or nest on the ground, where they generally sit during the day. Hares have a curious habit of stamping with their hind feet when they are alarmed or excited. They are very timid. This family is represented in nearly all parts of the world.

The Genus *Lepus* is characterized by molars $\frac{6-6}{5-5}$, large ears, short and bushy tail, hind legs powerful and much longer than the fore legs. About twenty Old-World species have been described, and rather less than that number of North American. Although the name is applied to several of our species, it is probable that there is no genuine North American *Rabbit*, of which the European Rabbit, *L. cuniculus*, may be taken as a type; but our species of this genus are Hares, which are mainly solitary in their habits, and do not construct burrows.

The Polar Hare, *L. glacialis*, Leach, of Northern and Arctic America, is twenty to twenty-five inches from the nose to the tail, and is one of the largest of the hares.

The Northern Hare or White Rabbit, *L. americanus*, Erxl., of Virginia to Labrador and westward, is nineteen to twenty inches long; the color cinnamon-brown in summer, and in winter white, but showing yellowish-brown between the tips of the long hairs. This species lives in the thickest woods, rarely or never goes into holes when pursued, but depends for its safety upon its fleetness and its windings and doublings among the thick cover. It follows the same paths year after year, both in winter and summer.

The Red Hare or Washington Hare, *L. Washingtonii*, Baird, of Washington Territory, is smaller than the preceding, and in summer of a rich cinnamon-red color.

The Prairie Hare, *L. campestris*, Bach., of the Upper Missouri region and northward and westward, is larger than the White Rabbit, the tail as long as the head, and the color in winter white, with a yellowish tinge, and in summer brownish-gray.

The Mule Rabbit, *L. callotis*, Wagler, is found in the southwestern parts of North America.

The California Hare, *L. californicus*, Gray, is twenty to twenty-five inches long, the general color above is mixed black and light cinnamon-red; the under parts cinnamon.

The Gray Rabbit, *L. sylvaticus*, Bach., common throughout a large part of the United States, is fifteen to sixteen inches in length, the general color yellowish-brown with a tinge of reddish, the lower parts pure white. It does not turn white in winter. When first started, the Gray Rabbit runs with great swiftness, but soon stops to listen. It is well known to hunters that they can stop it, when first started, by whistling. If pursued, and if the woods be open, it enters the first hole it can find. It often falls a prey to the weasel, as well as to other larger enemies. Its flesh is excellent food.

The Sage Hare, *L. artemisia*, Bach., found west of the Missouri, is smaller than the Gray Rabbit.

Bachman's Hare, *L Bachmani*, Waterh., of the Lower Rio Grande region, is also smaller than the Gray Rabbit.

Audubon's Hare, *L. Audubonii*, Baird, of California, is a little smaller than *L. sylvaticus*, the ears are longer than the head, and the tail rather long. The color above is mixed yellowish-brown and black, beneath pure white.

Trowbridge's Hare, *L. Trowbridgii,* Baird, of California, is smaller than Audubon's Hare, and the tail very short.

The Water Rabbit, *L. aquaticus,* Bach., of the Lower Mississippi region, is larger than *L. sylvaticus,* is common in wet grounds, often takes to the water when pursued, and swims and dives with facility.

The Marsh Rabbit, *L. palustris,* Bach., of South Carolina to Florida, is about the size of *L. sylvaticus,* the head and incisors disproportionately large, tail very short.

All the domestic varieties of Rabbits are supposed to have sprung from the European Rabbit, *L. cuniculus,* Linn., which lives in troops, and constructs burrows.

The Genus *Lagomys* — Pikas — is characterized by the molars, which are $\frac{5-5}{5-5}$, the short and rounded ears, short hind legs, and the absence of a visible tail. Its members are confined to the Northern hemisphere, and mostly to elevated regions. They are all small, the largest not exceeding in size the Guinea Pig. In Siberia they are called Pikas. Only one species is found in North America.

The Little Chief Hare, *L. princeps,* Rich., of the South Pass of the Rocky Mountains and northward, is about eight inches in length, and appears in color like a young rabbit, and utters a low bleat.

SUB-SECTION VIII.

THE ORDER OF EDENTATA, OR EDENTATES.

THE Order of Edentata comprises all Mammals which are destitute of incisor teeth, and some members of this order have no teeth. Wagner recognizes three families, — Bradypoda or Sloth Family, Effodientia or Armadillo Family, and Biclaviculata or Monotremata.

BRADYPODA, OR SLOTH FAMILY. — This Family comprises animals which have canine and molar teeth, anterior limbs very long, much exceeding the posterior, mammæ pectoral, tail wanting or very short, and the hair long and coarse. The fingers are united by the skin, and only marked by enormous, compressed, crooked nails, which when at rest are always bent towards the palm of the hand or the sole of the foot. By their whole structure, these animals are fitted to pass their lifetime on trees, and it is said they never remove from a tree until they have stripped it of its leaves. With their long arms and long claws, they cling firmly around the branches, and it is an interesting fact that they almost always keep on the under side of the branch. In this position they move and repose in perfect security. On the ground they move awkwardly and with difficulty. Observing this, and not knowing that the sloths are strictly arboreal, some of the earlier zoölogists regarded their structure as unfortunate. Two species, one of the size of the domestic cat and another larger, inhabit the hot portions of South America.

The Megatherium, having a skeleton eighteen feet long and eight feet high, only found fossil, is allied to the sloths. The bones of this animal are of colossal dimensions, the femur being three times as thick as that of the elephant. The Megalonyx and Mylodon are also huge extinct sloth-like animals, whose remains, like those of the Megatherium, are found in the superficial deposits of South America, and also to some extent in those of the United States, especially in South Carolina and Georgia.

Effodientia, or Armadillo Family. — This Family comprises Armadillos, Ant-eaters, and Pangolins, and is characterized by a long, pointed muzzle.

Armadillos or Tatous are at once distinguished from all other Mammals by their bony or horny armor. This armor is not a consolidated framework, but is composed of several parts, which are so arranged as to allow freedom in the bending of the body. One large shield covers the head, another the shoulders, and another the rump, and between the two last there are several parallel movable bands of the same material. The tail in some cases is covered with successive rings, and in others, as the legs, with mere horny tubercles. All this armor is attached to the skin of the body; and it is made up of numerous many-sided plates placed together as in inlaid work. The Armadillos have a pointed muzzle, slightly extensible tongue, and powerful claws. They inhabit the warm and hot parts of America, dig burrows, and live upon vegetables, insects, and worms. The Genus *Dasypus*, as limited by Linnæus, included all the species, but they are now distributed among several genera.

Fig. 66.

Nine-banded Armadillo, *D. novem-cinctus*, Linn.

The Nine-banded Armadillo, *D. novem-cinctus*, Linn., of Texas to Paraguay, is eighteen inches long to the tail, which is about eight inches, and the body has nine bands between the shield over the shoulders and that over the rump. Other species have respectively three, six, seven, and twelve intermediate bands.

The Giant Armadillo, *D. gigas*, Cuv., is about three feet long without the tail, and has twelve or thirteen intermediate bands.

The Genus *Chlamyphorus* includes *C. truncatus*, Harlan, of Chili, which is six inches long, and has the back only covered with a suit of transverse plates, and these are attached to the body only along the spine. The body is truncated behind.

The Glyptodon is a fossil Armadillo, found in South America, whose shield is compared to a huge cask, being five feet long, and the total length of the animal nine feet.

The Genus *Orycteropus* comprises the Earth-Pig of South Africa, celebrated for its unique teeth.

The Genus *Myrmecophaga* — Ant-eaters — is characterized by a long muzzle, toothless mouth, filiform tongue capable of great extension, and used to penetrate ant-hills and nests of termites, whence the insects are withdrawn, being entangled in the viscid saliva which covers it. The body is covered with much hair, and the claws of the fore feet are strong and trenchant, and suited to tearing open ant-nests. These animals inhabit the warm and hot parts of South America.

The Giant Ant-eater, *M. jubata*, Buff., is more than four feet in length, grayish-brown with an oblique black band edged with white upon each shoulder. Its tongue can be elongated more than two feet. Other species are much smaller than this, and one is no larger than a rat.

The Genus *Manis* — Pangolins or Scaly Ant-eaters — of the Eastern hemisphere, differs from the last genus in having the body, limbs, and tail clothed with large trenchant scales arranged like tiles, which they elevate when they roll themselves into a ball, as they do when they would ward off the attacks of an enemy.

The Short-tailed Pangolin, *M. pendactyla*, Linn., of the East Indies, is three or four feet in length.

Monotremata, or Ornithorhynchus Family. — This Family comprises animals which vary widely from all other Mammals, having their organic structure in many respects much like that of Birds. They have double clavicles, and well-developed marsupial bones, — Waterhouse places them at the end of the Marsupialia; the males have a peculiar spur on the hind feet, besides the ordinary nails, and they have no external conch to their ears, and their eyes are very small. They belong to Australia.

The Genus *Echidna* comprises those which have a long, slender muzzle and extensible tongue, like the Ant-eaters, and which are covered with spines.

The Genus *Ornithorhynchus*, or *Platypus*, is characterized by an elongated, enlarged, and flat muzzle, presenting the closest external resemblance to the bill of a duck, and the more so as its edges are similarly furnished with small transverse laminæ. These animals have no teeth except at the bottom of the mouth, and these are without roots, with flat crowns, and composed of little vertical tubes. There is a membrane to the fore feet which not only unites the toes, but extends far beyond the nails; in the hind feet the membrane terminates at the root of the nails, and the tail is flat. The whole body is covered with short, brown fur. These animals live in ponds and quiet streams, and dig burrows in the banks, in which they rear their young. Only one or two species are known. They are less than two feet in length.

Fig. 67.

Ornithorhynchus, or Platypus.

SUB-SECTION IX.

THE ORDER OF PACHYDERMATA, OR PACHYDERMS.

THE Order of Pachydermata comprises, according to Cuvier, three families, — Proboscidiana, Pachydermata Ordinaria, and Solipedes; or, for each of these substituting family names drawn respectively from a prominent genus, — or two genera in the second case, — we may consider the families as Elephantidæ, or Elephant Family, Rhinoceridæ or Rhinoceros Family, Suidæ or Swine Family, and Equidæ or Horse Family. They are all herbivorous

ELEPHANTIDÆ, OR ELEPHANT FAMILY. — This Family comprises animals of colossal size, — the largest and the most powerful of all the land animals, — with the nose extended into a very long prehensile snout, upper incisors developed into enormous tusks, head short and expanded above by large sinuses, neck and body short and thick, limbs long, without angles or bends, and the toes five and united to the hoofs. Their gigantic proportions, their peculiar organization, and their intelligence and sagacity, combine to make them objects of great interest to the common observer, as well as to the naturalist. One of the remarkable features of the elephant is the proboscis or trunk. This long and cylindrical organ is composed of several thousand muscles variously interlaced, is extremely flexible, and endowed with the most exquisite sensibility, and is terminated by an appendage which serves as a sort of finger. This trunk, agile and powerful, is at the same time the organ of smell, of touch, of prehension, and of defence. With it its possessor seizes everything he wishes to convey to his mouth, drink as well as food; thus obviating the necessity of a long neck, which would be inconsistent with the enormous head and heavy tusks, the latter weighing sixty to one hundred pounds each. Elephants at the present day are confined to the warm regions of the Eastern hemisphere. They are seven to ten feet high, and ten to fifteen feet in length, and covered with thick, nearly naked skin. One distinctive characteristic of these animals is found in the grinders, the crown of which is deeply divided into transverse vertical plates, each consisting of dentine coated by enamel, and this by a bone-like substance which fills the spaces between the plates and cements them together. The grinders succeed each other from behind forward, and there is never more than one, or two partially, on each side of both jaws at the same time; for the series is in constant process of shedding and replacement.

The total number of grinders which follow one another on each side of both jaws is seven, or at least six.

The Genus *Elephas* comprises the Elephants proper.

The Asiatic Elephant, *E. indicus*, Cuv., of India, is specially characterized by its oblong head, concave forehead, and the undulating sections of the laminæ which are seen on the crown of the grinders. This species has smaller ears than the next, and four nails to each hind foot. It has been used for a beast of burden from the earliest times.

The African Elephant, *E. africanus*, Cuv., of Southern Africa, is distinguished by its round head, convex forehead, large ears, and the lozenge-shaped figures on the crown of the grinders. Both species are hunted for their tusks, which furnish the world with ivory.

In both hemispheres the superficial deposits abound with skeletons and parts of skeletons of elephants which are now extinct. An elephant, covered with long, thick hair, and wholly unlike anything now living, was found encased in ice on the coast of Siberia. It was in such a state of preservation that dogs fed upon the flesh, although it is probable that it had been there thousands of years.

The Genus *Mastodon* comprises extinct Pachyderms, whose remains abound in the superficial accumulations of America, as well as in those of the Old World. In general appearance the Mastodon was much like the elephant, but differed from the latter in the grinders, the crowns of which are studded with large conical points. A skeleton of Mastodon dug up at Newburgh, New York, is seventeen feet long to the tail, which is six feet, and the tusks are nearly eleven feet in length, the whole weighing two thousand pounds. This splendid specimen is in the museum of the late Dr. Warren of Boston.

RHINOCERIDÆ, OR RHINOCEROS FAMILY. — This Family comprises very thick and naked skinned Pachyderms, which are distinguished from Elephantidæ by the absence of a proboscis, although the nose is much developed, by the existence of small canines instead of enormous tusks, and by incisors in both jaws. The feet are three- or four-toed, hoofs of unequal size, limbs short, body, neck, and head more or less elongated.

The Genus *Rhinoceros* comprises the largest of all land animals except the elephant. They are huge, bulky animals on short, stout legs, supported by broad three-toed feet, and the whole body is covered with an exceedingly tough hide, which appears in several species in large plaits or folds. From the upper surface of the muzzle, where the bones are very thick and strong and somewhat arched, there rises a horn composed of a solid mass of horny fibres. These animals inhabit Africa, Asia, and the Asiatic Archipelago. Seven species have been described. They are stupid and ferocious, frequent wet places, and feed upon herbs and tender branches.

The Indian Rhinoceros, *R. indicus*, Cuv., is about five feet high, nine feet long, and the largest individuals weigh six thousand pounds. Its horn is two to three feet in length.

The Genus *Hyrax* comprises the Damans of Africa and Asia, of the size of a rabbit, and which Cuvier calls rhinoceroses in miniature without the horn.

The Genus *Hippopotamus* is represented by only one species, *H. amphibius*, Linn., which inhabits the rivers of Africa. The body is massive, legs short, four hoofed toes to each foot, head enormous and terminated by a large inflated muzzle, eyes and ears small, and tail short. It feeds upon aquatic plants.

The Genus *Tapirus* is characterized by the nose, which resembles a small, fleshy proboscis, and by four toes to the fore and three to the hind feet. Two species are known.

Fig. 68.

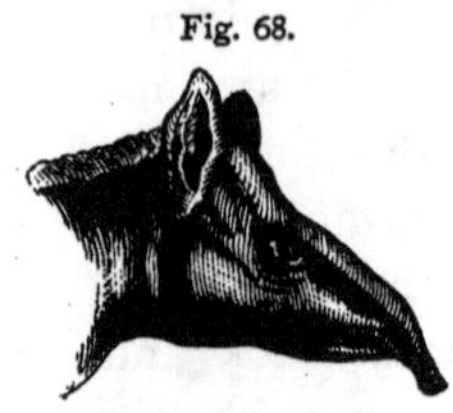

Head of the Tapir.

The American Tapir, *T. americanus*, Linn., of South America, is about the size of the ass, skin brown and nearly naked, and the neck fleshy, forming a sort of crest on the nape. The Indian Tapir, *T. indicus*, Fark., of Sumatra and adjacent regions, is larger than the American species. The remains of extinct Tapirs are found in Europe, one of which must have nearly equalled the elephant in size.

The Genus *Paleotherium* and allied genera comprise extinct, more or less tapir-like Pachyderms, the remains of which abound in the Paris basin, and in other parts of France, as well as in America. Both Rhinoceridæ and the next family are well represented in the fossils found in Nebraska,—a region rich in the remains of extinct Pachyderms and other animals.

Suidæ, or Swine Family.—This Family comprises much smaller Pachyderms than those before described, with a more compressed body, pointed head, large ears, blunt snout, slender legs, hoofs in pairs, and the skin covered with bristles.

The Genus *Sus*—Hogs—has the feet with two large middle toes armed with strong hoofs, and two much shorter lateral ones that hardly touch the ground. The number of incisors is variable, the canines project from the mouth and curve upwards, and the muzzle terminates in a sort of truncated button fitted for turning up the earth.

The Wild Boar, *S. scropha*, Linn., is the parent stock of all the varieties of the domestic hog. It has a short and thick body, straight ears, black bristly hair, and angular tusks which curve outwards and upwards. The young are striped black and white. When wounded, the wild boar is ferocious and formidable. These animals defend themselves from the attacks of wild beasts by forming a circle, and showing a savage front to the enemy. They are indigenous to Europe, Asia, and Africa, where in the wild regions they are still common.

The Babiroussa, *S. babiroussa*, Buff., of the Indian Archipelago, is of a lighter build, and is characterized by its long slender tusks, the under ones turned vertically upwards, and the upper ones inclining spirally backwards.

The Genus *Phacochærus*—Wart-bearing Hogs—is characterized by a fleshy lobe on each cheek, and enormous tusks which incline upward. These hogs inhabit Africa.

The Genus *Dicotyles* — Peccaries — differs from the preceding genera in its canines, which are directed like those of animals in general, the hind feet three-toed, the tail rudimentary, and there is a peculiar gland upon the loins. Peccaries inhabit the warmer regions of America.

The Texas or Collared Peccary, or Mexican Hog, *D. torquatus*, Cuv., is about three feet long, weighs fifty or sixty pounds, is of a general gray color with a whitish band stretching obliquely from the angle of the lower jaw over the shoulders.

The White-lipped Peccary, *D. labiatus*, Cuv., is larger than the preceding one, and is very destructive to vegetation.

EQUIDÆ, OR HORSE FAMILY.—This Family, called Solepedes by Cuvier, comprises animals which have only one apparent toe and a single hoof to each foot; although under the skin, on each side of their metatarsus and metacarpus, there are spurs representing two lateral toes. The Genus *Equus* comprises all the species.

The Horse, *E. caballus*, Linn., is indigenous to the Old World, but has accompanied man, and become established in every quarter of the globe. This noble animal is the most beautiful, graceful, and the most useful to man of all the Pachyderms. The Horse no longer exists in a wild state, except in those countries where horses once in a state of domestication have been set at liberty, as in Tartary and in America. Here they live in troops, each of which is led by an old male.

The varieties of the Horse are exceedingly numerous; but these varieties are all regarded as one species. The Arab Horses are the most beautiful and the fleetest; the largest and strongest are from the coasts of the North Sea; and the smallest from the North of Sweden and from Corsica.

In South Carolina are found the fossil remains of a horse which was probably indigenous to this country.

The Ass, *E. asinus*, Linn., is distinguished by its long ears, and the tuft which terminates the tail. It is indigenous to the great deserts of Central Asia, where it still exists in the wild state; but, like the horse, it has been domesticated, and is used for a beast of burden and for draught in all parts of the world. It is noted for its patience and great power of endurance. The hoarseness of its bray depends upon two small cavities situated at the bottom of the larynx. The Mule is the offspring of this species and the horse.

The Zebra, *E. zebra*, Linn., Southern Africa, has the general form of the ass, and is regularly marked throughout with black and white transverse stripes.

The Dzigguetai, *E. hemionus*, Pall., is intermediate, in size and appearance, between the horse and the ass. Its home is the dry regions of Central Asia.

The Quagga, *E. quaccha*, Gm., of Africa, resembles both the horse and zebra, but differs from both in specific characteristics. The neck and shoulders are brown striped transversely with white, the rump reddish gray, and the tail and legs whitish. Its voice resembles the barking of a dog.

The Onagga, *E. montanus*, Burchell, of Africa, is smaller

than the ass, of the general form of the quagga. Its color is bay, with black stripes; legs and tail white.

The so-called Herbivorous Whales which Cuvier grouped with the Cetaceans are now placed with the Pachyderms, with which they undoubtedly belong. They are whale-like in general appearance, but their teeth have flat crowns, and they have corresponding herbivorous habits. They frequently leave the water, and crawl upon shore, and feed upon the vegetation. Such are the Manati or Cow Whales, and the Dugong. The former are about fifteen feet long, and inhabit the warm parts of the Atlantic; and the Dugongs are found in the Pacific, and much resemble the Manati.

SUB-SECTION X.

THE ORDER OF RUMINANTIA, OR RUMINANTS.

THE Order of Ruminantia is one of the best defined of the Mammalia. It comprises all those animals which masticate their food, which is wholly vegetable, the second time. In accordance with this singular faculty, the stomach, with few exceptions, is composed of four different compartments, each having a special function. The food, being hastily and partially chewed, is passed into the largest stomach or paunch, and thence into the second, called the honey-comb. This second stomach, small and globular, seizes the food, moistens and compresses it into little pellets, which afterwards ascend to the mouth to be rechewed. The animal is at rest during this process, which lasts until all the food first taken into the paunch has been thus remasticated. The remasticated food descends directly to the third stomach, called the leaflet; thence to the fourth, or caillette, which is the true organ of digestion, analogous to the simple stomach of Mammals generally.

The feet of Ruminantia are terminated by two toes and two hoofs, appearing like a single hoof which has been cleft. Hence they are often called cloven-footed animals. Behind the hoofs and higher up are generally to be found two rudimentary toes. The two bones of the metatarsus and metacarpus are generally united into one, called the cannon. With few exceptions, the head of the males, and in many cases of the females also, is armed with horns. Excepting Camelidæ, the Ruminantia have no incisors in the upper jaw, but in nearly all cases eight in the lower, which shut against a callous pad above. Between the incisors and the molars there is a vacant space, which in some cases contains one or two canines. There are six molars in each side of both jaws, which have their flat crowns marked with two double crescents, the convexity of which is turned inwards in the upper, and outwards in the lower ones. Of all animals the Ruminants seem to be the most useful to man, furnishing him with flesh and milk for food, and hides for leather; and many of them are used for beasts of burden and for draught. The Ruminantia may be divided into three great groups: those with solid and usually deciduous horns, as the Deer, and called the Cervidæ, or Deer Family; those with permanent horns, consisting of an exterior hollow horn encasing a bony process of the skull, as the Antelopes, Goats, Sheep, and Oxen, and called the Cavicornia Family; and those which have no horns, as the Camels and Llamas, and called the Camelidæ, or Camel Family.

Cervidæ, or Deer Family. — This Family, as stated above, comprises all Ruminants which have the horns solid, and, excepting the Giraffe, deciduous. These prominences, or horns, are at first covered with skin similar to that upon the rest of the head. At their base is a ring of bony tubercles, which, as they enlarge, compress and obliterate the bloodvessels of that skin, and the latter becomes dry and peels off, leaving the horns bare. At length the horn separates from the cranium and falls. Others, however, and larger ones, take their places, and these in turn are subject to the same changes. Thus the horns of these animals are shed and renewed periodically. Such horns are called *antlers*. The dental formula is incisors $\frac{0}{8}$, canines $\frac{1-1}{0-0}$ or wanting, and molars $\frac{6-6}{6-6}$. This Family is represented in almost every region of the globe.

Fig. 69.

Virginia Deer, *C. virginianus*, Boddaert.

The Genus *Alce* — Moose — is characterized by very broadly palmated horns, found only on the male, and the nose wholly covered with hair except a small spot between the nostrils.

The Moose, *A. americanus*, Jardine, is the largest member of the deer family, quite equalling the horse in bulk, and standing very high; and its broad antlers weigh from fifty to seventy pounds. The muzzle is very broad and prolonged, the ears long and hairy, the neck short and thick, the latter and the shoulders covered by a sort of mane, and the throat with long hair. The general color is grayish-brown, and the hair is coarse and brittle. The movements of the moose are rather heavy, but its speed is great. It does not leap, but strides along without apparent effort over fallen trees, fences, and other like obstructions. It is common in the unsettled parts of Maine and New York, thence westward in corresponding latitudes, and northward to the frozen regions. It frequents wooded hillsides in winter, and the borders of lakes in summer. Moose are hunted for their flesh, which is excellent. They sometimes turn against the hunters before being wounded or even shot at. Their usual mode of defence consists in striking with their fore feet.

The Elk of the North of Europe is so nearly like our moose, that the two have been regarded by most authors as one species.

The Great Irish Elk, *Megaceros hibernicus*, Owen, ten feet high to the top of the horns, whose tips are ten feet apart, is an extinct species found in marl at the bottom of the peat bogs of Ireland.

The Genus *Rangifer* — Reindeer — has the horns broadly palmated at the tip, and present in both sexes; the nose wholly hairy, and the hoofs suboval and dilated.

The Reindeer, *R. tarandus*, Linn., of Northern Europe, is about four feet and a half long and three feet high, and is celebrated for the services it renders to the Laplanders, who possess large herds of them, and use them as beasts of burden and for draught, their milk and flesh for food, and their skins for clothing and covering for sledges. The reindeer is very hardy, and draws the sledge of its owner with great speed. In summer it feeds upon the tender portions of shrubs, but in winter it scrapes the snow from the ground, and feeds upon the so-called reindeer-moss. The hair is brown in summer, white in winter.

The Woodland Caribou or Reindeer, *R. caribou*, Aud. & Bach., of Maine and New Brunswick and westward to Lake Superior, is believed by some to be identical with the European species.

The Barren Ground Caribou, *R. groenlandicus*, Baird, is found in the Arctic regions of America and Greenland, beyond the limit of trees.

Fig. 70.

American Reindeer, or Woodland Caribou, *R. caribou*, Aud. & Bach.

The Genus *Cervus* — Common Deer — has the horns more or less rounded, cylindrical or conical, sometimes partly flattened, the nose tapering, naked, and moist.

The American Elk, or Wapiti, *C. canadensis*, Erxl., of the northern and northwestern portions of the United States, and northward to the fifty-seventh parallel, is about the size of the horse, the horns five to six feet long and much branched, the color in summer light chestnut-red, and in winter grayish.

The Virginia Deer, *C. virginianus*, Bodd., (Fig. 69,) of the United States east of the Missouri River, is one of the most beautiful and graceful of all the deer. It is very timid, and, when alarmed, bounds through the forest and over the plains with almost incredible velocity. The weight of an adult is about two hundred pounds. The color, light fawn in summer, reddish-gray in winter, the under part of the throat and tail always white. The Vir-

Fig. 71.

American Elk, or Wapiti, *C. canadensis*, Erxl.

ginia Deer is hunted for its flesh, which is considered one of the luxuries of the table during the winter months.

The White-tailed Deer, *C. leucurus*, Douglas, is found from the Upper Missouri and Platte to the Columbia River and Washington Territory.

The Mule Deer, *C. macrotis*, Say, is found from the Upper Missouri to Oregon.

The Columbia Black-tailed Deer, *C. columbianus*, Rich., is confined to the Pacific coast of North America.

The Stag or Red Deer, *C. elaphus*, Linn., inhabits the forests of all Europe, and of the temperate parts of Asia. Its weight is about two or three hundred pounds. The Stag-hunt has always ranked as the most fashionable of field and forest sports.

The Daim or Fallow Deer, *C. dama,* Linn., originally from Barbary, but now common throughout Europe, is smaller than the stag, and is the species common in parks of the wealthy, especially in England.

The Axis Deer, *C. axis*, Linn., indigenous to India, but domesticated in Europe, is about the size of the Fallow Deer, and is always of a rich fawn color spotted with white.

The European Roebuck, *C. capreolus,* Linn., is a very small deer weighing only about sixty pounds, and inhabiting the high mountains of the temperate parts of Europe.

The Genus *Moschus*—Musk Deer—is characterized by the absence of horns, and by having a long canine tooth on each side of the upper jaw. The members of this genus are light and elegant in their appearance.

The Musk Deer, *M. moschiferus,* Linn., is the most celebrated species, being the one which furnishes the well-known musk of commerce. This animal is about the size of the common goat, has scarcely any tail, and is covered with coarse and brittle hairs. It inhabits Thibet and the adjacent countries.

Fig. 72.

Musk Deer.

The Genus *Camelopardalis*—Giraffe—is characterized by both sexes having conical horns, which are always covered with a hairy skin, and which are never shed. Only one species, the Giraffe, *C. girafa,* F. Cuv., of the deserts of Africa, is known. It is one of the most remarkable animals in existence in respect to the great length of the neck and the disproportionate length of the fore legs. Its head in some cases is eighteen feet from the ground. Its hair is short and gray, sprinkled with fawn-colored angular spots, and it has a small fawn-colored and gray mane. It is gentle in disposition, and feeds upon leaves of trees.

CAVICORNIA, OR HOLLOW-HORNED RUMINANT FAMILY. — This Family, as stated above, comprises all the Ruminants which have the horns permanent, hollow, and enclosing a process of the frontal bone. The Cavicornia may be divided into three sub-families, which are here presented together, that their resemblances and differences may be seen at a glance: —

Antilopinæ, or Antelopes, characterized by horns rounded or conical, without sharp angles, variously curved, annulated or wrinkled, and black; the muzzle elongated, attenuated, generally hairy, and the end of the upper lip with a shallow groove.

Ovinæ, or Sheep and Goats, characterized by horns more or less angular and compressed, usually twisted and curved backwards, wrinkled, and generally dull yellowish-brown. The muzzle is broader than in the antelopes, generally hairy, and with a shallow groove.

Bovinæ, or Oxen, characterized by horns rounded, muzzle broad, usually naked, and without a vertical furrow at the end.

1. Antilopinæ, or Antelopes, are very numerous in species, no less than ninety having been described, varying in size from the light and graceful gazelle and chamois to those as large as the largest horse. Two of these belong to North America, two to Europe, and the rest to Southern Asia and to Africa, but mainly to Africa.

The Genus *Antilocapra* is characterized by erect horns, the base compressed, with a flattened process in front, the end conical and recurved. The nose is sheep-like, entirely hairy at the end except a narrow central line; the tail is very short, and there are no false hoofs behind the large ones.

Fig. 73.

Pronghorn Antelope, *A. americana*, Ord.

The Prong-horn Antelope or Cabree, *A. americana,* Ord, of the plains west of the Missouri River, from the Lower Rio Grande to the Saskatchawan, and westward to the Cascade and Coast Range of the Pacific slope, exceeds in size the domestic sheep, and has longer legs and a longer and more erect neck. The hair is very coarse and thick; the color above yellowish-brown, the entire under parts and a square patch on the rump white; the horns, hoofs, and naked parts of the nose black. About half-way up the horns on their anterior face there is a branch or prong, from which the animal gets its popular name.

The Genus *Aplocerus* is characterized by horns which are small, conical, nearly erect, slightly inclined, recurved at the tip, and ringed at the base.

Fig. 74.

Mountain Goat, *A. montanus*, Rich.

The Mountain Goat, *A. montanus*, Rich., of the Rocky Mountains, is an antelope. Its jet-black, polished, slender, and conical horns are much like those of the chamois. It is covered with long and pendent hair, and the color is white.

The Genus *Antilope* comprises Antelopes proper.

Fig. 75.

Gazelle, *A. dorcas*, Linn.

The Gazelle, *A. dorcas*, Linn., of the North of Africa, is a beautiful and graceful antelope about the size of the roebuck, with large black horns, and of a fawn color above and white beneath, with a brown band along each flank. The soft expression of the eye of the gazelle furnishes numerous images to the Arabian poets.

The Springbok, *A. euchore*, Forster, found in large herds in South Africa, is an antelope larger than the gazelle, but of the same form and color, and is remarkable for a fold of the skin of the croup, which opens and expands at every bound of the animal, disclosing the brilliant white hair with which the fold is lined. It gets its name from its habit of jumping upward whenever it is excited. In seasons of drought these beautiful animals are seen in herds of ten to twenty thousand, wandering over the country in search of pasturage.

The Saiga, *A. saiga*, Pall., of Poland and Russia, resembles *A. dorcas*, but is larger, and its horns are transparent.

Fig. 76.

Chamois, *A. rupicapra*, Linn.

The Chamois, *A. rupicapra*, Linn., of the middle regions of the high mountains of Western Europe, is about the size of a goat, of a deep brown color, and its horns towards the summit are bent abruptly backward like a hook. The chamois is exceedingly shy, and on the slightest alarm bounds away with a speed that is truly wonderful, over rocks, glaciers, along the brinks of dizzy heights, and up and down precipices where it would seem no animal could get a foothold, — often leaping upon a shelf of rock of scarcely more than a hand's breadth, or just large enough to receive its four feet placed close together.

The Long-horned Antelope or Oryx, *A. oryx*, Pall., of Central and Southern Africa, is as large as the stag, and has straight, slender, round and pointed horns two or three feet long, with the lower third obliquely annulated. The tail is long and blackish, and the hairs of the spine are directed towards the neck. It is often called Gemsbok.

The Canna or Eland, *A. oreas*, Pall., of South Africa, is an antelope which attains the weight of eight hundred or a thousand pounds, and has horns very long and straight, and with a spiral ridge.

The Koodo, *A. strepciseros*, Pall., is another very large antelope of the same region as the preceding one.

The Gnu, *A. gnu*, Gm., is a curious animal which Cuvier describes among the antelopes, and which is one of the most extraordinary forms of life to be found among the Ruminantia. Its head and horns remind us of the Cape Buffalo; the body, mane, and tail resemble those of a horse, and its feet are as light as those of a stag. The muzzle is large, flattened, and encircled with projecting hairs, and the general color is brownish.

2. Ovinæ, or Sheep and Goats.

The Genus *Ovis* — Sheep — is characterized by horns which are directed backwards, and then incline more or less spirally forwards; the chanfrin is convex, and there is no beard on the chin, as in the goats.

The Mountain Sheep or Big Horn, *O. montana*, Cuv., of the Rocky Mountains, is much larger than the domestic sheep, with very large horns. The female has smaller horns similar to those of the goat. A large individual of this species weighs about three hundred and fifty pounds. The Argali of Siberia, *O. ammon*, Linn., is

Fig. 77.

Mountain Sheep, or Big-horn, *O. montana*, Cuv.

regarded by Cuvier as identical with the Big Horn. The Mouflon of Sardinia, *O. musimon*, Pall., differs in being smaller, and in the smallness or deficiency of the horns of the female.

The Mouflon of Barbary, *O. tragelaphus*, Cuv., has soft and reddish hair, with a long mane under the neck. It is from the Mouflon and Argali that our numerous domestic varieties are supposed to have sprung. Of these the Merino from Spain is one of the most noted, on account of the length and the fineness of its wool. Persia, and countries of Asia farther east, furnish a variety whose tail is a double globe of fat. Syria and Egypt have a variety whose tail is so long, and so loaded with fat, that it attains a weight of fifty to one hundred pounds.

The Genus *Capra* — Goats — is characterized by horns directed upwards and backwards; and the chin is generally furnished with a long beard, and the chanfrin is generally concave. This genus is not represented in America, the so-called Rocky Mountain Goat being considered an antelope, as before stated. Goats are exceedingly active, and the wild species inhabit high and rugged parts of the mountains, where they subsist upon coarse grass, and leaves, and shoots of low shrubs which such localities afford. They are sure of foot, and bound along the verge of dizzy heights with great rapidity, and with an air of conscious security, which shows how well they are adapted to the regions which they inhabit.

The Wild Goat, *C. ægragus*, Gm., is found in herds on the mountains of Persia, where it is called Paseng, and on other mountains of the Eastern hemisphere. This is regarded as the parent stock of all the numerous domestic varieties.

The Angora Goat of Asia Minor is noted for furnishing the softest and most silky hair, which is largely manufactured by the inhabitants of Angora, no less than thirteen million pounds of fabrics and yarns being exported by them annually.

The Cashmere Goat of Thibet is the most celebrated of all for its fine wool. This goat is covered with long silky hair, under which is a delicate gray wool, about three ounces of which are obtained from a single individual; and it is of this wool that the renowned Cashmere shawls are made.

The Ibex, *C. ibex*, Linn., of the high mountains of the Old World, is distinguished from all the preceding by its large horns, square in front, and marked with transverse and prominent ridges.

The Caucasian Ibex, *C. caucasica*, Guldenst., is distinguished by its large triangular horns, but not square in front.

3. Bovinæ, or Oxen.

The Genus *Ovibos* is characterized by horns curving outward and downwards, hairy muffle except between the nostrils, tail very short, hoofs broad and inflexed at the tips, and the hair long and pendent.

The Musk Ox, *O. moschatus*, Blainville, of the Barren Grounds of Arctic America, is about the size of a two-

Fig. 78.

Musk Ox, O. *moschatus*, Blainville.

year-old cow; the horns united on the summit of the head, flat, broad, bent down against the cheek, with the points turned up. The color is brownish-black.

The Genus *Bos* is characterized by horns curving outwards and upwards, broad naked muzzle, wide space between the nostrils, large ears, rather long tail, and broad hoofs. It comprises about ten species.

The Common Ox, *B. taurus*, Linn., so serviceable to man, is too well known to need description. Its varieties are numerous. The Zebu is a variety inhabiting India, which has a large hump of fat upon the back between the shoulders. The male is known as the Brahmin Bull, and is held sacred by the Hindoos.

The American Buffalo, *B. americanus*, Gm., formerly inhabiting nearly all North America, but now only the Western plains, is the largest quadruped of America, being of the size of a large domestic ox, and characterized by a large head, which is carried low, broad forehead, broad full chest, large hump between the shoulders, narrow loins, and comparatively slender legs. The horns, set far apart, are thick at base, and taper rapidly to a sharp point. The Buffalo is covered with a thick coat of hair; that upon the head, neck, hump, shoulders, and fore legs to the knees, is very long and shaggy. The horns, hoofs, and hair — except the middle of the back, which is brownish — are black. The Buffalo is found in herds from a score to several thousand in number.

Though naturally timid, they are furious and formidable when wounded by the hunter.

Fig. 79.

American Buffalo, *B. americanus*, Gm.

The Aurochs, or Bison of the ancients, *B. urus*, Gm., formerly an inhabitant of all Europe, but now found only in the forests of Lithuania and of the Caucasus, is closely related to the American Buffalo.

The Buffalo of Southern Europe, *B. bubalus*, Linn., introduced from India, is related to the Arni, *B. arni*, Shaw, of India, whose enormous horns are ten feet from tip to tip.

The Cape Buffalo, *B. caffer*, Sparm., of South Africa, is characterized by its large horns, which are so wide at the base that they nearly cover the forehead. It is a very large animal, with a very ferocious disposition.

The Grunting Cow, or Yak, of Tartary, *B. grunniens*, Pall., is smaller than any of the preceding, with a tail resembling that of a horse, and a long mane upon the back. It makes a grunting noise similar to that of a hog.

Fossil remains of extinct Bovinæ are found in various parts of North America; also in the Old World.

Camelidæ, or Camel Family. — This Family comprises the Camels of the Old World and the Llamas of the New.

The Genus *Camelus* comprises the Camels, which have the two toes united below nearly to the point by a common sole, and the back furnished with humps of fat. They are natives of Central and Southwestern Asia, and from earliest times have been celebrated for their important services to the inhabitants of the arid regions of the East. Possessed of great strength and power of endurance, capable of subsisting on the coarsest and most scanty vegetation, able to travel for days without drinking, having feet suited to walking over sand, and withal gentle and obedient, the Camel is as indispensable to the merchant and traveller for traversing the deserts of Asia and Africa, as are vessels for crossing the ocean. The Camel can bear from five hundred to one thousand pounds during a long journey. It kneels to receive and to be relieved of its load. With an intuitive knowledge of its own powers, it obstinately refuses to rise when a greater load is put upon it than it can comfortably bear. The power of resisting thirst is due to the large number of cells on the walls of the paunch, in which is stored an extra supply of water. Camels have canine teeth in both jaws, two pointed teeth in the incisor bone, six incisors in the lower jaw, and eighteen to twenty molars; peculiarities unknown among all other Ruminants.

The Two-humped Camel, *C. bactrianus*, Cuv., is originally from Central Asia. This species is ten feet long, and eight feet high between the humps.

The One-humped Camel, *C. dromedarius*, Linn., smaller than the last, has spread from Arabia into Persia, Syria, and Africa. The Dromedary is a variety of this species.

The Genus *Auchenia* — Llamas — differs from the preceding in having the two toes separate, and in the absence of humps. The Llamas are confined to South America, and chiefly to the Andes, and are the American representatives of the Camels.

The Llama, or Guanaco, *A. llacma,* Linn., is about the size of the Stag, and covered with coarse, chestnut-colored hair. This species was early domesticated and extensively used as a beast of burden, in which capacity it is still employed. The Alpaca is a variety with long, woolly hair, which furnishes material for the best of fabrics.

Fig. 80.

Llama.

The Paco or Vicuna, *A. vicunna,* Linn., is of the size of a sheep, and covered with fawn-colored wool, which is also used in the manufacture of valuable fabrics.

Remains of extinct Camelidæ are found in the tertiary rocks of Nebraska.

SUB-SECTION XI.

THE ORDER OF CETACEA, OR CETACEANS.

THE Order of Cetacea comprises Mammalia which are formed for an exclusive residence in the water. They have no hind feet, two small bones suspended in the flesh being the only vestiges of posterior extremities; their anterior members closely resemble fins; and, excepting the tail, which spreads horizontally, their general appearance is decidedly fish-like. They are, however, genuine Mammals, have warm blood, respire by means of lungs, come frequently to the surface of the water to take in fresh supplies of air, — though some can remain beneath the water for half an hour or more, — and produce and nourish their young in the same manner as other Mammals. They are destitute of hair, and covered with a smooth skin, under which is a thick layer of fat called blubber. They propel themselves with rapidity by the downward and upward movement of the tail. In the most prominent members of the Cetacea the breathing-hole, which corresponds to the nostrils of other animals, is situated on the top of the head, and through this the water which has been taken into the mouth is spouted to a great height, and this spouting or blowing may be seen at great distances, and often serves to reveal these animals when they would otherwise be unobserved. The species are numerous; and there are at least three families,—Balænidæ or Right-Whale Family, Physeteridæ or Sperm-Whale Family, and Delphinidæ or Dolphin Family. The first two have the head excessively large.

BALÆNIDÆ, OR RIGHT-WHALE FAMILY.—This Family comprises Whales which have no real teeth, but the two sides of their upper jaw, which is keel-shaped, are furnished with rows of vertical horny plates, called whalebone, formed of a sort of fibrous horn, and which are fringed on their inner edges. This arrangement is adapted to the nature of the food of these whales, which consists of small marine zoöphytes, mollusks, and crustaceans. Swimming through schools of these little animals, the Whale engulfs myriads of them at once in its enormous mouth; and the water taken with them is strained off through the fringes, and all the animals, even the smallest, retained and swallowed.

Fig. 81.

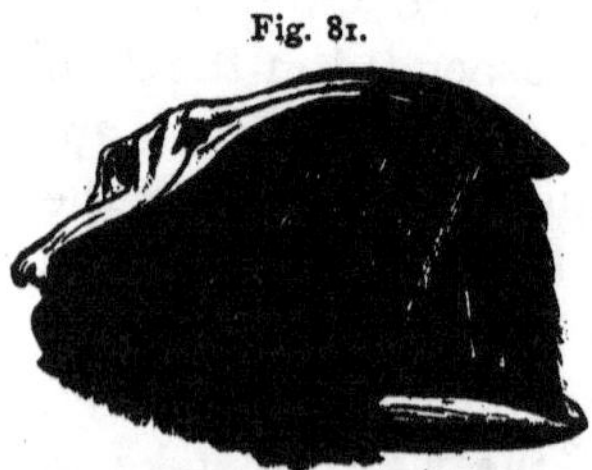

Skull of Right Whale, showing the whalebone.

The Genus *Balæna* comprises the Right Whales proper.

The Great Greenland, or Right Whale, *B. mysticetus*, Linn., attains the length of seventy feet, and is sometimes furnished with blubber two or three feet in thickness. This species supplies the black, flexible whalebone, in slabs of eight to ten feet in length, an individual yielding six to nine hundred strips or slabs on each side of the palate. It also furnishes more oil than any other whale, — a single individual yielding, in some cases, one hundred and twenty tuns. The Right Whale is confined to the frigid regions, and is common to the North Atlantic and North Pacific, but is never found in the tropics. It has the seven cervical vertebræ consolidated into one.

B. australis is a species confined to the Antartic regions, and is smaller than the Greenland Whale.

Fig. 82.

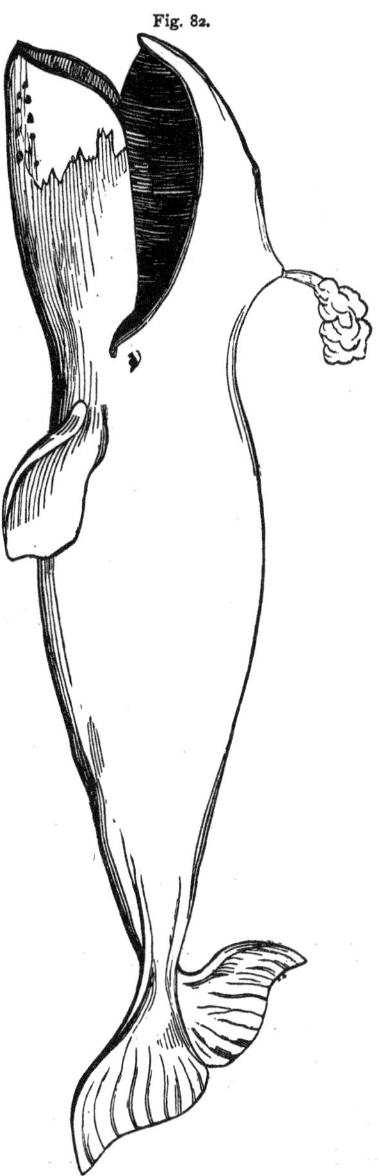

Right Whale, *B. mysticetus*, Linn.

The Genus *Balænoptera* comprises Whales with a dorsal fin and short baleen, and known under the names of Fin-backs, Razor-backs, and Rorquals. They equal and in many cases much exceed the *B. mysticetus* in length,—some have been seen one hundred feet long,—but yield far less oil. They are exceedingly powerful and rapid in their movements, and are captured with the greatest difficulty and danger. One or two species are common on the North Atlantic coast of America.

PHYSETERIDÆ, OR SPERM-WHALE FAMILY.—This Family comprises Whales with excessively enlarged heads, and whose upper jaw has neither teeth nor whalebone, and whose lower jaw is narrow, elongated, and corresponds to a furrow in the upper one, and is armed on each side with a range of cylindrical or conical teeth, which, when the mouth is closed, fit corresponding cavities in the upper jaw. The upper portion of the head consists mainly of large cavities, separated and covered by cartilages, and filled with an oil which becomes fixed as it cools, and is known under the name of spermaceti. The body yields sperm oil. The substance known under the name of ambergris is a concretion formed in the intestines of Sperm Whales. These animals inhabit deep, tropical, and temperate seas, and never enter the Polar regions.

Fig. 83.

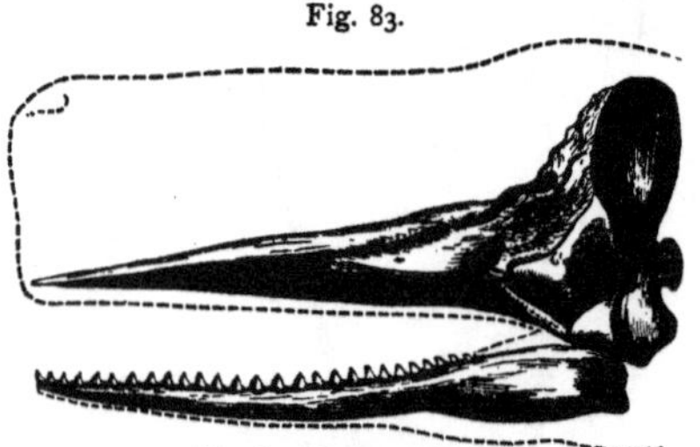

Head of Sperm Whale.

The Genus *Physeter* comprises the Sperm Whales proper.

The Great Sperm Whale or Cachelot, *P. macrocephalus*, Shaw, is the largest and most important species, equalling the Right Whale in size, averaging sixty feet in length. The largest specimen recorded was seventy-six feet in length, and thirty-eight feet in girth. The head constitutes one third of the whole animal. In this species the atlas is separate, but all the rest of the cervical vertebræ are consolidated into one. The Sperm Whale is usually found in companies of twenty to one hundred or more, and these companies are composed of females and their young and an old male. The Sperm-Whale fishery has employed at one time six hundred American vessels and fifteen thousand American seamen.

DELPHINIDÆ, OR DOLPHIN FAMILY. — This Family comprises Cetaceans which are included in the Linnæan genus *Dolphinus*, and whose head bears the usual proportion to the body, and whose jaws are both armed with simple and generally conical teeth, which they shed more or less with age. They live in communities, and are the most rapacious of the whole order.

The Genus *Delphinus* — Dolphins proper — is characterized by a convex forehead and a beak-like muzzle. The species are quite numerous, varying from six to fifteen feet in length, and are celebrated for their great velocity of movement.

The Common Dolphin, *D. delphis*, Linn., of all seas, is six to ten feet long, with from forty to forty-seven slender, arcuate, and pointed teeth on each side, both above and

Fig. 84.

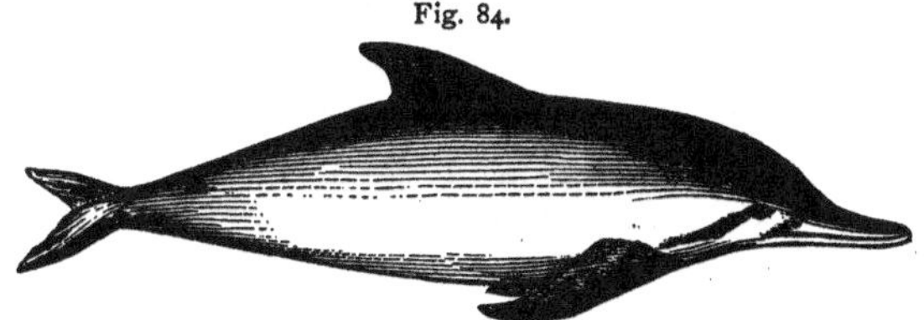

Common Dolphin, *D. delphis*, Linn.

below; the color black above and white beneath. This is the Dolphin of the ancients, so celebrated for its alleged docility and fondness for music.

The Genus *Phocæna* — Porpoises and Grampuses — is distinguished from the Dolphins proper by a short and convex muzzle. The members of this genus are from four to twenty feet in length, and, like dolphins, are often seen in large herds.

The Common Porpoise, *D. phocæna*, Linn., is the smallest of the Cetaceans, being only four or five feet long; the color blackish above, whitish below.

The Grampus or Killer, *D. orca*, Cuv., is twenty to twenty-five feet long, and is said to attack the whale.

The Blackfish or Round-headed Grampus, *D. globiceps*, Cuv., is twenty feet long; shining bluish-black above, lighter below.

The White Grampus or White Whale, *D. leucus*, Gm., *Beluga borealis*, Lesson, of the Northern regions, is ten to twenty or more feet in length, with the dorsal fin small,

Fig. 85.

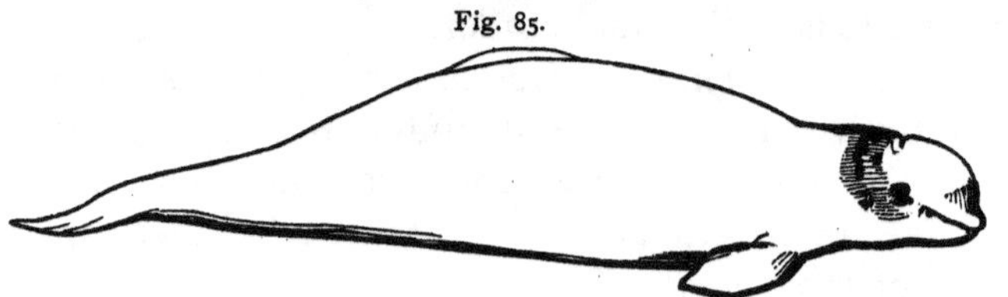

White Whale, *D. leucus*, Gm.

and for a Cetacean with an unusually distinct neck, the vertebræ of which are separate, and move freely upon one another. It often ascends rivers for a considerable distance, and is common in the St. Lawrence. A specimen of this animal, about ten feet long, and weighing about seven hundred pounds, was kept in a tank in the Aquarial Gardens, Boston, for about two years. He was quite docile, learned to recognize his keeper, and would come and take food from his hand. He was trained to the harness, and drew a young lady in a car prepared for the purpose.

The Genus *Monodon* — Narwhal — has no true teeth,

but a long, straight, pointed tusk implanted in the intermaxillary bone, and directed in the line of the body. The developed tusk is on the left side; an undeveloped one exists on the right side.

The Narwhal, *M. monoceros*, Linn., resembles a porpoise except in its spirally furrowed tusk, sometimes ten feet long. These animals inhabit the Arctic seas, and are pursued and skilfully captured by the Esquimaux.

Remains of extinct Cetacea occur in various parts of North America, as well as in the Old World. The Zeuglodon was a Cetacean seventy feet long, whose remains abound in the Southern States, especially in Alabama.

SECTION II.

THE CLASS OF BIRDS.

THE Class of Birds comprises all oviparous vertebrates which are clothed with feathers, furnished with a bill, and organized for flight. They have warm blood, and a complete double circulation. They are all bipeds; the body is inclined before their feet, the thighs are directed forward, and the toes elongated, forming a broad supporting base. The head and neck are more or less prolonged, the latter very flexible, and generally containing twelve or more vertebræ. The length of the neck and its great flexibility enable these animals to touch every part of the body with the bill. The trunk, serving as a point of support for the extended locomotive members, has little flexibility, the vertebræ of this portion being more or less firmly joined together. The pelvis is much lengthened to furnish points of attachment for the muscles of the thighs; and the sternum is of great extent, to bear the extensive muscles for moving the wings in flight. The sternal as well as the vertebral parts of the ribs are

Fig. 86.

Skeleton of a Bird.

hd, head; *cv*, cervical vertebræ; *p*, pelvis; *sc*, scapula; *cl*, clavicle; *cd*, corocoid bone, formerly regarded as the clavicle; *st*, sternum; *h*, humerus; *u*, ulna; *r*, radius; *c*, carpus; *mc*, metacarpus; *ph* and *th*, phalanges, *th* being the thumb; *fe*, femur; *ft*, fibula and tibia, more or less united; *t*, tibia, where the fibula is no longer seen, or only faintly indicated; *ts*, tarsus; *mt*, metatarsus more or less consolidated with the tarsus; *ps*, phalanges, or toes.

ossified in order to give greater strength to the trunk; and a small bone is attached obliquely across each rib, as a sort of cross-beam, which also contributes to the same result. The shoulders, which would otherwise be brought together by the effort of flying, are kept apart by two bony braces. Regarding the wrist as part of the hand, the wings are each made up of three sections, — the arm, forearm, and hand, — thus corresponding to the anterior extremities of man and other Mammals. The wings are furnished throughout their whole length with a range of quills, thus presenting a great surface to the air. The quills attached to the hand are called *primaries*, and are the largest and firmest; those attached to the fore-

Fig. 87.

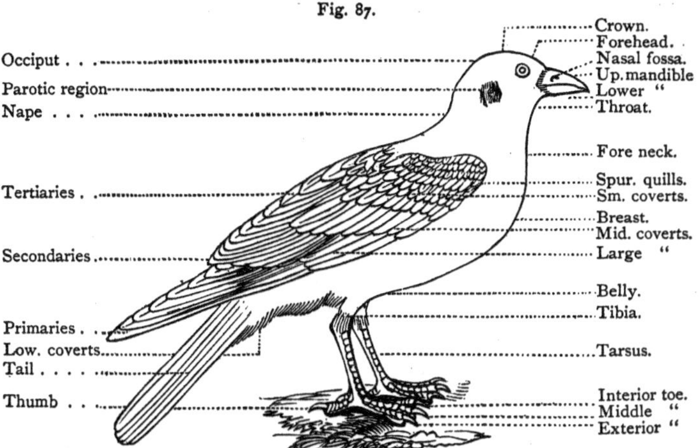

Showing the names of some of the principal parts of a Bird.

arm, *secondaries;* to the humerus, *tertiaries.* Ranges of shorter feathers cover the base of the quills above and below, and are called *coverts.* The feathers that grow from the shoulder are called *scapulars;* those from the thumb, *spurious quills.* The bony part of the tail is very short, but, like the wings, has a range of long quills, with upper and lower coverts, which serve both for ornament, and to aid in supporting the animal in the air. Both the quills and feathers consist of two parts, the *shaft* and the *vane;* the former is the axis, and the latter the expanded portion. The vane consists of laminæ, which are connected by minute barbs along their edges, and thus rendered firm to resist the air. There are, however, on every bird downy feathers, or such as do not have the laminæ united. The feathers of Birds do not grow from the entire surface of the body, but are symmetrically and systematically arranged in rows and patches, with bare intervening spaces; and the arrangement is different in different groups, and may yet be made an important element in classification. The hind locomotive members of birds are each composed of a femur; a tibia and fibula, the last two more or less united; a tarsus and metatarsus, the latter more or less merged in the former; and generally of three toes before and a thumb behind, the latter, however, sometimes wanting. The tibia and fibula are connected with the femur by an articulation with a spring, which keeps up the extension without any effort on the part of the muscles. And here it may be stated as an interesting fact, that there is a series of muscles reaching from the pelvis to the toes, and so arranged that the mere weight of the bird flexes the toes, thus enabling it to sleep in perfect security, even though perched on one foot. The number of joints in the thumb is two, in the next toe three, in the next four, and in the outer one five. The bones of birds are hollow; hence very light in comparison to their size and strength.

The respiratory system of birds is extensive; the air filling not only the lungs, but cells in other parts of the body, and even the cavities of the bones. Thus the lightness of the bones and the numerous air-cavities combine to diminish the specific gravity of the animal, and to adapt it to the medium in which it moves. The rings of the trachea are entire. At its bifurcation there is a glottis, generally furnished with peculiar muscles, and called the inferior larynx; and this is the point where the voice of birds is produced. The superior larynx is very simple. It enters the inferior, but has little to do with the voice.

The horny covering upon the mandibles performs the functions of teeth, and in some species the edges are so notched as to resemble real teeth. The stomach is composed of three parts: the *crop*, which is an enlargement of the œsophagus; a *membranous stomach*, in whose walls are numerous glands which furnish juices to moisten the food; and the *gizzard*, where the food is finally digested.

The brain of birds is comparatively very large. The eyes are so constructed, that with equal facility they can distinguish objects near or remote. Besides the two ordinary eyelids, there is a third one placed at the inner angle, and which can instantly be drawn over the eye like a curtain. It is called the nictitating membrane, and serves a most important office in the protection of the eye. The ear has but a single small bone, formed of one branch that adheres to the tympanum, and of another terminating in a plate that rests upon the fenestra ovalis; the cochlea is a slightly arcuated cone, but the semicircular canals are large. Nocturnal birds alone have an external conch. The organ of smell is concealed in the base of the bill. The tongue has but little muscular substance, and the taste is probably not very delicate.

The plumage of birds is rendered water-proof by the oil with which they dress their feathers, and which is furnished by a special gland at the hind part of the body.

Birds moult their feathers twice a year. In some, the winter plumage differs in its colors from that of summer. In a majority of cases the colors of the male are more brilliant than those of the female; and when this is the case, the young of both sexes resemble the adult female. When the adult male and female are of the same color, their young have colors peculiar to themselves.

Birds lay eggs, and sit upon them to hatch them. The egg in the ovary consists merely of the part we call yolk; it imbibes the external fluid called the white in the upper part of the oviduct, and becomes covered with a shell at the bottom of the same canal. The young bird of every kind has a horny point at the extremity of the bill, with which it breaks the shell, and which falls off a few days after it is hatched. This may be seen by every one on the bill of the young chicken. Most birds build nests in which to lay their eggs, and it is an interesting fact that all individuals of a species build alike, and, in a given locality, of the same kinds of material. Their skill and industry are in the highest degree remarkable, but cannot be dwelt upon here. Their ability to anticipate atmospheric changes is truly wonderful, and caused the ancients to attribute to them the power of divination. Cuvier says that on this faculty depends the instinct which acts upon birds of passage, prompting them to seek the sunny climes on the approach of winter, and their old homes as summer comes again.

The longevity of birds is regarded as about ten times as great as the period which they require to come to maturity or full growth. Domestic fowls live to the age of twenty years; parrots, thirty years; geese, fifty; while swans, ravens, and eagles are said to live a century.

The Class of Birds comprises ten to twelve thousand species, and is divided into seven Orders*: the Order of Raptores or Raveners, the Order of Scansores or Climbers, the Order of Insessores or Perchers, the Order of Rasores or Scratchers, the Order of Cursores or Runners, the Order of Grallatores or Waders, and the Order of Natatores or Swimmers.

SUB-SECTION I.

THE ORDER OF RAPTORES, OR BIRDS OF PREY.

THE Order of Raptores comprises all Birds of Prey, or those which, with few exceptions, pursue and capture birds and other animals for food. They are mostly of

* Bonaparte and some others divide the Class of Birds into two Sub-classes, called ALTRICES and PRECOCES, the former comprising those whose young hatch in a very feeble condition, and require to be fed for a considerable time from the bill of the parent, and the latter those whose young are able to run about and pick up food for themselves as soon as hatched. Each of these Sub-classes is divided into orders which stand in parallel series.

large size, and are characterized by strong, hooked bills, sharp claws, great extent of wing, and very powerful muscles; and the females are generally larger than the males. The plumage of the young and of the immature differs greatly from that of the adult. They live in pairs, and choose their mates for life. The Raptores comprise three families,—Vulturidæ or Vulture Family, Falconidæ or Falcon Family, and Strigidæ or Owl Family.

VULTURIDÆ, OR VULTURE FAMILY. — This Family comprises rapacious birds which have the eyes on a level with the sides of the head, which is partially naked or only sparsely covered with downy feathers. The talons are comparatively slender, and but little curved. Vultures are cowardly, seldom capturing prey unless forced to do so by hunger, preferring to feed on dead and decaying animals which they find, and which either by scent or sight, or both together, they discover at great distances. Of all animals, they are probably the most gluttonous and loathsome in their manner of feeding. They are found in nearly all countries of both hemispheres.

The Genus *Vultur* — Vultures proper — comprises the Vultures of the Old World, and the King of Vultures, and the Condor of South America.

The Condor, *V. gryphus*, Linn., of the Andes, is the largest of all the birds of prey, being four feet long, and having a spread of wing of nine feet; and it is said in some cases the spread is fourteen feet. This gigantic bird perches upon the high cliffs of the mountains till impelled by hunger, when it soars away, sometimes at the immense height of six miles, keenly surveying the surrounding country for its accustomed food.

The Genus *Cathartes* comprises the North American Vultures.

The Turkey Buzzard or Turkey Vulture, *C. aura*, Illig.,

of all North America except the Arctic regions, is thirty inches long, and the wing twenty-three inches.

Fig. 88.

California Vulture, *C. californianus*, Shaw.

The California Vulture, *C. californianus*, Cuv., of Western North America, is the largest rapacious bird in America, except the Condor, being from forty-five to fifty inches long, and the wing thirty inches; the color black, with a white transverse band upon the wings, the head and neck orange yellow and red.

The Black Vulture or Carrion Crow, *C. atratus*, Lesson, of the Southern States to Chili, is about twenty-three inches long, and is abundant even in the cities, where it performs the important office of scavenger.

Burrough's Vulture, *C. Burrovianus*, Cassin, of Mexico and Lower California, is the smallest of the genus, being about twenty-two inches in length.

The Genus *Gypætos* comprises the Læmmergeyer of the Alps and other high mountains of the Eastern hemisphere, which is the largest bird of prey in the Old World, being but little smaller than the Condor. It builds its nests upon inaccessible acclivities, attacks lambs, goats, and the chamois, and it is asserted that even infants have been carried off by it.

FALCONIDÆ, OR FALCON FAMILY. — This Family comprises birds of prey which have the head completely covered with feathers, the eyes more or less sunken, and exceedingly sharp talons and powerful muscles. They are adapted, both by their organization and courage, for the capture of living prey, although in these respects they differ greatly among themselves.

The Genus *Falco* — Falcons — has the form robust and compact; the bill short and strongly curved from the base to the point, near which is a distinct and generally prominent tooth; the nostrils circular, with a central tubercle; wings long and pointed; tail long and wide; tarsi short and covered with circular or hexagonal scales, and the middle toe long; claws large, strong, curved, and very sharp. The falcons are remarkable for exceedingly rapid flight, and great boldness in attacking their prey. They are the birds used in falconry, and several of the species are extremely docile, being readily trained to pursue game, and return at call.

The Duck Hawk or Peregrine Falcon, *F. anatum*, Bonaparte, of North America east of the Rocky Mountains, is eighteen to twenty inches long, the wings fourteen to fifteen inches; the upper parts bluish-cinereous, with transverse bands of brownish black; under parts yellowish white, with heart-shaped and circular spots of black on the breast and abdomen, and transverse bands of black upon the sides, under tail coverts, and tibiæ; quills and tail brownish black. The frontal band is white, the cheeks with a patch of black, bill light blue, and the legs and toes yellow. Younger specimens have the upper parts brownish black, the under parts darker than in the adult, and with longitudinal stripes of brownish black; the tarsi and toes of a bluish lead-color, frontal band obscure, and a large black patch on the cheek. This falcon pursues its prey with almost inconceivable velocity through all its turnings and windings, and when within a few feet of the quarry protrudes its powerful legs and talons to their full extent, almost closes its wings for a moment, and the next instant grasps the prize, and bears it away to a secluded place, or, if too heavy, forces it obliquely to the ground, and devours it on the spot. The Duck Hawk not only pursues ducks and other birds upon the wing, but, sweeping over the water, it catches up ducks and teal and other swimming-birds. One has been known to come, at the report of a gun, and bear away a teal not thirty paces from the hunter who had shot it.

Fig. 89.

Duck Hawk, or Peregrine Falcon, *F. anatum*, Bonap.

The Common Peregrine Falcon, *F. peregrinus*, Gm., of Europe, closely resembles the preceding. It was formerly much used in falconry.

The Black-headed Falcon, *F. nigriceps*, Cass., of Western North and South America, is closely related to the two preceding, but is smaller.

The Pigeon Hawk, *F. columbarius*, Linn., of the warm and temperate parts of America, is twelve to fourteen inches long, and the wing eight to nine inches; the male ten to eleven inches, and the wing seven and a half to eight inches. In the adult the upper parts are bluish-slate, every feather with a longitudinal black line; forehead and throat white; and the other under parts pale yellowish or reddish white, every feather with a longitudinal line of brownish black. The bill is blue, cere and legs yellow, quills black, tipped with ashy white, tail light bluish-ashy tipped with white, and with a broad subterminal black band, and several other narrower transverse bands of the same color. This spirited little falcon preys upon pigeons, teal, and most of the smaller birds. In the latter part of summer and early autumn it is a constant attendant upon the flocks of birds that are assembling for, or making, their southward migrations, and fattens upon those it chooses to select. It generally flies low, skimming over fields, and along the hedges and skirts of woodlands, searching for its favorite prey.

The Orange-breasted Hawk, *F. aurantius*, Gm., of Texas and South America, is somewhat smaller than the pigeon hawk.

F. femoralis, Temm., of New Mexico and South America, is somewhat larger than the pigeon hawk.

The Prairie Falcon, *F. polyagrus*, Cass., of Western North America, is eighteen to twenty inches long, and the wing thirteen to fourteen inches; the plumage above brown; a narrow frontal band, a line over the eye, and under parts, white; the breast and abdomen with longitudinal stripes and spots of brown, which color also forms a large spot on the flank.

The Gerfalcon, *F. candicans*, Gm., of Northern North America and Greenland, is about twenty-four inches long, and the wing sixteen inches, the plumage white, the upper parts with irregular confluent bands and large subterminal sagittate or hastate spots of ashy-brown, and the under parts with a few narrow stripes of brown. Younger specimens have the brown predominating, of a lighter shade than in the adult, and barred and spotted with white, the under parts whitish, with longitudinal stripes of brown.

The Gerfalcon or Iceland Falcon, *F. icelandicus*, Sabine, of Northern North America and Greenland, can only be distinguished from the preceding by the markings on the upper surface of the body, the brown, transverse bands in the present species being regular and very distinct.

The Gerfalcon of Iceland and the North of Europe is undoubtedly the same species as the Iceland falcon. The Gerfalcons are the most highly esteemed by falconers.

Fig. 90.

Sparrow Hawk, *F. sparverius*, Linn.

The Sparrow Hawk, *F. sparverius*, Linn., of the entire continent of America, is eleven to twelve inches long, and the wing seven to seven and a half inches; the top of the head, neck behind, back, rump, and tail, light rufous or cinnamon color; the under parts generally of a paler shade of the same rufous as the back, and always with circular or oblong spots of black. The frontal band, and space including the eyes and throat, are white; a spot on the neck behind, two on each side of the neck, and a line running downwards from before the eye, black. The Sparrow Hawk feeds upon small birds, mice, and other small animals, and never attacks poultry. It becomes greatly attached to a particular locality, and may be seen day after day on the same tree, stump, or stake, watching for prey.

The Genus *Astur* — Goshawk and allies — has the upper mandible lobed, but not toothed, the form somewhat long, wings rather short, tail broad and long, tarsi long and covered in front with wide transverse scales. Twelve species are known which belong to this genus, only one of which is found in North America.

The Goshawk, *A. atricapillus*, Bonap., of North America, is twenty-two to twenty-four inches long, and the wing about fourteen inches; the male about twenty inches, and the wing twelve and a half inches. The head above, neck behind, and a stripe from behind the eye, black; the other upper parts dark ashy-bluish. There is a conspicuous white stripe over the eye, and the entire under parts are mottled with white and light ashy-brown. Young specimens have the upper parts dark-brown, and the under parts white, every feather with a longitudinal stripe terminating in an ovate spot of brown. The Goshawk spends much of the time upon the wing. It pursues birds with great swiftness, and sometimes with meteor-like velocity it glides into the forest and emerges with a hare or squirrel which its quick eye had singled out. The nest is large, and placed on a tree near its trunk; and the eggs are bluish-white, sparingly spotted with light reddish-brown.

The Genus *Accipiter* has a more slender form than the preceding, but is otherwise similar, having also the lobed upper mandible. About twenty species are known, three of which belong to North America.

Cooper's Hawk, *A. Cooperii*, Bonap., of all temperate North America, is eighteen to twenty inches long, the wing ten to eleven inches; the male sixteen to eighteen inches long, and the wing nine and a half to ten inches. The upper parts are dark ashy-brown, the head above brownish-black, and an obscure rufous collar on the neck behind. The throat and under tail coverts are white, and the other under parts transversely barred with light rufous and white. The tail is dark cinereous tipped with white, and crossed by four wide bands of brownish-black. This hawk attacks poultry, grouse, hares, and squirrels.

The Blue-backed Hawk, *A. mexicanus*, Sw., of Western North America, is intermediate between the preceding and the following species.

The Sharp-shinned Hawk, *A. fuscus*, Bonap., of the whole of North America, is twelve to fourteen inches long, the wing seven and a half to eight inches; the male ten to eleven inches, and the wing six inches to six and a half. The upper parts brownish-black tinged with ashy; throat and under tail coverts white, the former with lines of black on the shafts of the feathers; the other under parts light rufous, deepest on the tibiæ, and with transverse bands of white; the tail ashy-brown tipped with white, and the secondaries and tertiaries with large partially concealed spots of white. Younger specimens are dull umber-brown above, tinged with ashy; under parts white, with stripes and spots of reddish-brown. The slender legs and toes of this species will generally be sufficient to distinguish it. Its flight is swift, but irregular and vacillating.

The Genus *Buteo* — Buzzards — has the bill short and wide at the base, the edges of the upper mandible lobed, the wings long and wide, the fourth and fifth quills usually longest, the tarsi moderate, robust and with transverse scales before and behind, and with hexagonal scales on the sides. About thirty species are known.

Swainson's Buzzard, *B. Swainsoni*, Bonap., of Northern and Western North America, is about twenty inches in length.

Baird's Buzzard, *B. Bairdii*, Hoy, of Northern and Western North America, is eighteen to twenty inches long, and wing fifteen inches.

The Red-tailed Black Hawk, *B. calurus*, Cass., of Western North America, is twenty-one inches long, and the wing sixteen and a half inches.

The Brown Hawk, *Buteo insignatus*, Cass., of Western North America, is nineteen and a half inches long, and the wing sixteen inches; the male seventeen inches, and the wing fourteen and a half inches.

Harlan's Buzzard, *B. Harlani*, Bonap., of Western North America, is twenty-one inches long, and the wings sixteen inches.

The Red-tailed Hawk, *B. borealis*, Vieill., of North America east of the Rocky Mountains and southward to the West Indies, is twenty-two to twenty-four inches long, the wing fifteen to sixteen inches in length; the male nineteen to twenty-one inches long, and the wing fourteen inches. The upper parts are dark umber-brown, the tail bright rufous tipped with white, and with a subterminal band of black; the under parts pale yellowish-white, with lines and spots of reddish-brown; and the under surface of the tail silvery white. This hawk is powerful; its flight is firm and protracted, and generally accompanied with a mournful cry. When it espies an intended victim, it alights upon a tree and watches for a short time, and then, with wings partly closed, descends swiftly upon the prey. When scanning a region, it sometimes sweeps around in broad circles, and thus rises to such a height as to be scarcely visible. Sometimes this hawk takes its stand upon a tall tree, and watches silently for hours for a good opportunity, which it seldom fails to improve. Poultry, hares, and squirrels fall an easy prey. This hawk builds its nest, which is large and flat, in the forked branches of one of the largest trees of the forest, and lays four or five eggs of a dull white color, with brown and black blotches.

The Western Red-tailed Hawk, *B. montanus*, Nutt., of Western North America, is closely related to *B. borealis*, but appears to be somewhat larger.

The Red-shouldered Hawk, *B. lineatus*, Jardine, of Eastern and Northern North America, is twenty-one to twenty-three inches long, the wing fourteen inches; the male eighteen to twenty inches long, and the wing twelve inches. This is one of the most common hawks of the region it inhabits, and is readily distinguished by its wing coverts, which, from the flexure to the body, are bright rufous. The upper parts are brown, the under parts paler orange-rufous, quills brownish-black with white spots on their outer webs, and with bars of a lighter shade of brown, and of white on their inner webs. The tail is brownish-black, with about five transverse bands of white, and tipped with white. This hawk prefers the forest, and generally hunts in pairs. During the spring, especially, its discordant notes may be heard daily. Its nest is made in the top of a large tree; eggs four to five, granulated, pale blue, faintly blotched with brownish-red at the smaller end.

The Red-bellied Hawk, *B. elegans*, Cass., of Western North America, is closely related to *B. lineatus*.

The Broad-winged Hawk, *B. pennsylvanicus*, Bonap., of Eastern North America, is seventeen to eighteen inches long, the wing eleven inches; the male sixteen inches, and the wing ten. The upper parts are umber-brown, throat white with lines of brown, the breast with a wide band of spots and bands of ferruginous tinged with ashy, and the other under parts white with numerous sagittate spots of reddish. The quills are brownish-black, widely bordered with white on their inner webs, and the tail dark brown, narrowly tipped with white, and with one wide band of white and several narrower bands near the base.

The California Hawk, *B. Cooperi*, Cass., of California, is about the size of *B. borealis*, and may be distinguished from all others of this genus in North America by its brighter colors.

The Genus *Archibuteo* is distinguished by tarsi densely feathered to the toes, but more or less naked behind, and covered with scales; wings long and wide, tail rather short and wide, and toes short.

The Rough-legged Hawk, *A. lagopus*, Gray, of temperate North America and Europe, is twenty-one to twenty-three inches long, the wing sixteen to seventeen inches; the male nineteen inches, and the wing fifteen to sixteen inches. The plumage is irregularly variegated with dark or light brown and white or whitish. It is one of the most widely diffused of all birds; rather sluggish in its habits, flies low, and frequents low grounds, where it sits for hours watching for birds and small quadrupeds.

The Black Hawk, *A. sancti-johannis*, Gray, of Eastern and Northern North America, is twenty-two to twenty-four inches long, the wing seventeen to seventeen and a half inches; the male twenty to twenty-two inches, the wing sixteen to sixteen and a half inches. The plumage is glossy black, with a brownish tinge. The tail has one transverse band of white, and is irregularly marked towards the base with the same color. Some specimens are dark chocolate-brown, with the head striped with yellowish-white and reddish-yellow; and the tail with several irregular transverse bands of white.

The California Squirrel Hawk, *A. ferrugineus*, Gray, of Western North America, is somewhat larger than either of the two preceding; the upper parts dark brown and light rufous; the under parts of the body white, with narrow longitudinal lines and spots on the breast of reddish-brown, and narrow irregular transverse lines of the same color, and of black, on the abdomen; the tibiæ and tarsi bright ferruginous, with transverse lines of black.

The Genus *Asturina* comprises *A. nitida,* Bonap., of Northern Mexico and South America.

The Genus *Nauclerus* is characterized by very long and pointed wings, and very long and forked tail, and by short bill, tarsi, and toes. Three species are known, two American and one African.

The Swallow-tailed Hawk, *N. furcatus,* Vigors, of the Eastern United States to the Mississippi and northward to Pennsylvania and Wisconsin, is twenty-three to twenty-five inches long, the wing sixteen to seventeen and a half inches; the head and neck and entire under parts white; the back, wings, and tail black, with a metallic lustre. The flight of this hawk is peculiarly graceful, and its motions very rapid. It glides along with gentle flappings, rises in circles, describes deep curves, and performs all kinds of evolutions in a manner that never fails to interest the beholder. It never attacks birds, but preys upon insects and reptiles, and always devours its prey while on the wing.

The Genus *Elanus* has the wings long and pointed, tail moderate and emarginate, but not forked.

The White-tailed Hawk, *E. leucurus,* Bonap., of the Southern and Western States and of South America, is fifteen to seventeen inches long; the head and tail and entire under parts white; the upper parts light cinereous, lesser wing coverts glossy black, inferior wing coverts white with a smaller patch of black.

The Genus *Ictinia* is characterized by a short and compact body, wings long and pointed, tail short and emarginated, and the tip of the bill emarginated.

The Mississippi Kite, *I. mississippiensis,* Gray, of the

Southern States, is about fifteen inches long, the upper parts dark lead-color, the head and under parts dark cinereous, the quills and tail brownish black.

The Genus *Rostrhamus* comprises the Black Kite, *R. sociabilis*, D'Orb., of Florida and southward, which is sixteen inches long, black except the tail at the base and under tail coverts, which are white; bill very long and slender.

The Genus *Circus* is characterized by a large head, short compressed bill, face partially encircled by a ring or ruff of projecting feathers, tarsi long and slender, and the claws rather slender and weak. This genus comprises fifteen species, only one of which is found in North America.

The Marsh Hawk or Harrier, *C. hudsonius*, Vieill., of North America and Cuba, is nineteen to twenty-one inches long, the wing fifteen and a half inches; the male sixteen to eighteen inches long, and the wing fourteen and a half inches. The upper parts and breast are pale bluish-cinereous, the upper tail coverts and under parts white, the latter with small cordate or hastate spots of light ferruginous.

About seventy species of eagles belong to the Falconidæ.

The Genus *Aquila* is characterized by a large and strong form, large, strong compressed bill, long and pointed wings, short and very strong tarsi feathered to the toes, and sharp, strong, and curved claws.

The Golden Eagle or Ring-tailed Eagle, *A. canadensis*, Cass., of all North America, is thirty-three to forty inches long, the wing twenty-five inches; the male thirty to thirty-five inches, the wing twenty to twenty-three inches. The head and neck behind light brownish-fulvous, tail at base white, terminal portion glossy black, all other parts rich purplish-brown, frequently nearly black on the under parts of the body. Younger specimens are lighter in plumage. The Golden Eagle has great power of flight, but not the speed of many of the falcons and hawks, and does not so readily pursue and capture birds upon the wing; but its keen sight enables it to spy an object of prey at a great distance, and with meteor-like swiftness and unerring aim it falls upon its victim. At times it soars to great heights, moving slowly and majestically in broad circles. The nest of the Golden Eagle is placed upon a shelf of a ragged and generally inaccessible precipice. It is flat and very large, and consists of dry sticks. The eggs are two in number, three and a half inches long, and two and a half inches through, and dull white with undefined patches of brown. The Golden Eagle preys upon fawns, hares, wild turkeys, and other large birds. It does not attain its full beauty of plumage till the fourth year. The so-called Ring-tailed Eagle is the present species before it has reached maturity. The European Golden Eagle is so nearly like the American one, that there is a question whether it is not the same species.

The Genus *Halietus* is characterized by large size, strong and very robust form, slightly lobed upper mandible, tarsi short and naked, or only feathered for a short distance below the joint of the tibia and tarsus, and the toes covered with scales. Ten or twelve species are known, all of which subsist mainly upon fishes.

The Northern Sea-Eagle, *H. pelagicus*, Sieb., of the northern parts of both continents, is the largest of all the eagles. The female is forty-five inches long, and the wing twenty-six inches. The frontal space, greater wing coverts, abdomen, and tail are white; all other parts of the plumage, dark brown; bill and legs, yellow.

The Washington Eagle, *H. Washingtonii*, Jard., of North America, is forty-three inches in length, the wing thirty-two inches, and the entire plumage dark brown mixed with fulvous.

The Gray Sea-Eagle, *H. albicilla,* Cuv., of Greenland and Europe, is thirty-five to forty inches long, the tail white, head and neck yellowish brown, and all the other plumage dark umber-brown. It is very common on the coast of Europe, and builds its nest upon high cliffs.

The Bald Eagle or White-headed Eagle, *H. leucocephalus,* Savig., of all temperate North America, Greenland, Iceland, and accidental in Europe, is thirty-five to forty inches long, the wing twenty-three to twenty-five inches; the male thirty to thirty-four inches, the wing twenty to twenty-two inches. The head, and the tail and its coverts, white; the remaining plumage brownish black; bill, feet, and irides yellow. Younger individuals have the entire plumage dark brown, bill brownish black, and irides brown. The term "bald" is unfortunate in its application to this eagle, for the white head is as densely feathered as any other part. When moving from one region to another, it flies by continued easy flappings. When searching for prey, it sails with wings extended, and occasionally allowing its legs to hang at their full length. It has the ability of ascending in circular sweeps without any apparent motion of the wings or of the tail, and it often rises in this manner until it disappears from view. When at an immense height, and as if observing an object on the ground, it sometimes closes its wings, and glides towards the earth with such velocity that the eye can scarcely follow it, causing a loud rustling sound like a violent gust of wind among the branches of the forest. The White-headed Eagle prefers the lowlands of the sea-shores and lakes, and the borders of large rivers, and is less frequently seen in mountainous regions. The nest, five or six feet in diameter, is placed on a tall tree, and is composed of sticks from three to

Fig. 91. Bald Eagle, *H. leucocephalus*, Savig.

five feet in length, together with turf, rank weeds, and lichens, and is occupied by the same pair year after year. The eggs are two to four, dull white, and equally rounded at both ends. The attachment of the parents to the young while unable to fly is very great, and for a man to ascend to the nest at such times is very dangerous; but when the young are able to fly, the old ones drive them from the nest, to which, however, they return at night for several weeks. This eagle preys upon fish, large birds, and various quadrupeds. During spring and summer, instead of fishing for itself, it watches the fish-hawk, and, as soon as the latter rises from the water with a fish, it rushes forth in pursuit, and the industrious bird is obliged to drop its well-earned prey in order to save its own life, when, with the quickness of thought, the eagle sweeps down and seizes the fish while it is yet falling, and bears it away. This eagle enjoys the honor of standing as our national emblem.

The Genus *Pandion* is characterized by a rather heavy form, very long wings, tarsi thick and strong and covered with small circular scales, claws large, curved, and very rough beneath. Three or four species are known.

The American Fish-Hawk or Osprey, *P. carolinensis*, Bonap., of all temperate North America, is twenty-five inches long; the wing twenty-one inches; the head and entire under parts white; a stripe through the eye, the top of the head, and upper parts of the body, wings, and tail, deep umber-brown; the bill and claws bluish black; the tarsi and toes greenish yellow. The Osprey preys wholly or mainly upon fish, and never attacks other birds. When searching for food, it flies with easy flappings at moderate heights, and when it spies a fish checks its course, seems to poise itself for a moment, and then plunges headlong and with great rapidity into the water to secure its prey. Rising, it mounts into the air, shakes the water from its plumage, squeezes the fish in its talons, and flies to feed its young, or to a tree to satisfy its own hunger. The Osprey makes its nest on a tall tree, generally in the vicinity of water. It is four feet across, and composed of sticks, weeds, and grasses. The eggs are three or four, yellowish-white, and densely covered with large irregular spots of reddish-brown. So mild is the disposition of this bird, that it suffers others to build their nests among the outer sticks of its own nest.

The Genus *Polyborus* is characterized by rather long bill, long, pointed wings, long and rather slender tarsi, and rather weak claws.

The Mexican Eagle or Caracara Eagle, *P. tharus*, Cass., of Southern North America and of South America, is sluggish in its habits, and walks on the ground with facility.

The Genus *Craxirex* has the edges of the upper mandible festooned.

Harris's Buzzard, *C. unicinctus*, Cass., of Southern North America and of South America, is twenty-two to twenty-four inches long, the body dark brown, shoulders, wing coverts, and tibiæ reddish-chestnut, and the tail white at the base and tipped with white.

Strigidæ, or Owl Family. — This Family comprises all the nocturnal birds of prey. They are characterized by a short heavy form, large head, large eyes directed forward, curved bill nearly concealed by bristle-like feathers, large ear-cavities, and face encircled by a disk of short rigid feathers, which, with the large eyes, give to these birds an expression very much like that of a cat. This family is represented in all parts of the world. About one hundred and fifty species are known, forty of which belong to America.

The Genus *Strix* is characterized by rather small eyes, and conspicuous facial disc. It contains twelve species.

The Barn Owl, *S. pratincola*, Bonap., of all temperate North America, sixteen inches in length and the wing thirteen, is our only representative of this genus. It is found near the borders of the forest, and frequently resorts to old buildings in its search for rats and mice.

The Genus *Bubo* — Great Horned or Cat Owls — is characterized by large size, robust and powerful form, conspicuous ear-tufts, and very large eyes. There are about fifteen species of this genus.

Fig. 92.

Great Horned Owl, *B. virginianus*, Bonap.

The Great Horned Owl, *B. virginianus*, Bonap., of all North America, is twenty-one to twenty-five inches long, the wing fourteen and a half to sixteen; the male eighteen to twenty-one inches, the wing fourteen to fifteen inches. Its large size and conspicuous ear-tufts are sufficient to distinguish it from all our other owls. Its plumage is exceedingly various. This owl makes great havoc among poultry, wild turkeys, and grouse. The nest is usually on a large branch, and not far from the trunk of the tree; the eggs three to six, almost globular, and white. There are several varieties of this species.

The Genus *Scops* — Mottled or Screech Owls — is characterized by small size, conspicuous ear-tufts, imperfect facial disk, short bill nearly covered, toes long and generally partly covered with hair-like feathers.

The Mottled Owl or Screech Owl, *S. asio*, Bonap., of all temperate North America, is nine and a half to ten inches long, the wing seven inches, and the male nearly of the same size. The upper parts are pale ashy-brown with longitudinal lines of brownish black, and irregularly mottled with the same and with cinereous; the under parts ashy white, with longitudinal stripes of brownish black, and with transverse lines of the same color. Younger individuals have the upper parts pale brownish-red with longitudinal lines of brownish black, and the tail rufous with bands of brown. This owl preys upon mice, small birds, and beetles. Its notes are uttered in a tremulous, doleful manner, and may be heard several hundred yards. It often comes to the farm-houses, and alights upon the roof. The nest is in a hollow tree; the eggs are four or five, white, and nearly globular.

The Western Mottled Owl, *S. McCallii*, Cass., of Western North America, is closely related to the preceding one, but is smaller.

The Genus *Otus* — Long-eared Owls — is characterized by a longer and more slender form than the preceding genera, moderate-sized head, ear-tufts long and erectile, facial disk well marked, and eyes small and surrounded by radiating feathers. Ten or twelve species are known, only one of which belongs to North America.

The Long-eared Owl, *O. Wilsonianus*, Lesson, of temperate North America, is fifteen inches long, the wing eleven, and readily distinguished by its long ear-tufts. It lingers about mountain streams, perching on a low tree or shrub. When disturbed, it does not fly, but bounds into the thicket, and makes off by long leaps. Its cry is prolonged and plaintive, consisting of two or three notes repeated at intervals. It rears its young in nests which it finds, seldom making one for itself.

The Genus *Brachyotus* — Short-eared Owls — is characterized by short and inconspicuous ear-tufts.

The Short-eared Owl, *B. Cassinii*, Brewer, of temperate North America and Greenland, is fifteen inches long, and the wing twelve inches.

The Genus *Syrnium* — Gray Owls — is characterized by large size, large head, absence of ear-tufts, rather small eyes, tail usually rounded at the end, and the prevalent colors of the plumage gray and cinereous.

The Great Gray Owl, *S. cinereum*, Aud., of Northern North America, is the largest of our owls, being twenty-five to thirty inches long, and the wing eighteen. In winter it wanders over a large part of the United States.

The Barred Owl, *S. nebulosum*, Gray, of Eastern North America, is about twenty inches long, and the wing thirteen to fourteen inches; the upper parts light ashy-brown, with transverse narrow bands of white; the breast with transverse bands of brown and white; the abdomen ashy white, with longitudinal stripes of brown. It destroys poultry, hares, and birds, and its cry is very loud and discordant. It lays four to six globular white eggs in a hollow tree.

The Genus *Nyctale* comprises small owls with small eyes, and ear-tufts which are only observable when erected. Four species belong to America.

The Sparrow Owl, *N. Richardsonii*, Bonap., of Northern North America, is about ten and a half inches long, and the wing seven and a half inches; the upper parts pale reddish-brown tinged with olive, with partially concealed spots of white; the under parts ashy white, with longitudinal stripes of reddish brown; the quills brown, with small spots of white on their outer edge, and large spots of white on their webs; the tail brown, every feather with about ten pairs of white spots.

Kirtland's Owl, *N. albifrons*, Cass., of Northern North America, is about eight inches long; the head, upper part of the breast, and entire upper parts, chocolate-brown; the forehead, throat, and a line running downwards from the base of the under mandible, white; the other under parts of the body reddish ochre-yellow.

The Acadian or Saw-whet Owl, *N. acadica*, Bonap., of temperate North America, is seven and a half to eight inches long; the upper parts brown tinged with olive; the under parts ashy white, with longitudinal stripes of pale reddish-brown. Its notes bear a strong resemblance to the noise made in filing the teeth of a large saw.

The Genus *Athene*—Burrowing Owls—is characterized by small size, rather long legs thinly covered with short feathers, and nearly or quite naked toes.

The Burrowing Owl, *A. hypugæa*, Bonap., is about nine inches long, and is found from the Mississippi to the Rocky Mountains. It lives in the holes of the Prairie Dog.

The Burrowing Owl, *A. cunicularia*, Bonap., is about ten and a half inches long, and is found west of the Rocky Mountains and in South America.

The Genus *Glaucidium*,—Pygmy Owls.

The Pygmy Owl, *G. gnoma*, Cass., of Oregon and California, is seven inches long, and the wing three inches and three fourths, being the smallest owl in North America.

Fig. 93.

Snowy Owl, *Nyctea nivea*, Gray.

The Genus *Nyctea* comprises the Snowy or White Owl, *Nyctea nivea*, Gray, of the northern regions of both continents, which is twenty-four to twenty-seven inches long, the wing sixteen to seventeen inches, and the entire plumage white, frequently with spots or irregular bars of dark brown. In winter it wanders over a great part of North

America and Europe. Unlike those before described, it hunts in the daytime as well as in twilight. It preys upon quadrupeds, birds, and fishes. It captures ducks and other birds upon the wing, striking them much after the manner of falcons.

The Genus *Surnia* comprises the Hawk Owl or Day Owl, *S. ulula*, Bonap., of the northern regions of both continents, which is sixteen to seventeen inches long, the wing nine inches, the upper parts brown, the throat white, a large brown spot on each side of the breast, and the other under parts with transverse stripes of pale ashy-brown. It has the general appearance and habits of both an owl and a falcon, and is mainly a diurnal bird.

SUB-SECTION II.

THE ORDER OF SCANSORES, OR CLIMBERS.

THE Order of Scansores comprises all birds which have their toes in pairs, two in front and two behind, the outer anterior one being usually directed backwards, — an arrangement which especially facilitates climbing. This order comprises five families, — Psittacidæ or Parrot Family, Ramphastidæ or Toucan Family, Trogonidæ or Trogon Family, Cuculidæ or Cuckoo Family, and Picidæ or Woodpecker Family.

PSITTACIDÆ, OR PARROT FAMILY. — This Family comprises birds which have a stout, thick, rounded bill, hooked at the tip, and the base covered with a soft skin or cere, as in the hawks. The tongue is thick and fleshy, the inferior larynx complicated and furnished with three muscles on each side, and their jaws are set in motion by a greater number of muscles than is found in other birds. Most of them are adorned with varied and gorgeous plumage, which, together with the facility with which many of them may be trained to imitate the human voice, has ever made them objects of attention. They are numerous, and confined to the tropical and subtropical regions of both hemispheres, mainly to the former, and each region has its peculiar species. They are known as Parrots, Macaws, Cockatoos, Paroquets, and are often tamed and kept as pets.

The Genus *Conurus* is characterized by feathered cheeks, and a long, conical pointed tail.

The Carolina Parrot, *C. carolinensis*, Kuhl., of the Southern and Southwestern States, is the only representative of its genus in the United States. It is thirteen inches long, the wing about eight inches, the head and neck gamboge-yellow, forehead and sides of the head brick-red, body and tail green, edge of wing yellow, and bill white. It is generally found in flocks.

Fig. 94.

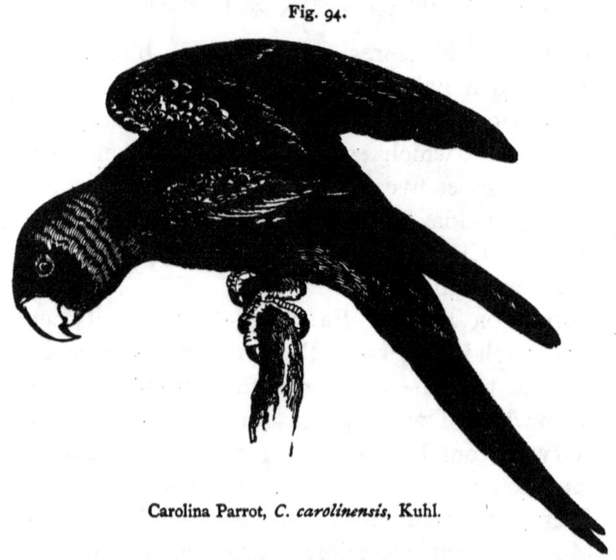

Carolina Parrot, *C. carolinensis*, Kuhl.

RAMPHASTIDÆ, OR TOUCAN FAMILY. — This Family comprises birds which are distinguished from all others by their enormous bill, which is almost as thick and long as their body, and which is light and cellular internally, arcuated near the end, and irregularly indented along its edges. Their tongue is long, and fringed with barbs on both sides. They inhabit the warm regions of South America. Contrary to what might be supposed, they are graceful in their movements, and in obtaining their food show a use for their long bills. They dip them down into the deep and hanging nests of other birds, and extract the eggs and young for food. They also feed upon fish, insects, and fruit. They throw their food into the air, and catch it as it descends, and thus swallow it with greater facility. About twenty species are known.

TROGONIDÆ, OR TROGON FAMILY.—This Family comprises birds with a broad short bill, the tip hooked and dentate, and the base surrounded by long stiff bristles, the wings short and rounded, tail elongated, legs small, tarsus short and hidden in the plumage. They live upon insects, build their nests in hollow trees, and are but little upon the wing. They are found in both continents.

The Genus *Trogon* — Trogons — has the bill broad, both mandibles serrated, and the anterior toes united beyond the first joint.

The Mexican Trogon, *T. mexicanus*, Sw., is the only representative of its family in North America. It is ten and three fourths inches long, the wing five and a quarter inches; the color golden green above, and on the neck all round; under parts carmine; forehead, chin, and side of the head black.

CUCULIDÆ, OR CUCKOO FAMILY.—This Family comprises birds with a compressed, gently curved, and generally lengthened bill, lengthened tarsi, rather short toes, long and soft tail, with eight to twelve feathers.

The Genus *Crotophaga* has the bill shorter than the

head, high and much compressed; tail feathers eight, and the plumage black.

The Black Parrot, *C. rugirostris*, Sw., of Florida to Brazil, is about fifteen inches long, and the wing six inches; and the bill with faint transverse wrinkles on the gently decurved culmen.

The Ani, *C. ani*, Linn., of the Atlantic coast of Southeastern North America and southward, is twelve inches long, the wing over six inches, the bill smooth, and the culmen abruptly decurved.

The Genus *Geococcyx* has the bill longer than the head, loral feathers stiff and bristly, and the tail feathers ten. Birds of this genus live on the ground.

The Paisano, Road Runner, or Chaparral Cock, *G. californianus*, Baird, of Texas to California, the only representative in the United States, is twenty to twenty-three inches long, and the wing six inches; all the feathers of the upper parts and wings of a dull metallic olivaceous-green, broadly edged with white near the end; under parts whitish. The legs are very long, and it can run faster than a fleet horse. It frequents the highways.

The Genus *Coccygus* — Cuckoos — comprises birds which have the bill shorter than the head, decurved, slender, and attenuated towards the end; loral feathers soft, tarsi shorter than the toes, and the tail feathers ten.

The Yellow-billed Cuckoo, *C. americanus*, Bonap., of the United States east of the Missouri plains, is twelve inches long, the wing about six inches; the upper parts metallic greenish-olive, under parts white. The upper mandible and tip of the lower are black; the rest of the lower mandible and edges of the upper, yellow. The flight of this bird is silent, rapid, and horizontal. It is shy, seeking the thickest foliage, where it sits by the hour uttering its unpleasant notes, which may be represented by *cow cow* eight or ten times repeated. It feeds on in-

Fig. 95.

Yellow-billed Cuckoo, *C. americanus*, Bonap.

sects, and upon eggs, which it steals from the nests of other birds. It builds its nest of dry sticks and grass, on a horizontal branch. The eggs are four or five, bright green.

The Black-billed Cuckoo, *C. erythrophthalmus*, Bonap., of the same region, and of about the same size and general appearance and habits as the preceding one, may at once be distinguished by its entirely black bill, and the naked red skin around the eye.

The Mangrove Cuckoo, *C. minor*, Cab., of Florida, is readily distinguished from both the preceding by its fulvous under parts and dark ear coverts.

PICIDÆ, OR WOODPECKER FAMILY.—This Family comprises birds characterized by a straight, rigid, and sharp bill, which is specially adapted to cutting into bark or wood; and by a long, acute tongue, armed towards the tip with barbs, and capable of great extension. They have stout feet, clothed before with broad plates, long wings, ten primaries, and twelve tail feathers, the exterior being small and concealed. Woodpeckers feed upon the larvæ of insects, which they secure by introducing their extensible tongue under the bark of trees, or into crevices, or into holes which they themselves have made, and then transfixing the larvæ with the barbed point; or the larvæ adhere to the viscid glue with which the tongue is covered. They are very common in both hemispheres, and about twenty-five species are found in North America.

The Genus *Campephilus* comprises Woodpeckers which have the bill considerably longer than the head, feet large, tail long and cuneate, wings long and pointed, the color black with white patches, and the head with a pointed crest.

The Ivory-billed Woodpecker, *C. principalis*, Gray, of the Southern States, is twenty-one inches long, the wing ten inches, the bill ivory-white, the body glossy-black, a stripe on each side of the neck and markings upon the wings white, and the crest scarlet. The female is without red upon the head, and with two spots of white on the end of the outer tail feather. This bird moves from the top of one tree to that of another by a single sweep, forming a most elegant curve. Its notes are clear and loud, yet plaintive. It makes its nest in the trunk of a live tree, in a hole which the male and female excavate by their mutual labor.

The Imperial Woodpecker, *C. imperialis*, of Mexico and Central America, is very similar to the preceding one, but larger.

The Genus *Picus* comprises small Woodpeckers which have the bill about equal in length to the head, or a little longer, and the lateral ridges conspicuous; the colors black and white. The four first mentioned below have the middle of the back longitudinally streaked with white, the under parts white, and a narrow red nuchal band; and the first two have the outer tail feathers pure white.

The Hairy Woodpecker or Sapsucker, *P. villosus*, Linn., of North America, is eight to eleven inches long, and, in addition to the above characteristics, has the wing coverts and innermost secondaries conspicuously spotted with white.

Harris's Woodpecker, *P. Harrisii*, Aud., from the eastern slope of the Rocky Mountains to the Pacific, is nine and a half inches long, and, in addition to the above characteristics, has the wing coverts uniform black, without spots.

The next two have the outer tail feathers white, with black transverse bands.

The Downy Woodpecker, *P. pubescens*, Linn., of Eastern North America, is six and a quarter inches long, and is a miniature of the Hairy Woodpecker.

Gairdner's Woodpecker, *P. Gairdneri*, Aud., from the eastern base of the Rocky Mountains to the Pacific, is six and three fourths inches long, with wing coverts and innermost secondaries nearly uniform black.

The two following have the middle of the back banded transversely with white and black, the under parts white with black spots upon the sides.

Nuttall's Woodpecker, *P. Nuttalli*, Gambel, of California, is about seven inches long, and, in addition to characteristics mentioned above, has the entire crown black, nape red, both spotted with white, and the feathers at the base of the bill white.

The Texas Sapsucker, *P. scalaris*, Wagler, of the Rocky Mountains, is six and a quarter inches long; crown and nape red, spotted with white.

The Red-cockaded Woodpecker, *P. borealis*, Vieill., of the Southern States, is seven and a quarter inches long, the wing four and a half inches; the upper parts, with top and sides of the head, black; the back, rump, and scapulars transversely banded with white; a silky patch on the side of the head, and the under parts generally, white.

The White-headed Woodpecker, *P. albolarvatus*, Baird, of Oregon and California, is about nine inches long, bluish-black, the head and a patch at the base of the primaries white.

The Genus *Picoides* comprises Woodpeckers with only three toes to each foot. The color generally is black above and white beneath ; the crown with a square yellow patch, a white stripe behind the eye, and another beneath it ; the quills spotted with white ; and the sides banded transversely with black.

The Black-backed Three-toed Woodpecker, *P. arcticus*, Gray, of Northern North America, is nine and a half inches long, the wing five inches, and is distinguished by its black back.

The Banded Three-toed Woodpecker, *P. hirsutus*, Gray, of the Arctic regions of North America, is about nine inches long, the wing four and three fourths inches, and the back transversely banded with white.

The Striped Three-toed Woodpecker, *P. dorsalis*, Baird, of the Rocky Mountains, is nine inches long, the wing five inches, and the back streaked longitudinally with white.

The Genus *Sphyrapicus* comprises Woodpeckers which have the lateral ridge of the bill very prominent, but terminating at the middle of the commissure ; the outer pair of toes longest, inner posterior one very short, wings long and pointed, fourth quill longest, the tail feathers very broad, abruptly acuminate, and with a long point. The colors are black and white, with a central yellow line on the belly.

The Yellow-bellied Woodpecker, *S. varius*, Baird, of North America east of the Rocky Mountains, is eight and a quarter inches long, the wing about four and three quarters inches, and, in addition to characteristics before mentioned, it has the crown red bordered with black, chin and throat red, a black patch upon the breast, and the outer and inner tail feathers varied with white.

The Red-breasted Woodpecker, *S. ruber*, Baird, of Western North America, is less than nine inches long, the wing five inches, the head, neck, and breast red.

Williamson's Woodpecker, *S. Williamsonii*, Baird, of

the Rocky Mountains, is nine inches long, the wing five inches; the head, neck, sides of the breast, and body, black; a stripe behind the eye white, a narrow line on the chin and throat red, tail feathers wholly black, and the back scarcely spotted.

The Brown-headed Woodpecker, *S. thyroideus*, Baird, of Western North America, is about nine inches long, the wing five inches; the head dark ashy-brown, the rest of the body encircled by transverse bands of black and brownish white, excepting a large, round black patch upon the breast; and, as in the preceding members of the genus, the central line beneath is yellow.

The Genus *Hylotomus* is represented by the Black Woodcock, *H. pileatus*, Baird, of North America generally, which is about eighteen inches long, the wing nine and a half inches; the general color dull greenish-black, a narrow white streak above the eye, a wider one under the eye and along the sides of the head and neck; the sides of the breast, under wing coverts, chin, and beneath the head, white tinged with yellow; and the entire crown, from the base of the bill to a well-developed occipital crest, scarlet. This bird is very shy, and, when followed by the hunter, goes rapidly from one tree to another, alighting upon the tallest, and generally keeping on the side farthest from the pursuer. Its notes are very loud and clear.

The Genus *Centurus* comprises Woodpeckers which are banded above transversely with black and white, the rump white, the head and under parts brown, and the belly with a red or yellow tinge.

The Red-bellied Woodpecker, *C. carolinus*, Bonap., of North America east of the Rocky Mountains, is nine and three fourths inches long, the wing five inches, and, in addition to the characteristics named above, it has the crown and nape red, forehead white tinged with red, and the middle of the belly red.

The Yellow-bellied Woodpecker, *C. flaviventris*, Sw., of the Rio Grande region, is nine and a half inches long, the wing five inches, with a square red patch on the crown.

The Gila Woodpecker, *C. uropygialis*, Baird, of the Lower Colorado region, is nine inches long.

The Genus *Melanerpes* comprises Woodpeckers which have the back black, with or without a white rump, and variable beneath, but without transverse bands.

Fig. 96.

Red-headed Woodpecker, *M. erythrocephalus*, Sw.

The Red-headed Woodpecker, *M. erythrocephalus*, Sw., of North America, east of the Rocky Mountains, is nine and three fourths inches long, the wing five and a half inches, and the head and neck all round crimson, margined with a narrow crescent of black upon the upper part of the breast; the back, primaries, and tail, black; the under parts, a broad band across the middle of the wing, and the rump, white. It excavates a hole for its nest in a decaying tree; eggs four to six, pure white.

The California Woodpecker, *M. formicivorus*, Bonap., of California and eastward, is about nine inches long, the wing five inches; above and on the anterior half of the body glossy black; the top of the head and a short occipital crest, red; forehead, rump, and belly, white; the sides of head, chin, and broad pectoral band, black; a collar on the throat passing up before the eyes into the frontal band, white tinged with yellow.

Lewis's Woodpecker, *M. torquatus*, Bonap., of Western North America, is ten and a half inches long, the wing six and a half inches; the color dark glossy green above; the breast, lower part of the neck, and a narrow collar all round, grayish white; base of the bill and sides of the head, dark crimson; belly red, streaked with whitish.

The Genus *Colaptes* comprises Woodpeckers with a slender bill, large feet, and long tail.

The Golden-winged Woodpecker, *C. auratus*, Sw., of North America east of the Rocky Mountains, is one of the most beautiful birds of this family. It is twelve and a half inches long, the wing six inches, the top of the head and the upper part of the neck bluish ash, a red crescent on the nape, the other upper parts, except the pure white rump, light olivaceous-brown with transverse bands of black, and a patch upon the cheeks black; the lower parts yellowish white tinged with brownish and ornamented with circular black spots, and with a black crescent upon the breast. The shafts and under surfaces of the wings and tail feathers are gamboge-yellow. On the first sunny days of spring the Golden-winged Woodpeckers appear on the tops of the decayed trees, and, as they hop about, striking with their bills here and there, make the woods resound with their loud, merry notes. Soon they are paired, and both male and female begin to excavate a hole in a tree for the nest. The female lays four to six beautiful white eggs for each brood, and two broods are reared in a season.

Fig. 97.

Golden-winged Woodpecker, *C. auratus*, Sw.

The Red-shafted Flicker, *C. mexicanus*, Sw., of Western

North America, is thirteen inches long, the wing over six and a half inches, the shafts and under surfaces of the wing and tail feathers orange-red, a red patch on each side of the cheek, throat and stripe beneath the eye bluish ash, and the back glossed with purplish brown. The female has no red on the cheek.

C. hybridus, Baird, is the name given to woodpeckers from the Upper Missouri, which combine characteristics common to both the preceding species.

SUB-SECTION III.

THE ORDER OF INSESSORES, OR PERCHERS.

THE Order of Insessores embraces far more species than any other in the whole class of birds, and those which in many cases seem widely different from one another; but they agree in many important respects, especially in their feet, which have three toes directed forward and one behind, the latter being on the same level with the others. This Order naturally divides into three groups, which may be called Sub-Orders, — Strisores, Clamatores, and Oscines.

The Sub-Order of Strisores comprises birds which have the hind toe versatile, or capable of being turned more or less laterally forward, thus making the bird appear to have four toes in front. They have ten primaries, and the tail feathers are never more than ten. The Strisores comprise three families, — Trochilidæ or Humming-Bird Family, Cypselidæ or Swift Family, Caprimulgidæ or Goat-sucker Family.

TROCHILIDÆ, OR HUMMING-BIRD FAMILY. — This Family comprises birds of the smallest size, and of the most gorgeous plumage to be found in the feathered race. The beauty and splendor of their colors are beyond description. One might as well attempt to describe the rainbow, as the hues of emerald, and ruby, and amethyst, and topaz, and burnished gold, which flash from these beautiful forms of life, as they glance among the foliage, or dart from flower to flower seeking their accustomed food. They belong exclusively to the continent and islands of America, and are the most numerous in the hot regions. Some species range north to the Arctic regions, and south to Patagonia ; and from the level of the sea to the cold heights of the Andes. The feet of humming-birds are very small, the wings very long and narrow, and the tail broad. Everything in their organization contributes to give them great power and rapidity of flight ; and they are able to balance themselves in the air, or beside a flower, with a facility that is truly wonderful, and which finds a parallel only among some of the insect tribes. The bill is awl-shaped, thin, sharp-pointed, straight, or curved ; in some cases as long as the head, and in others much longer. The mandibles are excavated to the tip for a lodgment of the tongue, and form a tube by the close fitting of their cutting edges. The tongue, which is split almost to its base, forming two hollow threads, is protruded at will, like that of the woodpeckers, and by the same sort of mechanism. The food consists of insects and honey, which are secured by extending the tongue into flowers without opening very wide the bill. About four hundred species are known ; and six or more are found in North America.

Fig. 98.

Humming-Bird, *T. colubris*, Linn.

The Black-throated Humming-Bird, *Lampornis mango*, Sw., of South America and perhaps northward to Florida, is four and a half inches long, and the wing two and six tenths inches.

The Ruby-throated Humming Bird, *Trochilus colubris*, Linn., of North America to Brazil, is three and a quarter inches long, and the wing one inch and six tenths.

The Black-chinned Humming-Bird, *T. Alexandri*, Bourc. & Muls., of California and southward, is three inches and three tenths long, and the wing one inch and seven tenths.

Fig. 99.

Humming-Bird's nest, *T. colubris*, Linn.

The Red-backed Humming-Bird, *Selasphorus rufus*, Sw., of Western North America, is three and a half inches long, and the wing over one inch and a half.

The Broad-tailed Humming-Bird, *S. platycercus*, Gould, of Mexico and Texas, is three and a half inches long, and the wing a little less than two inches.

The Anna Humming-Bird, *Atthis anna*, Reichenb., of California, is three and six tenths inches long, and the wing two inches.

The Ruffed Humming-Bird, *A. costæ*, Reichenb., of Southern California, is three and two tenths inches long, and the wing one inch and three fourths.

CYPSELIDÆ, OR SWIFT FAMILY. — This Family comprises small, dull-colored birds, which have the general appearance of swallows, but differ from the latter in many essential characteristics. The Swifts have a much smaller and shorter bill, with the edges greatly inflected; the nostrils are superior instead of lateral, and without bristles; the wing more falcate, and with ten primaries instead of nine; the tail with ten feathers instead of twelve; the feet are weaker, the hind toe more or less versatile, and the anterior toes usually lack the normal number of joints; and there are peculiarities in their vocal organs.

The Genus *Panyptila* comprises Swifts which have the legs thick, hind toe directed laterally, legs feathered to the claws, second primary longest, and tail forked.

The White-throated Swift, *P. melanoleuca*, Baird, of

New Mexico, is five and a half inches long, and the wing five inches.

The Genus *Nephœcetes* comprises Swifts with naked, slender legs, forked tail, and first primary longest.

The Northern Swift, *N. niger*, Baird, of Northwestern America to the West Indies, is six and three quarters inches long, with the wing of the same length. The general color is dark sooty-brown, with a greenish gloss.

The Genus *Chætura* is characterized by the even tail and stiffened shafts projecting as spinous points.

The Chimney Swallow, *C. pelasgia*, Steph., of North America east of the Rocky Mountains, is five and a quarter inches long, the wing over five inches.

The Oregon Swift, *C. Vauxii*, De Kay, of the Pacific coast, is less than five inches long, and the wing four and three fourths inches.

CAPRIMULGIDÆ, OR GOAT-SUCKER FAMILY. — This Family comprises birds with a short triangular bill, and soft, lax, owl-like plumage. They feed upon insects, which they capture while upon the wing.

The Genus *Antrostomus* is characterized by a bill with conspicuous bristles, rounded wings, broad graduated tail, and very lax plumage.

Chuck-will's Widow, *A. carolinensis*, of the South Atlantic and Gulf States, is twelve inches long, the wing eight and a half inches, the bristles of the bill with lateral filaments; general color pale rufous, top of the head reddish-brown streaked with black, and the terminal two thirds of the tail, except the four central feathers, rufous-white. The female is without the white upon the tail. In early spring the forests echo with the notes of this interesting bird.

The Whippoorwill, *A. vociferus*, Bonap., of Eastern United States to the Central Plains, is ten inches long, the wing six and a half inches, the bristles of the bill

Fig. 100.

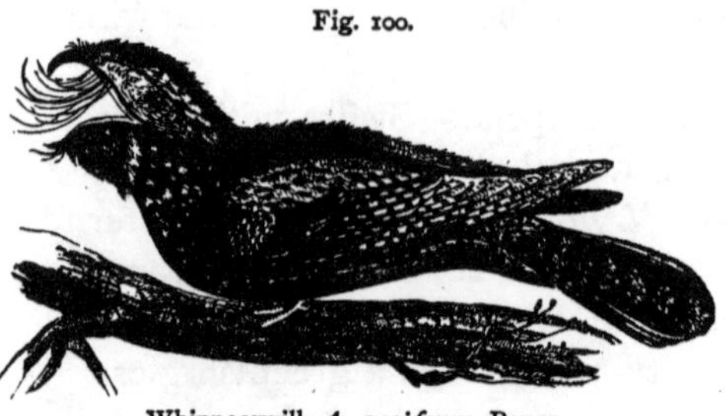

Whippoorwill, *A. vociferus*, Bonap.

without lateral filaments. The general color is similar to that of the preceding, the top of the head ashy gray, longitudinally streaked with black. Its notes are three, and have a fancied resemblance to the syllables *whip-poor-will*, and hence its name. It begins its song soon after sunset, and continues till late at night; then remains silent till near the dawn, when it resumes and continues till sunrise. During the day the Whippoorwill sleeps upon the ground, or on fallen trunks of trees, or on low branches, and may often be approached to within a few feet before it flies. It is said that it always sits with its body parallel to the branch on which it alights, and never across it. Its eggs are always two, short elliptical, much rounded, and nearly equal at both ends; the color greenish white, spotted and blotched with bluish gray and light brown. These are laid in May, on the bare ground or on dry leaves, and in the most secluded parts of the thickets.

Nuttall's Whippoorwill, *A. Nuttalli*, Cass., of the high central plains and westward to the Pacific coast, is eight inches long, the wing five and a half inches.

The Genus *Chordeiles* has the bill without bristles, or with very feeble ones, the wings very long and pointed, tail narrow, forked, and plumage rather compact.

The Night-Hawk, *C. popetue*, Baird, of North America generally, is nine and a half inches long, and the wing over eight inches, and is so well known as to require no further description here. Night-Hawks are not strictly nocturnal, as the name implies; but are often upon the wing throughout the entire day, especially if it be cloudy. They are generally most active just before night, and retire to rest at dark. Their loud, squeaking notes are familiar to all. The singular loud and half-booming sound which they make in plunging from a great height is said to be produced by the concussion caused by the new position of the wings at the moment when the bird passes the centre of its plunge and commences the ascent. The Night-Hawk makes no nest, but deposits its two oval, freckled eggs on the bare ground, or on a flat rock, in fields or in very open woods.

Fig. 101.

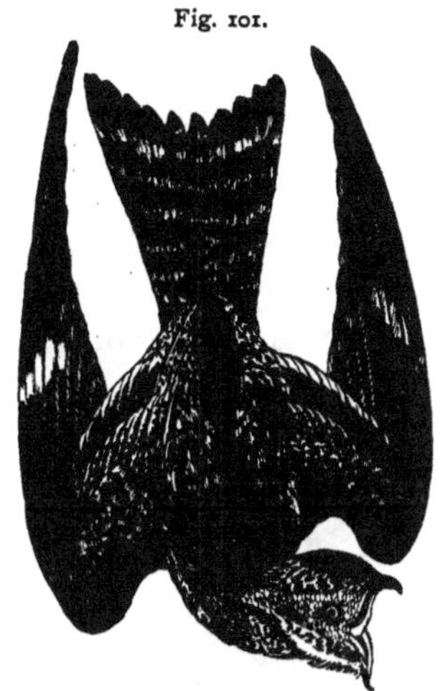

Night-Hawk, *C. popetue*, Baird.

Some persons suppose that the Night-Hawk and Whippoorwill are identical, but they do not even belong to the same genus. The Night-Hawk has the bristles of the bill hardly appreciable, wings sharp-pointed, longer than the tail, which is rather narrow, and forked or emarginate. The Whippoorwill has the mouth margined by long stiff bristles, the wings short, not reaching the end of the tail, which is short and rounded ; and they differ in their colors and markings.

The Western Night-Hawk, *C. Henryi*, Cass., of New Mexico, and the Texas Night-Hawk, *C. Texensis*, Lawrence, are other North American species.

The Sub-Order of Clamatores comprises birds which have three toes before and one behind, and the latter not versatile ; the primaries ten, the first nearly as long as the second ; and the tail feathers usually twelve. It comprises three Families, — Alcedinidæ or Kingfisher Family, Pronitidæ or Saw-bill Family, and Colopteridæ or Flycatcher Family.

ALCEDINIDÆ, OR KINGFISHER FAMILY. — This Family comprises birds with a large head, a long, straight, sub-pyramidal bill, very small tongue, short wings, small legs, very short tarsi, and the outer and middle toes united half their length.

The Genus *Ceryle* comprises two species.

Fig. 102.

Belted Kingfisher, *Ceryle alcyon*, Boie.

The Belted Kingfisher, *Ceryle alcyon*, Boie, of the entire continent of North America, is about thirteen inches long, the wing six inches, the head with a long crest; the color above, blue; beneath, and a spot anterior to the eye, white; a band across the breast, and the sides of the body, like the back; the primaries white on their basal half, and the tail with transverse bands and spots of white. Ponds and slow streams are its favorite resorts, near which it sits on a branch or decayed limb, and watches for fish, which constitutes its food. At the proper moment it plunges headlong into the water, seizes the fish, flies to the nearest tree, and swallows its victim in a moment, and is immediately on the watch for another. Its notes are harsh. The nest is made in a horizontal hole excavated in a high bank of a stream, by the mutual labor of the male and female to the depth of three to six feet; eggs six, pure white.

The Texas Kingfisher, *C. americana*, Boie, is much smaller, being only eight inches long, and the wing three and a quarter inches; and the head is only slightly crested. The upper parts and a pectoral and abdominal band of blotches, and a line on each side of the thorax, are glossy green; the under parts generally, a collar on the back of the neck, and a double series of spots on the quills, white.

PRONITIDÆ, OR SAW-BILL FAMILY.—This Family comprises birds with the bill as long as the head, gently decurved near the tip, but not hooked, and the cutting edges dentated; the tarsi rather long, feet large, the middle and outer toes connected for more than half their length. Three genera are known, only one of which is represented in North America.

The Genus *Momotus* has the bill as long as the head, both mandibles dentated, and the tail long.

The Saw-Bill, *M. cæruliceps*, Gould, of Mexico, is fifteen inches long, the wing five and a half, the general color yellowish-green, the top of the head and occipital crest bright blue encircled with black.

COLOPTERIDÆ, OR FLYCATCHER FAMILY.—This Family comprises small birds which connect the non-melodious birds with the Oscines, to be noticed hereafter. The bill in most cases is bent abruptly down at the tip, before which is a slight notch; the sides of the mouth are provided with stiff bristles; the wings of moderate length, the first primary always more than half the length of the second, usually nearly as long as the latter. This large family is represented in North America by about thirty species.

The Genus *Pachyrhamphus* has the Rose-throated Flycatcher, *P. aglaiæ*, Lafresn., of Mexico to the Rio Grande, as its only representative in North America. It is seven and a half inches long, and the wing three and three fourths inches.

The Genus *Milvulus* is characterized by a bill shorter than the head, tail twice as long as the wing and excessively forked.

The Fork-tailed Flycatcher, *M. tyrannus*, Bonap., of South America and accidental in the United States, is fourteen inches long, wing four and three fourths inches, the top and sides of the head glossy-black, the rump, wings, and tail almost black, the rest of the upper parts ash-gray, the under parts white, and the crown with a concealed patch of yellow.

The Scissor-tail, *M. forficatus*, Sw., of Texas to Mexico, is thirteen inches long, the wing four and three fourths inches, and is distinguished by the habit of closing and opening the long feathers of the tail, like the blades of a pair of scissors.

The Genus *Tyrannus* has the tail rather shorter than the wings, and nearly even or only moderately forked, the wings long and pointed, the outer primaries attenuated, and the head with a concealed patch of red.

The Kingbird or Bee Martin, *T. carolinensis*, Baird, of North America east of the Rocky Mountains, is eight and a half inches long, the wing less than five inches, the color above dark bluish-ash, the lower parts white, the sides of the throat and across the breast tinged with pale bluish-ash, the top and sides of the head bluish black, and the concealed crest vermilion in the centre, white behind, and before partially mixed with orange. Its favorite resorts are open fields or orchards. Perched upon a stake or tall weed

Fig. 103.

Kingbird, *T. carolinensis*, Baird.

or a low tree, it watches for insects, which it darts upon with unerring aim. It is very courageous, never hesitating to attack hawks, crows, and other large birds which it dislikes.

The Gray Kingbird, *T. dominicensis*, Rich., of the West Indies and accidental in the Southern States, is eight inches long, the wing less than four and three fourths inches.

The Arkansas Flycatcher, *T. verticalis*, Say, of Western North America, is eight and a quarter inches long, the wing four and a half inches, the general color ashy above, yellow beneath, tail nearly black, wings brown, pectoral band pale ashy, and the crest vermilion in the centre, and yellowish before and behind.

Cassin's Flycatcher, *T. vociferans*, Sw., of Mexico and Texas, is nearly nine inches long, and the wing five and a quarter inches.

Couch's Flycatcher, *T. Couchii*, Baird, of Mexico, is nine inches long, and the wing five inches.

The Genus *Myiarchus* has the bill wide at base; tarsus equal to, or not longer than, the middle toe; tail broad, long, even, or slightly rounded, and about equal to the wings, which scarcely reach to the middle of the tail; the head with elongated distinct feathers The general color above brownish olive, the throat ash, and the belly yellow.

The Great-crested Flycatcher, *M. crinitus*, Cab., of North America east of the Missouri, is eight and three fourths inches long, the wing four and a quarter inches, the general color dull greenish-olive above, sides of the head as high as the upper eyelid, and sides of the neck, throat, and fore part of the breast, bluish ashy; the other under parts bright sulphur-yellow; head with a depressed crest. This species delights in forests. Its flight is rapid and powerful. Seeing an insect, it sweeps downward, secures it, returns to the tree, swallows its victim, erects its crest, and immediately utters its harsh, squeaking note.

The Ash-throated Flycatcher, *M. mexicanus*, Baird, of California to Texas, closely resembles the preceding, but the bill is narrower and blacker, tarsi longer, wings less pointed, the throat and fore part of the breast white, and the sulphur-yellow of the under parts pale.

Cooper's Flycatcher, *M. Cooperi*, Baird, and Lawrence's Flycatcher, *M. Lawrencii*, Baird, of Mexico, are other North American species.

The Genus *Sayornis* is characterized by a depressed, moderate crest, rather narrow bill, tail long, broad, and slightly forked, and equal to the wings, which reach to the middle of the tail.

The Black Flycatcher, *S. nigricans*, Bonap., of Western North America, is about seven inches long, the wing about three and a half inches, the general color sooty-brown, under parts and edge of the tail white.

The Pewee or Phœbe Bird, *S. fuscus*, Baird, of Eastern North America, is seven inches long, the wing less than three and a half inches, the general color above brownish olive, crown darker, the under parts and edge of the tail yellowish. The Pewee lingers around bridges, old mills, and caves, in some secure part of which it makes its nest of mud, grass, and moss, with a soft lining within to receive the pure white eggs with reddish spots near their larger end.

Say's Flycatcher, *S. Sayus*, Baird, of Western North America, is seven inches long, the wing four and a third inches, the general color grayish brown, reddish cinnamon beneath.

The Genus *Contopus* is characterized by very short, stout tarsi, wings very long and much pointed, and reaching beyond the middle of the tail, the head moderately crested, the general color olive above, pale yellowish beneath, with a darker patch on the sides of the breast.

The Olive-sided Flycatcher, *C. borealis*, Baird, of Greenland and rare upon the Atlantic and Pacific coasts of North America, is seven and a half inches long, the wing less than four and a half inches; in addition to the above-mentioned characteristics, it has a silky white tuft on each side of the rump.

The Short-legged Pewee, *C. Richardsonii*, Baird, of North America, is over six inches long, the wing over three and a half inches.

The Wood Pewee, *C. virens*, Cab., of Eastern North America to the high central plains, is six and a quarter inches long, the wing three and a half inches; the upper parts, side of the head, neck, and breast, dark olivaceous-brown; the lower parts pale yellowish tinged with ash across the breast; a ring around the eye, and two narrow bands across the wing, white. The Wood Pewee loves the dark, quiet retreats of the forest. Here, sitting on a dry branch, it may always be found in summer and early autumn, watching for insects, and, with its wings quivering, uttering its low, melancholy notes. It makes its nest on a horizontal branch, constructing it of lichens and mosses without, and of fine grasses and hairs within. The eggs are four or five, light yellowish, and spotted with reddish on the larger end.

The Genus *Empidonax* is characterized by long tarsi, tail a little shorter than the wings, head moderately crested, the general color olivaceous above, yellowish beneath.

Traill's Flycatcher, *E. Traillii*, Baird, of the Eastern United States to Mexico, is six inches long, the wing two and nine tenths inches.

The Little Flycatcher, *E. pusillus*, Cab., of Western North America, is five and a half inches long, the wing two and eight tenths inches.

The Least Flycatcher, *E. minimus*, Baird, of the United States east of the Missouri plains, is five inches long, the wing two inches and six tenths.

The Small Green-crested Flycatcher, *E. acadicus*, Baird, of the United States east of the Mississippi, is five inches and six tenths long, and the wing three inches; the upper parts, with sides of the head and neck, olive green; throat and middle of the belly, white; the other under parts pale greenish-yellow.

The Yellow-bellied Flycatcher, *E. flaviventris*, Baird, of North America, is over five inches long, the wing less than three inches; and distinguished by the bright sulphur-yellow of the under parts.

Hammond's Flycatcher, *E. Hammondii*, Baird, of the vicinity of Fort Tejon, is five inches and a half long, the wing less than three inches; and at once distinguished by its very slender bill.

Wright's Flycatcher, *E. obscurus*, Baird, of the Rocky Mountains, is five and three fourths inches long, and the wing two and three fourths.

The Genus *Pyrocephalus* comprises the Red Flycatcher, *P. rubineus*, Gray, of the Rio Grande region, which is five and a half inches long, the wing three and a quarter inches, the head fully crested, crown and whole under parts bright carmine; the remaining upper parts dark brown.

The Sub-Order of Oscines comprises the true singing-birds, such as have the larynx provided with five pairs of peculiar muscles, which are used in the production of song. North America has twelve families,—Turdidæ or Thrush Family, Sylvicolidæ or Warbler Family, Hirundinidæ or Swallow Family, Bombycillidæ or Waxwing Family, Lanidæ or Shrike Family, Liotrichidæ or Mocking-Bird Family, Certhiadæ or Creeper Family, Paridæ or Titmouse Family, Alaudidæ or Lark Family, Fringillidæ or Finch Family, Icteridæ or Blackbird Family, and Corvidæ or Crow Family.

Turdidæ, or Thrush Family. — This Family comprises birds with the bill notched near the tip, wings rather long, primaries ten, of which the first is very short, the second nearly equal to the longest; tarsi usually rather long, and mainly without scutellæ, the lateral toes about equal, and the basal joint of the middle toe united by its basal two-thirds to the outer, and by its basal half to the inner toe.

The Genus *Turdus* has the bill shorter than the head and stout, culmen gently curved from the base, tarsi longer than the middle toe, lateral toes nearly equal, but the outer one longer, the wings pointed and longer than the tail, which is nearly even, or slightly emarginate.

The Wood Thrush, *T. mustelinus*, Gm., of the United States east of the Missouri and southward to Guatemala, is eight and one tenth inches long, the wing four and a quarter inches; the color above clear cinnamon-brown, the top of the head more rufous, and more olivaceous on the rump and tail; the under parts are clear white, sometimes tinged with buff before, and thickly marked with sub-triangular sharply-defined spots of blackish. The sides of the head are dark brown streaked with white, the legs yellow, bill brown except its yellow base beneath. The nest is built in a laurel, or other low shrub, and composed of leaves, grass, and mud, lined with fibrous roots; eggs four to five, of a uniform light blue. The Wood Thrush delights in deep shady glens where there is a brook or little stream, and in thick

Fig. 104.

Wood Thrush, *T. mustelinus*, Gm.

dark woods. Its soft, liquid, half-plaintive notes excel in sweetness those of any other American bird, and can only be approximated, never equalled, by those of the flute in the hands of a master. They are few in number, but possess a charm beyond description, touching the heart of every cultivated listener, and calling forth all the nobler feelings of our nature. Says Audubon, "How often, as the first glimpses of morning gleamed doubtfully amongst the dusky masses of the forest-trees, has there come upon my ear the delightful music of this harbinger of day, and how fervently on such occasions have I blessed the Being who formed the Wood Thrush, and placed it in those solitary forests, as if to console me amidst my privations, to cheer my depressed mind, and to make me feel, as I did, that man never ought to despair!"

The Hermit Thrush, *T. Pallasi*, Cab., of the United States east of the Mississippi, is seven and a half inches long, the wing over three and three quarters inches; the color above light olive-brown, passing into rufous on the rump, upper tail coverts, and tail, and with less intensity on the outer surface of the wings. The under parts white, scarcely tinged with buff across the fore part of the breast; the sides of the throat and the fore part of the breast with rather sharply defined subtriangular spots of dark olive-brown, and the sides of the breast with less distinct and paler spots of the same; and there is a whitish ring around the eye. Its song, as I have learned since the above was in type, rivals even that of the Wood Thrush.

The Dwarf Thrush, *T. nanus*, Aud., of the Pacific coast of North America, is very similar to *T. Pallasi*, but smaller, being only six and a half inches long, with the wing three and a half inches, and the white of the under parts is purer, and the sides are glossed with bluish ash instead of yellowish olive-brown, and the tail is tinged with purple.

Wilson's Thrush, *T. fuscescens*, Stephens, of North America east of the Missouri, is seven and a half inches long, the wing four and a quarter inches; the color above, and on the sides of the head and neck, nearly uniform light reddish-brown; beneath, white, the fore part of the breast and throat tinged with pale brownish-yellow; and the sides of the throat and the fore part of the breast are marked with small obscurely defined triangular spots of light brownish.

The Oregon Thrush, *T. ustulatus*, Nutt., of the Pacific coast, is seven and a half inches long, the wing three and three fourths inches.

The Olive-backed Thrush, *T. Swainsonii*, Cab., of Eastern North America, Greenland, and southward to Peru, and accidental in Europe and Siberia, is seven inches long, the wing over four inches; the color of the upper parts uniform olivaceous, with a decided shade of green; the fore part of the breast and throat pale brownish-yellow, and the rest of the lower parts white; the sides of the throat and fore part of the breast with somewhat rounded spots of well-defined brown.

The Gray-cheeked Thrush, *T. alciæ*, Baird, of the Mississippi region to the Missouri, is nearly eight inches long, the wing about four and a quarter inches; the color above dark olive-green, sides of the head ash-gray; the under parts white; the sides of the throat, and the breast, with arrow-shaped spots of dark plumbeous-brown.

The Common Robin, *T. migratorius*, Linn., of all North America to Mexico, is nine and three fourths inches long, and the wing nearly five and a half inches; and is so well known that it needs no further description. This is one of the most common and most interesting birds, coming to the temperate districts early in the spring, and

8

remaining late in the autumn. Some remain through the whole winter even in New England, but keep in the thick swamps and on the sunny sides of woods. The song of the Robin at the close of the early days of spring is among the sweetest that issues from our groves and orchards.

The Varied Thrush, *T. nævius*, Gm., of Western North America, and accidental on Long Island and near Boston, is nine and three fourths inches long, the wing five inches, and much resembles *T. migratorius*.

The Misle Thrush, *T. viscivorus*, Linn., of Europe, is brown above, the under parts of the wings white, and breast spotted. It is extremely fond of the mistletoe.

The Genus *Saxicola* comprises the Stone-Chats, small birds common in the Old World, and one species inhabits Greenland and is accidental in the northern portions of North America.

The Genus *Erythaca* comprises the Robin Redbreast, *E. rubecula*, Sw., *Motacilla rubecula*, Linn., of Europe, which is familiar to every one, by name. This pretty little bird is five and a half inches long, brownish gray above, the throat and breast red, and belly white. It delights in the presence of man, and often enters his dwelling. In the cold weather it sometimes takes up its abode in houses, and, selecting a perch, warbles its song when the day is clear or the fire burns brightly.

The Genus *Sialia* is characterized by a short, stout bill slightly notched at the tip, wings much longer than the tail, and the claws considerably curved.

The Blue-Bird, *S. sialis*, Baird, of North America east of the Rocky Mountains, is six and three quarters inches long, and the wing four inches; the color above uniform azure-blue; beneath reddish-brown, the abdomen and under tail coverts white. The female has the blue lighter, and tinged with brown on the head and back. The Blue-

Bird makes its nest in a hollow apple-tree or post, and lays four to six pale blue eggs. Two or three broods are raised in a season, and generally from one nest. While the female is sitting on the second set of eggs, the male takes charge of the first brood.

The Western Blue-Bird, *S. mexicana*, Sw., of Western North America, is six and a half inches long, the wing four and a quarter inches; the bill more slender, wings longer, and blue more intense, than in the preceding.

The Rocky Mountain Blue-Bird, *S. arctica*, Sw., of the Rocky Mountains, is six and a quarter inches long, the wing over four and a quarter inches; the color azure-blue; the belly and under tail coverts white.

The Genus *Regulus* has the bill slender, much shorter than the head, depressed at the base, moderately notched at the tip; the rictus well provided with bristles, and the nostril covered by a single bristly feather projecting forwards. The birds of this genus are very small, olive green above, and whitish beneath.

Fig. 105.

Ruby-crowned Wren, *R. calendula*, Licht.

The Ruby-crowned Wren, *R. calendula*, Licht., of North America, is four and a half inches long, and at once distinguished by the crown, which has a large concealed patch of scarlet feathers which are white at the base. Its song is clear, varied, and harmonious, charming all who hear it.

The Golden-crested Wren, *R. satrapa*, Licht., of the Northern United States from the Atlantic to the Pacific, is less than four inches long, the wing two and a quarter inches, and is distinguished by the black of the crown embracing a central patch of orange-red encircled by gamboge-yellow; the forehead, line over the eye, and space beneath it, white. It is exceedingly active, and may generally be found with other small birds gleaning among the foliage of trees and bushes in search of minute insects and larvæ.

Cuvier's Golden Crest, *R. Cuvieri*, Aud., differs mainly from the preceding in having two black bands on the crown anteriorly, separated by a whitish one.

The Genus *Hydrobata* comprises the Water Ouzels.

The American Dipper or Water Ouzel, *H. mexicana*, Baird, of the Rocky Mountains, from British America to Mexico, is seven and a half inches long, and the wing four inches; the color dark plumbeous above, and paler beneath. A closely related species is found in Europe.

Fig. 106.

Water Ouzel, *H. mexicana* Baird.

The Ouzel frequents mountain streams, into which it walks or dives, and moves about in search of aquatic insects and other small animals, which constitute its food.

Sylvicolidæ, or Warbler Family. — This Family comprises a large number of very small, but exceedingly beautiful and interesting birds. They are characterized by a conical, slender, or depressed bill usually half the length of the head, nine primaries, the first nearly as long as the second and third, the tarsi distinctly scutellate anteriorly, lateral toes nearly equal and shorter than the middle one, and the basal joint of the middle one free nearly to its base externally, and united for half the length interiorly. This family is numerously represented in all parts of the world. Many species are generally found in the same locality, and may be seen a great part of the day gliding among the thick foliage, busily engaged in catching the minute insects which lurk beneath the leaves and in the buds and blossoms, and which for the most part escape the sight of other and larger birds. Some of the warblers and some of the thrushes are so nearly related, that each is often placed in the group of the other, according to the importance attached to different characters by different writers. Some of the warblers are among the sweetest of the feathered songsters.

Fig. 107.

Nightingale, *P. luscinia*, Sw.

The Genus *Philomela* comprises the Nightingale, *P. luscinia*, Sw., of Europe, the sweet and celebrated songster of the night. It is about six inches long, reddish brown above, whitish gray beneath.

More than fifty species of Warblers are found in the United States.

The Genus *Anthus* has the bill slender, much attenuated, and distinctly notched, the wing very long, the first primary nearly equal to the longest, and the tertials almost as long as the primaries.

The Tit-Lark, *A. ludovicianus*, Licht., of North America generally and accidental in Europe, is six and a half inches long, the wing three and three fourths inches; the color above olive brown, beneath dull buff or yellowish brown, with a series of dark brown spots and streaks across the breast and along the sides. This is one of the few species of its family which frequent open fields.

The Genus *Neocorys* is closely related to the preceding, but is stouter, and the tail shorter.

The Missouri Skylark, *N. Spragueii*, Sclater, of Nebraska, is our only species. It is five and three fourths inches long, the wing less than three and a half inches.

The Genus *Mniotilta* comprises the Black and White Creeper, *M. varia*, Vieill., of North America east of the Missouri, which is five inches long, the wing less than three inches; the color black, the feathers broadly edged with white; the head black; a median, superciliary, and maxillary stripe, the middle of the belly, two bands upon the wings, outer edges of tertials, inner edges of wing and tail feathers, and a spot on the inner webs of the outer two tail feathers, white. Its notes are few, being a series of rapidly repeated *tweets*, the last one much prolonged. It flies only from one tree to another which is nearest, and which it ascends or descends in a spiral direction, searching for insects and larvæ.

The Genus *Parula* comprises the Blue Yellow-backed Warbler, *P. americana*, Bonap., of North America east of the Missouri, which is four and three quarters inches long, the wing about two and one third inches; the color blue above, with a patch of yellowish green upon the back; under parts yellow before and white behind; two white bands across the wings, a small white spot on either eyelid, and a conspicuous white spot on the outer two tail feathers. This species utters a soft prolonged twitter, its only song.

The Genus *Protonotaria* comprises the Prothonotary Warbler, *P. citrea*, Baird, of the Southern and Western States, which is less than five and a half inches long, and the wing less than three inches; the head, neck, and under parts rich yellow; back dark olive-green, with a tinge of yellow; rump, tail above, and wings, bluish ash.

The Genus *Geothlypis* has the bill distinctly notched, bristles short or wanting, wings short and rounded, tail long, the general color olive green above, yellow below.

Fig. 108.

Maryland Yellow-throat, — female, *G. trichas*, Cab.

The Maryland Yellow-throat, *G. trichas*, Cab., of North America, is five and a half inches long, the wing two and four tenths inches; and readily distinguished by a band of black on the forehead, cheeks, and ear-coverts.

The female is without the black band. This Warbler builds its nest upon the ground, and lays four to six white eggs speckled with light brown.

The Gray-headed Warbler, *G. velatus*, Cab., of the West Indies, and perhaps found in the United States, is four and three fourths inches long, and distinguished from the preceding by a narrower black frontal band, and by the dark ash of the crown.

The Mourning Warbler, *G. philadelphia*, Baird, of Eastern North America, is five and a half inches long, the wing less than two and a half inches; and distinguished by the ashy gray of the head and neck, and the black patch on the fore part of the breast.

Macgillivray's Warbler, *G. Macgillivrayi*, Baird, of Western North America, is five inches long, the wing less than two and a half inches, the head and neck ash, a narrow frontlet and space around the eye black, the feathers of the forward under parts really black, but appearing gray from the ashy tips of the feathers; the rest of the upper parts dark olive-green, and of the lower, yellow.

The Genus *Oporornis* has the bill rather compressed, wings elongated, tail slightly rounded, tarsi elongated, and claws large; above olive green beneath yellow.

The Connecticut Warbler, *O. agilis*, Baird, of the Eastern United States, very rare, is six inches long, the wing three inches.

The Kentucky Warbler, *O. formosus*, Baird, of the Eastern United States, is five inches long, the wing less than three; and distinguished by the yellow throat and superciliary stripe, and by the top of the head and streak beneath the eye, which are black.

The Genus *Icteria* comprises the Chats.

The Yellow-breasted Chat, *I. viridis*, Bonap., of the United States east of the Missouri, is nearly seven and a half inches long, the wing three and a quarter inches; the color of the upper parts olive-green, the forward half of the under parts, including the inside of the wing, gamboge-yellow; the rest of the under parts white.

The Long-tailed Chat, *I. longicauda*, Lawr., of Western North America, is seven inches long, and very similar to the preceding, but with a longer tail.

The Genus *Helmitherus* has the bill large, stout, compressed, about as long as the head, with neither notch nor bristles; tarsi short, wings long.

The Worm-eating Warbler, *H. vermivorus*, Bonap., of the United States east of the Missouri, is five and a half inches long, the wing three inches; the upper parts clear olive-green, the head with four black stripes and three brownish-yellow ones; the under parts pale brownish-yellow, tinged with buff across the breast.

Swainson's Warbler, *H. Swainsonii*, Bonap., of the Southern States, is over five inches and a half long, the wing less than three inches; above dull olive-green, a superciliary stripe and the under parts of the body white, tinged with yellow.

The Genus *Helminthophaga* has the bill elongated, very acute, without a notch, wings long and pointed, the first quill nearly or quite the longest, tail nearly even and rather slender, and the tarsi longer than the middle toe.

The Blue-winged Yellow Warbler, *H. pinus*, Baird, of the United States east of the Missouri, is four and a half inches long, the wing two and four tenths inches; above olive-green, the wings and tail bluish-gray, the crown and under parts rich orange-yellow. Its nest is elongated and attached by its upper edge to several stout stalks of grass; eggs four to six, pure white, with a few pale red spots at the larger end.

The Golden-winged Warbler, *H. chrysoptera*, Baird, of the United States east of the Missouri, is about five inches long, the wing less than two and three fourths inches; the upper parts bluish gray, the head above and a large patch upon the wings yellow; the throat and fore part of the breast black, the rest of the under parts white.

Bachman's Warbler, *H. Bachmani*, Cab., of the Southern States, is four and a half inches long, the color olive green above; the throat, fore part of the breast, and band across the crown, black; the forehead, lesser wing-coverts, chin, and under parts, yellow.

The Nashville Warbler, *H. ruficapilla*, Baird, of the United States east of the Missouri, is over four and a half inches long, the wing less than two and a half inches; the color olive green above, under parts deep yellow.

The Orange-crowned Warbler, *H. celata*, Baird, of Western North America, is less than five inches long, the wing two and a quarter inches, the color above olive green, the under parts greenish yellow, and there is a concealed patch of pale brownish-orange on the crown.

The Tennessee Warbler, *H. peregrina*, Cab., of the United States east of the Missouri, is four and a half inches long, the wing two and three fourths inches; the color above olive green, beneath dull white, the top and sides of the head ash gray.

The Genus *Seiurus* has the bill compressed, distinctly notched, and with very short bristles; the wings longer than the tail, which is slightly rounded, and the feathers acuminate; the color above olivaceous, beneath whitish.

The Golden-crowned Thrush, or Oven-Bird, *S. aurocapillus*, Sw., of North America east of the Missouri, is six inches long, the wing three inches; the middle of the crown brownish orange bordered by black. Its nest is on the ground, and is oven-shaped; eggs white, spotted with reddish brown near the larger end.

The Water Thrush, *S. noveboracensis*, Nutt., of the United States east of the Missouri and southward, is over six inches long, and the wing over three inches, the bill small, a superciliary stripe brownish yellow, the under parts streaked with olivaceous brown, and the breast almost black.

The Large-billed Water-Thrush, *S. ludovicianus*, Bonap., of the United States east of the Missouri, is six and one third inches long, the wing three and a quarter inches; and distinguished by its large bill, and superciliary white stripe.

The Genus *Dendroica*, formerly Sylvicola, has the bill attenuated, depressed at the base, compressed from the middle, bill distinctly notched, bristles short but distinct, tarsi long, the hind claw long, the wings long and pointed, the second quill usually a very little longer than the first, tail slightly rounded, and always with a white spot. More than twenty species belonging to this genus are found in the United States. By adopting the synopsis of the species as given by Baird, they may readily be defined.

A. Those belonging to this group have the chin, throat, and fore part of the breast black bordered by lighter, the back streaked, two white bands upon the wing, and the outer tail-feathers mainly white.

The Black-throated Green Warbler, *D. virens*, Baird, of the United States east of the Missouri, is five inches long, the wing over two and a half inches, the crown and back olive, the forehead, superciliary and maxillary stripes yellow.

The Western Warbler, *D. occidentalis*, Baird, of the Pacific coast, is four and seven tenths inches long, the wing two and seven tenths inches, the top and sides of the head yellow, the back ash conspicuously streaked.

Townsend's Warbler, *D. Townsendii*, Baird, of the Pacific coast, is five inches long, the wing two and six tenths inches, the crown blackish, back olive, superciliary and maxillary stripes yellow.

The Black-throated Gray Warbler, *D. nigrescens*, Baird, of the Pacific coast, is four and seven tenths inches long, the wing two and three tenths, the crown black, back ash, superciliary and maxillary stripe white.

B. This group comprises those which have the sides and under parts of the head black.

The Black-throated Blue Warbler, *D. canadensis*, Baird, of the United States east of the Missouri, is five and a half inches long, the wing two and six tenths, blue above, white beneath, primaries with a white patch at base, and the sides and under parts of the head black.

C. Those in this group have a central longitudinal yellow patch on the crown.

The Yellow-rump Warbler, *D. coronata*, Gray, of Eastern North America to the Missouri plains, is five and six tenths inches long, the wing three inches, the color slate-blue above, throat white, breast blackish, the sides and rump with a yellow patch.

Audubon's Warbler, *D. Audubonii*, Baird, of Western North America, is five and a quarter inches long, the wing nearly three and a quarter inches, the throat yellow, and one large white patch upon the wing.

The Blackburnian Warbler, *D. Blackburniæ*, Baird, of North America east of the Missouri, is five and a half inches long, the wing less than three inches, the back black, throat bright orange, and a patch on the wing and outer tail-feathers white.

Fig. 109.

Blackburnian Warbler, *D. Blackburniæ*, Baird.

D. In this group the sides and throat are chestnut, and the back streaked.

The Bay-breasted Warbler, *D. castanea*, Baird, of North America east of the Missouri, is five inches long, the wing three inches, the crown chestnut, sides of the head black, and belly white.

E. Those in this group have the under parts white, back streaked with black, and the wings with white bands.

The Chestnut-sided Warbler, *D. pennsylvanica*, Baird, of the United States east of the Missouri, is five inches long, the wing two and a half inches, the crown yellow encircled with white, sides of the head black, enclosing a white patch behind, and sides of the body chestnut.

The Blue Warbler, *D. cærulea*, Baird, of the United States east of the Missouri, is four and a quarter inches long, the wing over two and a half inches, blue above and across the breast,. the sides of the crown and body streaked with black.

The Black Poll Warbler, *D. striata*, Baird, of North America east of the Missouri, is five and three quarters inches long, the wing three inches, the crown and streaks upon the sides black, the cheeks below the eye white.

F. Those of this group have the throat uniform yellow, separated from the belly by a series of pectoral streaks, and the sides are streaked.

The Pine-Creeping Warbler, *D. pinus*, Baird, of North America east of the Missouri, is five and a half inches long, and the wing three inches, olive green above, and yellow beneath ; two dull white bands upon the wings.

The Blue Mountain Warbler, *D. montana*, Baird, of the Blue Mountains, Virginia, is four and three quarters inches long, olive green above, forehead and under parts yellow.

The Yellow Warbler, *D. æstiva*, Baird, of the United States, is five and a quarter inches long, the wing over two and a half inches, the general color yellow, back olivaceous, ventral streaks brownish red.

The Black and Yellow Warbler, *D. maculosa*, Baird, of the United States east of the Missouri, is five inches long, the wing two and a half inches, the crown blue, rump yellow, sides of the head and back black, spots on the central third of the tail and a large patch upon the wing white, and large black streaks on the under parts.

Kirtland's Warbler, *D. Kirtlandii*, Baird, of Ohio, is five and a half inches long, the wing two and eight tenths inches, blue above streaked with black, the sides of the head and inferior streaks black, and a white patch at the end of the tail.

The Yellow-Red Poll, *D. palmarum*, Baird, of the United States east of the Mississippi, is five inches long, the wing nearly two and a half inches, the crown, sides of the head, and inferior streaks rufous, rump greenish-yellow, a white spot on the end of the tail, superciliary streak and under parts yellow.

The Cape May Warbler, *D. tigrina*, Baird, of the United States east of the Mississippi, is five and a quarter inches long, the wing nearly three inches, the bill acute and decurved, the color olive above, rump and under parts yellow, crown blackish, sides of the head chestnut.

G. Those of this group have the throat yellow, and not separated from the belly by pectoral streaks.

The Carbonated Warbler, *D. carbonata*, Baird, of Kentucky, is four and three quarters inches long, olive green above, spotted with black, crown black, beneath yellowish, a band on the wing and edge of the tail whitish.

The Yellow-throated Warbler, *D. superciliosa*, Baird, from Pennsylvania to the Missouri, is over five inches long, the wing over two and a half inches, the back slate-colored, under parts white, crown and sides of the head and neck black, superciliary stripe yellow, changing to white behind.

The Prairie Warbler, *D. discolor*, Baird, of the Atlantic States, is nearly five inches long, the wing two and a quarter inches; olivaceous above, the back streaked with red; under parts and superciliary stripe yellow, and a black mark on the side of the head.

The Genus *Sylvia* comprises the Tailor-Bird, *S. sutoria,* Lath., of the East Indies, which is about five inches long, and celebrated for the ingenious way in which it prepares a place for its nest. Picking up a leaf, it sews its edges, with a thread which it makes or finds, to a living leaf, leaving an opening above; and thus a pouch is formed which is suspended by the leaf-stalk of the living leaf. In the bottom of this the nest is made. Sometimes it sews together two contiguous living leaves.

The Genus *Myiodioctes* is characterized by a depressed bill notched at the tip, the gape with long bristles, wings longer than the tail, and the latter rounded or graduated; the colors olive or plumbeous above, and yellow beneath.

The Hooded Warbler, *M. mitratus,* Aud., of the United States east of the Missouri, is five inches long, the wing two and three quarters inches, the head and neck black, back olive green, the front, cheeks, and under parts yellow, and the tail with white on the outer feathers.

The Small-headed Flycatcher, *M. minutus,* Baird, of the Atlantic States, is five inches long, olive above, yellow beneath, and the wings with two white bands.

The Green Black-cap Flycatcher, *M. pusillus,* Bonap., of the United States, is four and three quarters inches long, the wing two and a quarter inches, the upper parts olive, the forehead, line over the eye, and under parts bright yellow, the crown with a black patch.

The Canada Flycatcher, *M. canadensis,* Aud., of the United States east of the Mississippi, is about five and a third inches long, the wing about two and two thirds inches, the back bluish; streaks upon the crown, stripe on the side of the head and neck, and collar of streaks upon the breast, black.

Bonaparte's Flycatcher, *M. Bonapartii,* Aud., of Louisiana, is over five inches long, and closely resembles the preceding species.

The Genus *Cardellina* comprises the Vermilion Flycatcher, *C. rubra*, Bonap., of Northern Mexico, which is over five and a half inches long, the color dark crimson.

The Genus *Setophaga* has the bill depressed, the tip abruptly decurved and much notched, the rictus with long bristles, wings rounded, and tail short.

The Redstart, *S. ruticilla*, Sw., of the United States east of the Missouri, is five and a quarter inches long, the wing two and a half inches, the general color black, the sides of the breast and base of the quills and tail reddish orange, and the abdomen white. This is one of the handsomest and liveliest birds of our forests. It is almost constantly hunting insects along the branches, and with every movement it opens and shuts its beautiful tail, then flirts it from side to side, and at the same time utters its pleasing *tetee whee*. The nest is built upon a low bush, and appears to hang to the twigs; eggs four to six, white, sprinkled with ashy gray and blackish dots.

The Painted Flycatcher, *S. picta*, Sw., of Northern Mexico, is five and a quarter inches long, the wing two and a half inches, the color black, belly red, patch on the wings and outer tail-feathers white.

The Genus *Pyranga* has the bill rather straight, notched at the tip, the wings elongated, the color of the male scarlet, of the female yellowish.

The Scarlet Tanager, *P. rubra*, Vieill., of the United States east of the Missouri, is seven and four tenths inches long, the wing four inches, the color bright scarlet, wings and tail black. Its notes are not very musical, and have been represented by Wilson by the syllables *chip churr*. Its nest is on a low branch of a tree; eggs three to five, dull greenish-blue, speckled with reddish brown and light purple.

The Summer Red Bird, *P. æstiva*, Vieill., of the Southern Atlantic and Gulf States, is seven and one fifth inches long, the wing three and three quarters inches; the color light red, back more dusky.

The Rocky Mountain Tanager, *P. hepatica*, Sw., is eight inches long, and dark scarlet-red tinged with ashy on the back and sides.

The Louisiana Tanager, *P. ludoviciana*, Bonap., of Western North America, is seven and a quarter inches long, the wing over three and a half inches; the general color of the interscapular region, wings, and tail, black; head and throat tinged with scarlet, and the wings with two whitish bands.

The Genus *Euphonia* comprises the Blue-headed Tanager, *E. elegantissima*, Gray, of Mexico, which is about four and two thirds inches long, the wing two and three quarters inches, bluish black above, yellowish below, the top of the head blue.

HIRUNDINIDÆ, OR SWALLOW FAMILY. — This Family comprises birds with a very short, depressed, and triangular bill, very long wings, very short tarsi, and tail generally forked.

The Genus *Hirundo* has the tail more or less forked.

The Barn Swallow, *H. horreorum*, Barton, of North America, is six and nine tenths inches long, the wings five inches, and the tail excessively forked; the color steel-blue above, forehead and throat chestnut brown, belly reddish white.

The Cliff Swallow, *H. lunifrons*, Say, of North America, is five inches long, the wing less than four and a half inches, the tail emarginate, the crown and back steel-blue, throat and sides of the head dark chestnut, breast fuscous, belly white.

The White-bellied Swallow, *H. bicolor*, Vieill., of North America, is six and a quarter inches long, the wing five inches; glossy metallic green above, white beneath.

The Violet-Green Swallow, *H. thalassina*, Sw., of Western North America, is four and three quarters inches long, the wing four and a half inches, the tail acutely emarginate; the color green above, pure white below.

The Edible Swallow-nest Swallow, *H. esculenta*, Linn., is a very small species of the East Indies, whose nests, composed of a whitish gelatinous substance, are held in high estimation by epicures, and which constitute an important article of commerce.

The Genus *Cotyle* is distinguished from Hirundo by the slightly forked tail, very slender toes, and dull color.

The Bank Swallow, *C. riparia*, Boie, of North America, is four and three quarters inches long, the wing four inches; the upper parts grayish brown, under parts white, with a band across the breast like the back.

The Rough-winged Swallow, *C. serripennis*, Bonap., of North America, is four and a half inches long, the first primary with the outer web much stiffened; the color sooty brown above, grayish beneath.

The Genus *Progne* has the bill strong and short, toes long and strong, and the size is the largest of the family.

The Purple Martin, *P. purpurea*, Boie, of North America, is seven and three tenths inches long, the wing less than six inches; the color glossy steel-blue, with purple and violet reflections.

It need scarcely be stated, that the notion which some entertain, that swallows spend the winter at the bottom of ponds, is entirely erroneous.

Bombycillidæ, or Waxwing Family. — This Family comprises birds with the bill short, broad, much depressed, and the gape opening to the eyes; both mandibles notched, the upper with a tooth behind the notch, the outer lateral toe the longest, and the head generally crested.

The Genus *Ampelis* — Waxwings — has the tail even, and some of the quills with horny appendages that look like sealing-wax.

The Bohemian Chatterer or Waxwing, *A. garrulus*, Linn., of the northern parts of both continents, is nearly seven and a half inches long, the wing four and a half inches; the general color brownish ash, primaries and tail-feathers plumbeous black, the tail with a terminal band of yellow, the head and throat marked with black, the wings with white, and the secondaries have red horny tips.

The Cedar Bird, *A. cedrorum*, Baird, of North America, is seven and a quarter inches long, the wing over four inches; the general color reddish olive passing into yellow below, and posteriorly above into ashy; the forehead, space below the eye, and a line above it, intense black; the quills and tail dark plumbeous and dusky, the tail tipped with yellow; and the secondaries have red horny tips. It is impossible to describe the plumage of this beautiful bird, it is so silky, and its tints so delicate in their shadings. The Cedar Bird builds her nest in low trees or bushes, and lays four purplish white eggs marked with black spots.

The Genus *Myiadestes* comprises Townsend's Flycatcher, *M. Townsendii*, Cab., of Western North America, which is eight inches long, the wing four and a half inches, the tail deeply forked, the general color bluish ash.

LANIDÆ, OR SHRIKE FAMILY.—This Family comprises birds with a strong compressed bill, the tip abruptly hooked, both mandibles distinctly notched, the upper with a distinct tooth, the lower with the point bent upward, and the tarsi longer than the middle toe and strongly scutellate. This family comprises the Shrikes and the Vireos.

The Genus *Collyrio* has the bill shorter than the head, the tip of the lower mandible bent upward, rictus with long bristles, legs stout, wings rounded, and claws very sharp.

The Great Northern Shrike, or Butcher-Bird, *C. borealis*,

Baird, of North America, is nearly nine inches long, the wing four and a half inches, the color above light bluish ash, the under parts white, the breast with fine transverse lines; the wings and tail black. It preys mainly on insects, sparrows, and other small birds. It has the power of imitating the sounds of other birds, especially those indicating distress; and has the singular habit of impaling birds and insects upon the points of twigs and thorns; but for what object is not well understood.

Fig. 110.

Great Northern Shrike, *C. borealis*, Baird.

The Loggerhead Shrike, *C. ludovicianus*, Baird, of the South Atlantic and Gulf States, is nine inches long, and the wing nearly four inches.

The White-rumped Shrike, *C. excubitoroides*, Baird, of Western North America, is eight and three quarters inches long, and the wing nearly four inches.

The White-winged Shrike, *C. elegans*, Baird, inhabits Western North America.

The Genus *Vireo* has the bill short, compressed, the tip bent downward, wings rather long and pointed, tail nearly even, and tarsi longer than the middle toe.

The Red-eyed Vireo or Flycatcher, *V. olivaceus*, Vieill., of North America east of the Missouri, is six and a half inches long, the wing three and a half inches; olive green above, white below, the crown dark ash, and iris red; a whitish line from the bill over the eye, a dark line between this and the ashy crown, and a dusky line through the eye. This is one of the earliest singers of spring, and latest of autumn. Its notes are loud, clear, and melodious, and are heard throughout the day among the taller trees. The nest is generally suspended from forked twigs; eggs four to six, spotted with reddish brown at the larger end.

The Yellow-green Vireo, *V. flavoviridis*, Cass., of Northern Mexico and southward, closely resembles the preceding, but the colors are more strongly marked.

Bartram's Vireo, *V. virescens*, Vieill., of Central and South America and possibly of the Atlantic United States, resembles *V. olivaceus*, but is smaller.

The Whip Tom Kelly, *V. altiloquus*, Gray, of Florida and the West Indies, is very similar to *V. olivaceus*.

Fig. 111.

Warbling Flycatcher, *V. gilvus*, Bonap.

The Warbling Flycatcher or Vireo, *V. gilvus*, Bonap., of North America, is about five and a half inches long, the wing about three inches; the color olive green above, beneath white, tinged with very pale yellow on the breast and sides. Its song is low, mellow, and sweet.

The Philadelphia Vireo, *V. philadelphicus*, Cass., of Pennsylvania to Wisconsin, closely resembles the preceding one, but is at once distinguished by the absence of the spurious primary.

Bell's Vireo, *V. Belli*, Aud., of Missouri River and Texas, is four and a quarter inches long, the wing two and a quarter inches, and is very similar to *V. gilvus*, but smaller, and the spurious primary is large.

The Black-headed Flycatcher, *V. atricapillus*, Woodh., of Texas, is four and three quarters inches long, the wing over two inches; olive green above, white beneath, the head and neck above and on the sides black.

The White-eyed Vireo, *V. noveboracensis*, Bonap., of North America east of the Missouri, is five inches long, the wing two and a half inches, the spurious primary half the length of the second; the upper parts bright olivaceous-green; under parts white, the sides of the breast and body yellow. The space around the eyes is greenish yellow, and the iris is white; the wings have two yellowish-white bands. This bird frequents the thickest bushes. It sings with great spirit, and often throughout the day. The nest is attached to the twigs of a low bush; eggs four to six, pure white, marked with dark spots near the larger end.

Hutton's Flycatcher, *V. Huttoni*, Cass., of Southern California, is about the same size as the preceding one, but has the bill much more slender.

The Blue-headed Flycatcher, *V. solitarius*, Vieill., of North America, is five and a half inches long, the wing two and four tenths inches, the spurious primary very small; olive green above, top and sides of the head and upper part of the neck dark bluish-ash; under parts white, the sides under the wings greenish yellow.

The Yellow-throated Flycatcher, *V. flavifrons*, Vieill., of North America east of the Missouri, is six inches long, the wing over three inches, and with no spurious primary; the color from the bill to the middle of the back, sides of the head, neck, and fore part of the breast, olive green; the rest of the upper parts ashy blue. The under parts, from the bill to the middle of the belly, with a ring around the eye, sulphur-yellow; the remaining under parts white. This bird prefers the taller trees, whose branches it ascends by regular short hops, searching every leaf in its way. Its notes are measured and plaintive. The nest is attached to the extremity of small twigs, and is sometimes five or six inches deep; it is an exceedingly interesting structure. The eggs are four to five, white, spotted with reddish brown.

LIOTRICHIDÆ, OR MOCKING-BIRD FAMILY.—This Family comprises birds with the bill slender, straight, or curved, as long or longer than the head, slightly notched, or not at all; the wings short, concave, and rounded, the tarsi long, and generally strongly scutellate. This is an extensive family, and embraces forms which at first seem to differ greatly, but which are now regarded as related in their most essential characteristics.

The Genus *Mimus* has the bill shorter than the head, decurved from the base, and distinctly notched.

Fig. 112.

Mocking-Bird, *M. polyglottus*, Boie.

The Mocking-Bird, *M. polyglottus*, Boie, of the Southern States, is nine and a half inches long, the wing four and a half inches; olive gray above, whitish beneath, the wing and tail black, the base of the primaries and the tip of the tail white. This bird imitates with ease the songs and notes of all the birds he hears. Audubon considers the singing of our Mocking-Bird superior to that of the Nightingale.

The Cat Bird, *M. carolinensis*, Gray, of the United States east of the Missouri, is less than nine inches long, the wing over three and a half inches; the general color dark plumbeous; the under tail coverts dark brownish-chestnut. In spring its song is exceedingly varied, mellow, and sweet. It also possesses a remarkable power of imitating the notes of other birds, and has been heard to imitate perfectly a strain of Yankee Doodle. Sometimes it mews or yawls like a cat, and in a most disagreeable manner, which greatly detracts from its proper estimation; because all do not know that at times it sends forth the sweetest music. The nest is generally built in low bushes, and composed of dry twigs and grass without, fibrous roots within; eggs four to six, glossy greenish-blue. Two broods are raised in a season.

The Genus *Oreoscoptes* comprises the Mountain Mocking-Bird, *O. montanus*, Baird, of Western North America, which is eight inches long, and the wing nearly five inches, and with the bill longer and more slender than in *Mimus*.

The Genus *Harporhynchus* has the bill as long as or longer than the head, no notch, the wings short, and tail long.

The California Thrush, *H. redivivus*, Cab., of California, is eleven and a half inches long, the wing over four inches, the bill much decurved and longer than the head. The color above brownish olive, beneath pale cinnamon, deepening into rufous on the under tail-coverts.

Five additional species of this genus are found in Mexico and California.

The Brown Thrush, *H. rufus*, Cab., of North America east of the Missouri, is over eleven inches long, the wing over four inches; the color above light cinnamon-red, beneath pale rufous-white, with longitudinal streaks of dark brown. In the pleasant spring mornings, this bird utters the sweetest melodies from the topmost twigs of some isolated tree. Later in the day, and at all times late in the season, it prefers low thick bushes. Its flight is low and heavy, and continued only a few rods at a time. The nest is made in a clump of low bushes a few feet from the ground; the eggs four to six, dull buff thickly sprinkled with dots of brown.

The Genus *Campylorhynchus* comprises *C. brunneicapillus*, Gray, Southwestern North America, which is eight inches long, the wing nearly three and a half inches, and is the largest Wren in the United States.

The Genus *Catherpes* comprises the White-throated Wren, *C. mexicanus*, Baird, of the Rio Grande region, which is six and a half inches long, the wing two and a half inches.

The Genus *Salpinctes* comprises the Rock Wren, *S. obsoletus*, Cab., of the Rocky Mountains, which is five and seven tenths inches long, the wing less than three inches.

The Genus *Thryothorus* has the bill about as long as the head, nearly straight to the tip, which is abruptly decurved. The wings are about equal to the tail, which is arched and nearly even, and the tarsus longer than the middle toe.

The Great Carolina Wren, *T. ludovicianus*, Bonap., of the United States north to Pennsylvania and west to the Missouri, is six inches long, the wing two and six tenths inches, the color reddish brown above, beneath pale yellowish-rusty.

Berlandier's Wren, *T. Berlandierii*, Couch, of Mexico, closely resembles the preceding, but is smaller.

Bewick's Wren, *T. Bewickii*, Bonap., of North America, is five and a half inches long, the wing two and a quarter inches; the color rufous brown above, plumbeous white below, wings and innermost tail-feathers barred with dusky; the remaining tail-feathers mostly black, marked with white.

The Genus *Cistothorus* has the tail much graduated and shorter than the wings, and the feet stout.

The Long-billed Marsh Wren, *C. palustris*, Cab., of North America, is five and a half inches long, the wing over two inches; the upper parts dull reddish brown; the crown, interscapular region, outer surface of tertials, and tail feathers, almost black; the under parts and streak over the eye white. This species lives among the rank vegetation growing around inlets to the sea. Here it builds its nest, and lays six or more eggs, of a deep chocolate color.

The Short-billed Marsh Wren, *C. stellaris*, Cab., of the United States east of the Platte River, is four and a half inches long, the wing one inch and three fourths; the bill is scarcely half the length of the head; the hind part of the crown, back, and rump, almost black, streaked with white; under parts white; the sides, upper part of the breast, and under tail-coverts, reddish brown. This species lives in marshy fresh-water meadows, and is very shy. When uttering its lively song, it stands on a tuft of sedge or a low bush, and its head and tail are alternately depressed and elevated as if the body were moving on a pivot.

The Genus *Troglodytes* has the bill nearly as long as the head, compressed, decurved; and the wings about equal to the tail.

The House Wren, *T. ædon*, Vieill., of North America east of the Missouri, is nearly five inches long, the wing over two inches; the color above reddish brown, barred with dusky, under parts brownish gray. This wren delights in being near the habitations of man, and builds its nest in any hole it finds in the timbers or walls of our buildings, or in a hollow tree of the orchard or garden. The nest is formed of dry twigs and grasses, and lined with soft materials; eggs five or six, pale reddish.

Parkman's Wren, *T. Parkmanni*, Aud., of Western North America, is very similar to the preceding, but the colors are grayer.

The Wood Wren, *T. americanus*, Aud., of Eastern United States, is four and a half inches long, the wing two inches, and is very similar to *T. ædon;* but the bill is shorter, tail more graduated, colors darker, and there is no light line over the eye.

The Winter Wren, *T. hyemalis*, Vieill., of North America, is four inches long, the wing over one inch and a half; the upper parts reddish brown, marked with transverse

Fig. 113.

Winter Wren,
T. hyemalis, Vieill.

bars of dusky and light, except on the head and upper part of the back; beneath pale reddish-brown, more or less spotted and barred. The scapulars, wing coverts, and outer web of primaries are spotted with white. The motions of the Winter Wren are exceedingly varied, rapid, and precise. It may be seen in a score of attitudes in the course of a few minutes. Now it is on one side of a brush-heap, and in a moment it has passed through and appears on the other. It reaches the upper branches of a small tree by hopping from twig to twig, and in the course of its passage presents each side in turn to you a dozen times; and when at the top it utters a delicate melody, and then dashes headlong, and is out of sight in a moment. Audubon says that the song of the Winter Wren excels that of any other bird of its size with which he is acquainted. The nest is long and bag-like, and attached to a rock or tree near the ground. It is made of moss and lichens, and lined with hair and feathers. The eggs are six or more, of the most delicate rose-color, dotted with reddish brown.

The Genus *Chamæa* comprises the Ground Tit, *C. fasciata*, Gambel, of California, which is six inches long; the upper parts olivaceous brown, the lower pale brownish-cinnamon.

Certhiadæ, or Creeper Family.—This Family comprises birds with the bill slender, as long or longer than the head, without a notch; and the entire basal joint of the middle toe united to the lateral ones.

The Genus *Certhia* has the bill as long as the head, and much compressed and decurved from the base.

The American Creeper, *C. americana*, Bonap., of North America, is five and a half inches long, the wing over two and a half inches; the color above dark brown, each feather streaked centrally with whitish, and the rump rusty; the under parts, and a streak over the eye, white; and the wings with a bar of reddish white across both webs.

Fig. 114.

American Creeper, *C. americana*, Bonap.

The Genus *Sitta* has the bill subulate, acutely pointed, compressed, and about as long as the head.

The White-bellied Nuthatch, *S. carolinensis*, Gm., of North America east of the Central Plains, is six inches long, the wing three inches and three quarters; the color ashy blue above, the under parts white, top of the head and neck black. It moves along trunks and branches with the greatest facility, and at a little distance is easily mistaken by the careless observer for a little woodpecker. The nest is made in a hole excavated in a decayed trunk or branch; eggs five or six, dull white, spotted with white at the larger end.

Fig. 115.

White-bellied Nuthatch, *C. carolinensis*, Gm.

The Slender-billed Nuthatch, *S. aculeata*, Cass., of the Pacific coast and eastward, is precisely similar to the preceding, but has the bill more slender.

The Red-bellied Nuthatch, *S. canadensis*, Linn., of North America, is four and a half inches long, the wing two and two thirds inches; the color ashy blue above, top of the head black, under parts brownish rusty. The nest is made in a low stump; eggs four, white, rose-tinged, and sprinkled with reddish dots. Like other species of its genus, the Red-bellied Nuthatch at night attaches its feet to the bark, and sleeps with its head downwards.

The Brown-headed Nuthatch, *S. pusilla*, Lath., of the South Atlantic States, is four inches long, the wing two and a half inches; the color above ashy blue, top of the head and upper part of the neck light brown, divided on the nape by white; beneath, dingy white.

The California Nuthatch, *S. pygmæa*, Vigors, of Western North America, is about four inches long, the wing two and four tenths inches, and closely resembles the preceding.

PARIDÆ, OR TITMOUSE FAMILY. — This Family comprises birds which have the bill short, straight, and conical, the wings short, and tail long.

The Genus *Polioptila* has the bill depressed at base, nearly as long as the head, notched at the tip, and with rictal bristles. The members of this genus are lead-color above, white beneath.

The Blue-gray Gnat-catcher, *P. cærulea*, Sclat., of North America, is four and three tenths inches long, the wing less than two and a quarter inches, and is distinguished by the two outer tail-feathers, which are entirely white, and a narrow black frontal line which extends back over the eye.

The Western Gnat-catcher, *P. plumbea*, Baird, of the valley of Colorado, has no black on the forehead, but a stripe over the eye; and the outer web only of the outer tail-feather is white.

The Black-tailed Gnat-catcher, *P. melanura*, Lawr., of the Rio Grande region, has the entire top of the head black, and the edge only of the outer tail-feather white.

The Genus *Lophophanes* has the crown with a conspicuous crest.

The Tufted Titmouse, *L. bicolor*, Bonap., of North America east of the Missouri, is six and a quarter inches long, the wing less than three and a quarter inches; the color above ashy black, forehead black, beneath dull whitish.

The Black-crested Tit, *L. atricristatus*, Cass., of the Rio Grande, is five and a quarter inches long, the wing three inches ; the color above plumbeous, crest long, pointed, and black.

The Gray Titmouse, *L. inornatus*, Cass., of California, is five inches long, the wing over two and a half inches, crest elongated ; the color above olivaceous ashy, beneath whitish.

Wollweber's Titmouse, *L. Wollweberi*, Bonap., of Southern Rocky Mountains, is four and a half inches long, the wing two and a half inches.

The Genus *Parus* has the head without a crest, body and head stout, the crown and throat black.

The Long-tailed Chickadee, *P. septentrionalis*, Harris, from the Missouri to the Rocky Mountains, is five and a half inches long, and the wing two and seven tenths inches.

The Black-cap Titmouse, or Chickadee, *P. atricapillus*, Linn., of Eastern North America, is five inches long, the wing two and a half inches, the back brownish ashy, under parts whitish, top of the head and throat black, and the sides of the head between white. The nest is generally made in a hole in a stump ; the eggs rarely exceed eight in number, the color white slightly dotted and marked with light reddish. Two broods are raised in a season.

Fig. 116.

Titmouse,
P. atricapillus, Linn.

The Western Titmouse, *P. occidentalis*, Baird, of the Pacific coast, closely resembles the preceding.

The Mexican Titmouse, *P. meridionalis*, Sclat., of Eastern Mexico, is five inches long, and the wing two and six tenths inches.

The Carolina Titmouse, *P. carolinensis*, Aud., of the Southern Atlantic States, is very similar to *P. atricapillus*, but is smaller.

The Mountain Titmouse, *P. montanus*, Gambel, of Western North America, is five inches long, the wing two and six tenths inches; the head and neck above, a line through the eye, and the under part of the head and the throat, glossy black; the forehead and a line over the eye, and one below it, white; the other parts ashy.

The Chestnut-backed Tit, *P. rufescens*, Towns., of Western North America, is four and three quarters inches long, the wing less than two and a half inches, and is readily distinguished by the dark brownish-chestnut of the back and sides.

The Hudsonian Titmouse, *P. hudsonicus*, Forster, of Northeastern North America, is five inches long, the wing two and four tenths inches; the color yellowish olive-brown, under parts white.

The Genus *Psaltriparus* has the bill very small and short, the outline much curved for its terminal half, the tail long and slender and much graduated, tarsi longer than the middle toe, and no black on the crown or throat.

The Black-cheeked Tit, *P. melanotus*, Bonap., of the Rio Grande region, is four inches long, the wing one inch and nine tenths, and is distinguished by a black patch on each cheek.

The Least Tit, *P. minimus*, Bonap., of the Pacific coast of the United States, is of the same size as the preceding; the color olivaceous cinereous.

The Lead-colored Tit, *P. plumbeus*, Baird, of the Southern Rocky Mountains, is four and one fifth inches long, the wing less than two and a quarter inches.

The Genus *Paroides* comprises the Verdin, *P. flaviceps*, Baird, of Texas, which is four and a half inches long, the wing less than two and a quarter inches; and distinguished from all the foregoing in the greater length of the quills; the color above cinereous, head yellow, under parts brownish white.

ALAUDIDÆ, OR SKYLARK FAMILY. — This Family comprises birds with a short conical bill, the first primary very short or wanting, tertiaries greatly elongated beyond the secondaries, tarsi scutellate before and behind, and the hind claw very long and nearly straight.

The Genus *Eremophila* has the first primary wanting, and the nostrils circular and concealed by a dense tuft of feathers.

The Skylark or Shore Lark, *E. cornuta*, Boie, of the plains and prairies of North America and in the Atlantic States in winter, is seven and three quarters inches long, the wing four and a half inches; the color above pinkish brown, the feathers of the back marked with dusky; a band across the crown and running back along the lateral tufts, a crescentic patch from the bill below the eye and along the side of the head, and a pectoral crescent, black; the frontal band and under parts white; chin and throat yellow. It sings sweetly while on the wing, but its song is short. It rises obliquely from the ground for about forty yards, begins and ends its song, then performs a few evolutions and returns to the ground, where it also sings, but less frequently, and with less fulness. The nest is built on the ground; the eggs are four or five, grayish, with numerous pale-blue and brown spots.

Fig. 117.

Skylark, *E. cornuta*, Boie.

The Skylark of Europe, *E. arvensis*, is brown above, whitish beneath, and spotted with deep brown. Next to the nightingale, this is the sweetest singer in Europe. It rises vertically, and when rising or descending utters its varied and powerful song.

FRINGILLIDÆ, OR FINCH AND SPARROW FAMILY.—This Family comprises birds with a short, robust, conical bill, nine primaries, tarsi scutellate anteriorly, the two sides with undivided plates meeting behind, forming a sharp ridge. This family comprises all the birds known as Grosbeaks, Finches, Buntings, Crossbills, Sparrows, and the like. Baird divides the family into four groups.

The first group comprises those which have the bill variable, very large to quite small, the base of the upper mandible with a closely pressed fringe of bristly feathers concealing the nostrils; the wings long, pointed, a half to a third longer than the forked or emarginate tail; and the tarsi short.

The Genus *Hesperiphona* has the bill enormously large.

The Evening Grosbeak, *H. vespertina*, Bonap., of Western North America, is seven and three tenths inches long, the wing four and three tenths inches, the forward half of the body yellowish olive, shading into yellow on the rump and under tail-coverts; the crown, wings, upper tail-coverts, and tail, black; frontal band yellow.

The Genus *Pinicola* has the bill much smaller than *Hesperiphona*, the culmen much curved, and tail nearly even.

The Pine Grosbeak, *P. canadensis*, Cab., of Arctic America and southward into the United States in winter, is eight and a half inches long, the wing four and a half inches; the general color carmine-red, wings marked with white, bill and legs black. The female is brownish above, greenish yellow beneath; the top of the head, rump, and upper tail-coverts brownish gamboge-yellow.

The Genus *Carpodacus* has the bill more or less curved above, and the tail moderately forked.

Fig. 118.

Purple Finch,
C. purpureus, Gray.

The Purple Finch, *C. purpureus*, Gray, of North America east of the Central Plains, is six and a quarter inches long, the wing three and a third inches, the color crimson; belly and under tail-coverts white; there are two reddish bands across the wings. The female is olivaceous brown, white beneath, the feathers streaked with brown. The song of the Purple Finch is prolonged and sweet. The nest is built in a tree a few feet from the ground; the eggs are four, bright emerald-green.

The Western Purple Finch, *C. californicus*, Baird, of the Pacific coast, is rather smaller than the preceding, and the purple of the head and rump much darker.

Cassin's Purple Finch, *C. Cassinii*, Baird, of the Rocky Mountains, is larger than *C. purpureus*.

The House Finch, *C. frontalis*, Gray, of the Rocky Mountains to the Pacific, is five and three quarters inches long, the wing three and a quarter inches; the forehead, superciliary stripe, throat, and upper part of the breast, crimson; the remaining upper parts grayish brown, and the under parts whitish.

The Genus *Chrysomitris* has the bill nearly straight, and the tail quite deeply forked; the general colors yellow and black.

The Black-headed Goldfinch, *C. magellanica*, Bonap., of South America and accidental in the United States, is four and a half inches long, the wing two and three quarters inches; the head is black all round.

Yarrell's Goldfinch, *C. Yarrelli*, Bonap., of California and Mexico, is four inches long, with the crown black.

The Yellow-Bird or Thistle-Bird, *C. tristis*, Bonap., of North America, is five and a quarter inches long, the wing three inches; the color gamboge-yellow, crown and wings black, tail and wings marked with white. The female is yellowish-brown, with no black upon the head. The nest is very handsome, made of lichens and fastened to a twig; eggs four to six, white tinged with bluish, and spotted with reddish brown at the larger end.

The Arkansas Finch, *C. psaltria*, Bonap., of the Southern Rocky Mountains to the coast of California, is four and a quarter inches long, the wing two and two fifths inches, the upper parts olive green, the head, wings, and tail black; beneath, bright yellow.

The Black Goldfinch, *C. mexicana*, Bonap., of Mexico near the Rio Grande, is over four inches long, the wing two and a quarter inches.

Lawrence's Goldfinch, *C. Lawrencii*, Bonap., of California, is four and seven tenths inches long, the wing two and three quarters inches; the crown, sides of the head anterior to the eye, chin, and throat, black; sides of the head and neck, upper part of the neck, back, and upper tail-coverts, ashy; under parts greenish yellow.

The Pine Finch, *C. pinus*, Bonap., of North America, is four and three quarters inches long, the wing three inches; olive brownish above, beneath whitish, every feather streaked with dusky; the concealed bases of tail feathers and quills and their inner edges sulphur-yellow; outer edges yellowish green; two brownish-white bands upon the wings, and the tail much forked.

The Genus *Pyrrhula* comprises the Bulfinches of the Eastern hemisphere, which have the bill very large and rounded. They are easily tamed, and sing well.

The Genus *Curvirostra* has the points of the mandibles greatly curved and overlapping. Crossbills live in flocks; and feed mainly upon seeds contained in the cones of pines and those of other allied trees, which, by the aid of their peculiar bill, they are able to secure with wonderful facility.

The Red Crossbill, *C. americana*, Wils., of North America, is six inches long, the wing over three and a third inches; the color dull red, the wings and tail dark blackish-brown. The female is dull greenish-olive, rump and crown bright greenish-yellow; beneath grayish.

The White-winged Crossbill, *C. leucoptera*, Wils., of North America, is six and a quarter inches long, the wing three and a half inches, and is readily distinguished by the white bands upon the wings.

Fig. 119.

White-winged Crossbill, *C. leucoptera*, Wils.

The Genus *Ægiothus* has the bill short and acutely conical, wings long, the second quill somewhat longer than the first and third, and the tail deeply forked; colors reddish.

The Lesser Red Poll, *Æ. linaria*, Cab., of North America, is five and a half inches long, the wing over three inches; the color above light yellowish, each feather streaked with dark brown, the crown crimson, upper part of the breast and sides of the body tinged with light crimson; the most of the remaining under parts white. Audubon says few birds exhibit a more affectionate disposition than this; and he enjoyed the pleasure of seeing several on a twig feeding each other by passing a seed from bill to bill, and one individual actually receiving food from two of his companions at the same time.

The Mealy Red Poll, *Æ. canescens*, Cab., of Greenland, is six inches long, and is further distinguished from the preceding by its white rump.

The Genus *Leucosticte* has the bill obtusely conical, and a conspicuous ridge on the side of the lower mandible.

The Gray-crowned Finch, *L. tephrocotis,* Sw., of the Northern Rocky Mountains, is over seven inches long, and the wing four and three tenths inches.

The Genus *Plectrophanes* has the bill always more or less curved or blunted, the wings one half longer than the tail, the hind claw much the largest ; colors black and white.

The Snow-Bunting, *P. nivalis,* Meyer, of Northern North America and south into the United States in winter, is six and three quarters inches long, the wing over four and one third inches ; the colors in full plumage black and white ; in winter, white beneath, head and rump yellowish brown, back brown. Snow-Buntings move in flocks, and keep mainly in open fields.

The Lapland Longspur, *P. lapponicus,* Selby, of Northern North America, is six and a quarter inches long, the wing nearly four inches ; the head all round and extending to the breast black ; the sides of the lower part of the neck and under parts white ; a chestnut collar on the back of the neck ; and the remaining upper parts brownish yellow streaked with brown.

Smith's Bunting, *P. pictus,* Sw., of Illinois in winter and northward in summer, is five and a half inches long, the wing three and a half inches, and is distinguished from all the preceding by the flesh-colored legs.

The Chestnut-collared Bunting, *P. ornatus,* Towns., of the Upper Missouri, is five and a quarter inches long, the wing three and one fifth inches, and is distinguished by a chestnut band on the neck.

The Black-shouldered Longspur, *P. melanomus,* Baird, of the eastern slope of the Rocky Mountains, closely resembles the preceding.

Maccown's Longspur, *P. Maccownii,* Lawr., of the eastern slope of the Rocky Mountains, is five and a half inches long, the wing over three and a half inches, and is distinguished by its large and stout bill.

The second group embraces all the plain sparrow-like birds marked with longitudinal stripes. The bill is small and conical, tarsi lengthened, and the lateral claws never reaching beyond the base of the middle claw.

The Genus *Centronyx* has the bill elongated, the wings reaching beyond the middle of the tail, tarsi elongated, the hind toe large and curved.

Baird's Bunting, *C. Bairdii*, Baird, of the Yellowstone region, is four and three quarters inches long, the wing two and four fifths inches.

The Genus *Passerculus* has the tarsus about equal to the middle toe, the lateral toes about equal, the wing reaching to the middle of the tail, the first primary longest, the tertiaries equal to the primaries; the tail is emarginate and slightly rounded, and the feathers acute.

The Savannah Sparrow, *P. savanna*, Bonap., of North America east of the Missouri, is five and a half inches long, the wing two and seven tenths inches; the upper parts streaked with dark brown, the crown with a median stripe of yellowish gray; a superciliary stripe, eyelids, and edge of elbow, yellow; the fore part of the breast streaked, and the rest of the under parts mainly white. It builds its nest on the ground at the foot of a tuft of grass or low bush; eggs four to six, pale bluish, softly mottled with purplish brown.

The Nootka Sparrow, *P. sandwichensis*, Baird, of Western North America, closely resembles the preceding.

The Spotted Sparrow, *P. anthinus*, Bonap., of the coast of California, is similar to *P. savanna*, but is only five inches long.

The Lark Sparrow, *P. alandinus*, Bonap., of California to Mexico, differs from *P. savanna* in being rather smaller, and in paler colors.

The Beaked Sparrow, *P. rostratus*, Baird, of the coast of California, is distinguished from all the preceding of its genus by its longer bill.

The Genus *Poœcetes* has the bill rather large, wings long and pointed, the second and third quills longest, tail short, forked, stiff, and its feathers acute.

The Grass Finch, or Bay-winged Bunting, *P. gramineus*, Baird, of the United States from the Atlantic to the Pacific is six and a quarter inches long, the wing three and one tenth inches; above yellowish brown, the feathers streaked abruptly with dark brown; beneath yellowish white, the breast, sides of neck, and body streaked with brown, and the wings with the shoulder light chestnut-brown. The Bay-winged Bunting sings sweetly, and, at times, for a half-hour without changing its place. Its nest is built in the grass, and partly sunk in the ground; the eggs four to six, bluish white, with undefined blotches of pale reddish-brown.

The Genus *Coturniculus* has the bill very large and stout, tarsus longer than the middle toe, wings short and rounded, tail short, narrow, graduated, but slightly emarginate, the feathers lanceolate and acute, but not stiffened.

The Yellow-winged Sparrow, *C. passerinus*, Bonap., of North America east of the Central Plains, is about five inches long, the wing two and two fifths inches; the feathers of the upper parts brownish rufous, margined narrowly with ash; the crown blackish with a central and superciliary stripe of yellowish tinged with brown, the bend of the wing bright yellow, the quill and tail-feathers edged with whitish. The lower parts are brownish yellow, nearly white on the middle of the belly; the feathers of the upper part of the breast and sides of the body with darker centres. Its nest is upon the ground, made of grass, and lined with fibrous roots and hair; eggs dingy white sprinkled with brown spots.

Henslow's Bunting, *C. Henslowi*, Bonap., of the Eastern United States as far north as Washington, and westward, is five and a quarter inches long, the wing less than two and a quarter inches; the upper parts yellowish brown, the crown with a broad black spotted stripe on each side, these spots continuing down to the back; two black maxillary stripes on each side of the head, and an obscure crescent of the same color behind the auriculars; under parts light brownish-yellow, the upper breast and sides streaked with black; edge of the wing yellow, wings and tail strongly tinged with chestnut.

Leconte's Bunting, *C. Lecontii*, Bonap., of the Yellowstone, is about four and one third inches long, the wing over two inches.

The Genus *Ammodromus* has the bill very long, slender, attenuated, considerably curved towards the tip above; the wings short and rounded, reaching only to the base of the tail; the latter is short and graduated, each feather stiffened, lanceolate, and acute; the legs and toes very long, reaching beyond the tip of the tail.

The Sharp-tailed Finch, *A. caudacutus*, Sw., of the United States on the Atlantic coast, is five inches long, the wing about two and three tenths inches; the upper parts brownish olivaceous, the head with black streaks upon the sides, and a central stripe of ashy; a superciliary and maxillary stripe and a band across the breast buff-yellow; a brown stripe on the side of the throat, the upper part of the breast and sides of the body streaked with black. The rest of the under parts are white, and the edge of the wing yellowish white. The nest is on the ground, near high-water-mark; eggs four to six, dull white dotted with light brown.

The Seaside Finch, *A. maritimus*, Sw., of the United States on the Atlantic coast, is about six inches long, the wing two and a half inches; olivaceous brown above, beneath white, the breast and sides yellowish brown; the sides of the head and body, a medial line on the head, and indistinct streaks on the breast, ashy brown; the superciliary stripe is bright yellow anterior to the eye, the remainder plumbeous; the edge of the wing yellow, and the bill blue. This bird may be seen, at any hour of the day during spring and early summer, upon the tops of tall plants that grow by the margin of tide-water, there uttering the few notes that compose its song. The nest is made of coarse grass without and fine within; eggs four to six, grayish white freckled with brown.

Samuel's Finch, *A. Samuelis*, Baird, of California, is five inches long, the wing less than two and a quarter inches.

The Genus *Chondestes* comprises the Lark Finch, *C. grammaca*, Bonap., from the Mississippi Valley to the Pacific coast. It is six inches long, the wing about three and one third inches; the upper parts grayish brown, the hood chestnut with a median and superciliary stripe of dingy white; the under parts white; a round spot on the breast, a maxillary stripe, and a short line from the bill to the eye, black. The tail is dark brown, broadly tipped with white.

The Genus *Zonotrichia* has the bill conical and slightly notched, wings not reaching to the middle of the moderately-rounded tail, the second and third quills longest, tarsus longer than the middle toe, lateral toes about equal, the hind toe longer than the lateral ones, the claws of the latter just reaching to the base of the middle one.

The White-crowned Sparrow, *Z. leucophrys*, Sw., of the United States east of the Rocky Mountains, is over seven inches long, the wing three and a quarter inches; the head above, the upper half of the loral region, and a line through and behind the eye, black; a longitudinal patch upon the crown, and a line from above the anterior corner of the eye, the two coming together on the occiput, white. Its song consists of six or seven notes, rather plaintive, the first of which is loud, clear, and sweet, the second broader but less firm, and the rest gradually diminishing in fulness and power. This song is repeated at short intervals throughout the day. The nest is made upon the ground, beautifully constructed, and found with difficulty; the eggs five, light sea-green, mottled towards the larger end with brownish spots and blotches, and a few spots of lighter tint are dispersed over the whole.

Gambel's Finch, *Z. Gambellii*, from the Rocky Mountains to the Pacific coast, is almost precisely like the preceding, only smaller.

The Golden-crowned Sparrow, *Z. coronata*, Baird, of the Pacific coast, is seven inches long, the wing about three and one third inches; the top of the head black; median stripe yellow anteriorly, and ashy posteriorly; the sides and under parts of the head and neck, and upper part of the breast, ash-color, passing into whitish beneath.

Harris's Finch, *Z. querula*, Gamb., of the Missouri region, is seven inches long, the wing less than three and a half inches, and distinguished by having the head all round, the neck, and the throat, black.

The White-throated Sparrow, *Z. albicollis*, Bonap., of the United States east of the Missouri, is seven inches long, the wing over three inches; the crown with two black stripes separated by one of white, a broad superciliary stripe yellow to the middle of the eye and white behind it, a broad black streak from behind the eye, the chin white, upper part of the breast dark ash, edge of wing and axillaries yellow, the back rufous brown streaked with dark brown, the belly and two bands across the wing coverts white.

The Genus *Junco* has the bill small and conical, the wings reaching over the basal fourth of the exposed portion of the tail, the tarsus longer than the middle toe, the lateral toes slightly unequal, the outer reaching to the base of the middle claw, and the tail slightly emarginate.

The Mexican Junco, *J. cinereus*, Cab., of Mexico, is six and two fifths inches long, the wing over three inches.

The Oregon Snow-Bird, *J. oregonus*, Sclat., of the Pacific coast to the eastern side of the Rocky Mountains, is six and a half inches long, wing three inches, the head and neck all round sooty black, the interscapular region and the wings dark rufous brown, a lighter tint of the same on the breast and below, the rump brownish ash, and the outer two tail-feathers white.

The Gray-headed Snow-Bird, *J. caniceps*, Baird, of the Rocky Mountains, is six inches long, the wing about three and a quarter inches.

The Black Snow-Bird, *J. hyemalis*, Sclat., of the United States east of the Missouri, is six and a quarter inches long; grayish or dark ashy-black, deepest before; the middle of the breast and belly, the under tail-coverts, and the first and second external tail-feathers white, and the third tail-feather white margined with black. These birds appear in flocks in winter, and are very tame.

The Genus *Poospiza* is represented by only two species in the United States.

The Black-throated Sparrow, *P. bilineata*, Sclat., of the Rio Grande, is nearly five inches long, the wing two and three quarters inches; the color above ashy gray, the under parts and superciliary stripe white, the chin and throat black.

Bell's Finch, *P. Belli*, Sclat., of Southern California, is six and a quarter inches long, the wing nearly three inches; the upper parts bluish ash, under parts pure white.

The Genus *Spizella* is distinguished from *Zonotrichia* by the smaller size, and a longer and forked tail.

The Tree Sparrow, *S. monticola*, Baird, of North America, is six and a quarter inches long, the wing three inches; the feathers of the back dark brown centrally, then rufous, and edged with pale fulvous; the hood, and a line from behind the eye, chestnut. The under parts are whitish, with a blotch of brownish on the breast.

The Field Sparrow, *S. pusilla*, Bonap., of North America east of the Missouri, is five and three quarters inches long, the wing two and one third inches; the bill red, the crown continuous rufous-red, the back similar, streaked with blackish; the under parts white, tinged before with yellowish. This sparrow builds upon the ground at the foot of a small bush, or on branches close to the ground; eggs four to six, light ferruginous.

The Chipping Sparrow, *S. socialis*, Bonap., of North America, is five and three quarters inches long, the wing nearly three inches; the bill black, crown continuous chestnut, the forehead black, separated in the middle by white, superciliary stripe white, a black stripe through the eye; the rump, sides of the head and neck, and back of the latter, ashy; and the interscapular space with black streaks margined with pale rufous. The under parts whitish, and two narrow white bands across the wing-coverts. This is one of the most common birds. Its song is six or seven notes, uttered with rapidity. Its nest is slender, formed of grasses, and lined with hair, and placed upon an apple-tree or some low bush, but never on the ground. The eggs are four or five, greenish blue marked with dark brown spots.

The Clay-colored Bunting, *S. pallida*, Bonap., of the Upper Missouri, is four and three quarters inches long, the wing over two and a half inches; brownish yellow above, the feathers of the crown and back conspicuously streaked with blackish; under parts whitish.

Brewer's Sparrow, *S. Breweri*, Cass., of the Rocky Mountains to the Pacific coast, is five inches long, the wing two and a half inches, and closely resembles the preceding.

The Black-chinned Sparrow, *S. atrigularis*, Baird, of Mexico near the Rio Grande, is five and a half inches long, the wing two and a half inches.

The Genus *Melospiza* has the bill conical, wings quite short and rounded, tail graduated and the feathers oval at the tips, tertiaries longer than the secondaries, the fourth quill longest. The crown and back are similar in color and streaked, the lower parts thickly streaked, and the tail unspotted.

Fig. 129.

Song-Sparrow, *M. melodia*, Baird.

The Song-Sparrow, *M. melodia*, Baird, of the United States east of the Rocky Mountains, is six and a half inches long, the wing over two and a half inches; the upper parts rufous brown streaked with dark brown and ashy gray. The crown is rufous with a median and superciliary stripe of dull gray. The interscapular region has the feathers dull brown in the centre, then rufous, then grayish on the margin. There is a light maxillary stripe bordered above and below by one of dark rufous-brown, with a similar one from behind the eye. The under parts are white, the breast and sides of the body streaked with dark rufous, and on the middle of the breast this color is concentrated into a spot. This species builds both on the ground and on trees. The nest is made of fine grass, lined with hair; eggs three to seven, light greenish-white, speckled with dark umber, the specks larger towards the larger end.

Heermann's Song-Sparrow, *M. Heermanni*, Baird, and Gould's Song-Sparrow, *M. Gouldii*, Baird, inhabit California.

The Rusty Song-Sparrow, *M. rufina*, Baird, of the Pacific coast, is six and three quarter inches long, the wing two and seven tenths inches, light rufous-brown above, beneath whitish.

The Mountain Song-Sparrow, *M. fallax*, Baird, of the Rocky Mountains, is very similar to *M. melodia*, but with smaller bill and longer wings and tail.

Lincoln's Finch, *M. Lincolnii*, Baird, of North America, is five and six tenths inches long, the wing two and six tenths inches; crown chestnut, with a median and superciliary ashy stripe; back streaked with black; beneath, white; a stripe behind the ear-coverts, band across the breast, and under tail-coverts, brownish yellow.

The Swamp Sparrow, *M. palustris*, Baird, of the United States east of the Missouri, is five and three quarters inches long, the wing two and two fifths inches; the middle of the crown chestnut, forehead black, superciliary streak, sides of the head, and back, ash, the latter broadly streaked with black; under parts whitish tinged with ashy across the breast. The nest is built at the foot of a tuft of grass; eggs four or five, dull white, speckled with reddish.

The Genus *Peucæa* has the upper mandible curved, very short and much rounded wings, tail long and much graduated, and the toes short.

Bachman's Finch, *P. æstivalis*, Cab., of Georgia, is six and a quarter inches long, the wing about two and a third inches, the feathers above dark brownish-red margined with bluish ash, under parts pale brownish-yellow, and edge of wing yellow.

Cassin's Finch, *P. Cassinii*, Baird, of Texas, is similar to the last, but paler.

The Brown-headed Finch, *P. ruficeps*, Baird, of California, is five and a half inches long, the wing over two and a third inches, ashy brown above, the crown and nape chestnut, superciliary stripe ashy; under parts pale yellowish-brown, chin with a line of black on each side.

The Genus *Embernagra* comprises the Texas Finch, *E. rufivirgata*, Lawr.

The third group comprises those which have the legs, toes, and claws very stout, and the lateral claws reaching nearly to the end of the middle one. They are sparrow-like species with triangular spots beneath.

The Genus *Passerella* has the body stout; wings long and pointed, reaching to the middle of the nearly even tail.

The Fox-colored Sparrow, *P. iliaca*, Sw., of the United States east of the Mississippi, is seven and a half inches long, the wing three and a half inches; the back dull ash, each feather with a blotch of brownish red; the top of the head and neck similar, but with smaller and less distinct blotches; the exposed surfaces of the wings, upper tail-coverts, and tail, bright rufous; the under parts white; the upper parts of the breast, sides of the body, and throat, with triangular spots of rufous; and on the middle of the breast a few smaller ones of blackish. This Sparrow lingers in clumps of bushes near the water, and patches of briers along the fences. Its flight is low, rapid, undulatory, and its notes sweet. The nest is made upon the ground in a tuft of grass, or under a low bush; eggs four or five, rather sharp at the smaller end, dull greenish sprinkled with blotches of brown.

P. obscura, Verrill, of Anticosti, is somewhat smaller than the preceding species.

Townsend's Sparrow, *P. Townsendii*, Nutt., of the Pacific coast of North America, is about seven inches long, the wing about three inches; the color above uniform dark olive-brown, with a tinge of rufous; the under parts are white, thickly covered with triangular blotches of the same color as the back.

The Slate-colored Sparrow, *P. schistacea*, Baird, of Western North America, is six and four fifths inches long, the wing over three inches; the color above uniform slate-gray; the upper surfaces of wings, tail, and coverts, dark brownish-rufous; under parts white, with arrow-shaped spots of slate-gray.

The fourth group comprises those which are bright-colored, usually without streaks, and with a large bill.

The Genus *Calamospiza* comprises the Lark Bunting or White-winged Blackbird, *C. bicolor*, Bonap., of the Central Plains, which is six and a half inches long, the wing three and a half inches.

The Genus *Euspiza* comprises the Black-throated Bunting, *E. americana*, Bonap., of the United States east of the Central Plains, which is six and seven tenths inches long, the wing three and a half inches; the sides of the head and neck and back of the neck, ash; crown yellowish green, streaked with dusky; superciliary stripe, middle of the breast, and edge of the wing, yellow; under parts white, with a black patch upon the throat.

The Genus *Guiraca* has the bill very large.

The Rose-breasted Grosbeak, *G. ludoviciana*, Sw., of North America east of the Missouri, is eight and a half inches long, the wing over four inches; the upper parts, head, and neck all round, glossy black; a broad crescent across the breast, axillaries, and under wing-coverts, carmine. The rest of the under parts, the rump, and upper tail-coverts, middle wing-coverts, spots on tertiaries and wing-coverts and the basal half of primaries and secondaries, and a large patch on the inner web of the outer three tail-feathers, pure white. The female is without black or carmine, or the white of the quills, tail, and rump. The song of this Grosbeak is loud, clear, and mellow.

Fig. 121.

Rose-breasted Grosbeak, *G. ludoviciana*, Sw.

The Black-headed Grosbeak, *G. melanocephala*, Sw., of Western North America, is eight inches long, the wing four and a quarter inches.

The Blue Grosbeak, *G. cærulea*, Sw., of the Southern United States, is seven and a quarter inches long, the wing three and a half inches, and at once distinguished by its brilliant blue color.

The Genus *Cyanospiza* has the upper outline of the bill considerably curved, wings long and pointed, tail rather narrow and nearly even, and the tarsus about equal to the middle toe. The species are all small and of very showy plumage.

The Blue Bunting, *C. parellina*, Baird, of the Rio Grande region, is five inches long, the wing two and a half inches, general color dark blue.

The Varied Bunting, *C. versicolor*, Baird, of Northeastern Mexico, is five and a half inches long, the wing two and three quarters inches; the colors reddish, blue, and black.

The Painted Bunting, or Nonpareil, *C. ciris*, Baird, of the Southern States, is five and a half inches long, the wing two and seven tenths inches; the head and neck ultramarine blue; a stripe from the chin to the breast, the under parts generally, and the rump, vermilion. The edges of the chin, loral region, greater wing-coverts, and interscapular region are green; and the tail, lower wing-coverts, and outer webs of quills, purplish blue. The female is dark green above, yellow below.

The Lazuli Finch, *C. amœna*, Baird, of the Central Plains to the Pacific, is five and a half inches long, the wing nearly four inches; the upper parts greenish blue, the upper part of the breast pale brownish-chestnut, the remaining under parts, except the blue throat, white.

The Indigo Bird, *C. cyanea*, Baird, of the United States east of the Missouri, is five and three quarters inches long, the wing three inches; the color bright ultramarine blue. The female is brown above, whitish beneath. This bird prefers the skirts of woodlands and detached thickets. Its song consists of eight or ten notes, and is generally uttered from the top of a low tree. It builds its nest of grass on the stalks of rank grass or other plants; eggs four to six, blue, with a spot or two of purple at the larger end.

The Genus *Spermophila* comprises the Little Seed-eater, *S. moreletii*, Puch., of Texas and southward.

The Genus *Pyrrhuloxia* has the bill very short, broad, and greatly curved.

The Texas Cardinal, *P. sinuata*, Bonap., of the Rio Grande region, is eight and a half inches long, the wing three and three quarters inches; the head with a long pointed crest; the upper parts generally pale ashy-brown; crest, wing, and tail, dark crimson; throat, breast, median line below, under tail-coverts, the edge and inner coverts of the wings, bright carmine; the bill yellowish.

The Genus *Cardinalis* has the bill enormously developed, and the head crested.

The Red Bird or Cardinal, *C. virginianus*, Bonap., of the more southern parts of the United States, is eight and a half inches long, the wing three and three quarters inches, and the general color vermilion; a band around the base of the bill and upper part of the throat black. The female with duller red, and less in extent.

The Genus *Pipilo* has the bill rather stout, feet large, the claws stout and curved, tail considerably longer than the wings; the upper parts generally black or brown, under parts white or brown.

The Ground-Robin, Towhee, or Chewink, *P. erythrophthalmus*, Vieill., of the United States east of the Missouri, is eight and three quarters inches long, the wing three and three quarters inches; the upper parts generally, head and neck all round, and the upper part of the breast, glossy black abruptly defined against the pure white below; the sides light chestnut, wings and tail black marked with white, and the iris red. The female has the black replaced with brown. This is one of the most common birds, and is usually seen upon the ground among low bushes. Every few moments its favorite *chewink* comes to our ears. Sometimes it ascends to the top of a small tree, and sings with a mellow sweetness which cannot fail to interest any one who hears it. The nest is made upon the ground, in a little hole scraped out for the purpose; eggs four to six, pale flesh-color, with dark spots. Two or three broods are raised in a season.

Fig. 122.

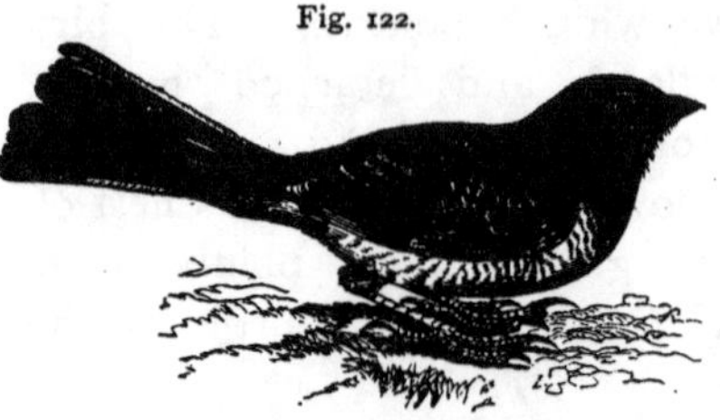

Chewink, *P. erythrophthalmus*, Vieill.

The Oregon Ground Robin, *P. oregonus*, Bell; the Arctic Towhee, *P. arcticus*, Sw., of the Central Plains; the Spurred Towhee, *P. megalonyx*, Baird, of California to the Rio Grande, with the claws enormously developed; Abert's Towhee, *P. Abertii*, Baird, of the Southern Rocky Mountains; the Brown Towhee, *P. fuscus*, Sw., of the coast region of California; the Canon Finch, *P. mesoleucus*, Baird, of the Rio Grande and westward,—are additional species.

Blanding's Finch, *P. chlorurus*, Baird, of the Rio Grande and Rocky Mountains, is seven inches long, the wing over three inches, the color above dull grayish olive-green, the crown chestnut, the upper part of the breast and sides of the body bluish ash, the exterior of the wings and tail bright olive-green, the edge and under surface of the former bright yellow.

ICTERIDÆ, OR BLACKBIRD FAMILY.—This Family comprises birds with a long bill, nine primaries, tail rather long and rounded, the legs stout, the basal joint of the middle toe free on the inner side, and united half-way on the outer.

The Genus *Ploceus* comprises the Weavers of Africa and India, celebrated for their curious nests woven of grass. Some species, like the Republican, *Loxia socia*, Lath., unite by hundreds, and construct a roof, beneath which they build their nests, each being separate and entered from below. The nest of others is a suspended sphere, from which, in some cases, there hangs down a long tube loosely woven of grass.

The Genus *Dolichonyx* has the middle toe very long, and the tail feathers with rigid acuminate points.

The Bobolink, Reed-bird, or Rice-bird, *D. oryzivorus*, Sw., of North America east of the Central Plains, is seven and seven tenths inches long, the wing less than four inches; the general color black, beautifully marked with cream-color and white. The female is yellowish beneath, the feathers above dark brown edged with brownish yellow. Late in the summer the male assumes the colors of the female. During summer this beautiful bird is seen in every meadow, and its jingling song is heard throughout a great part of the day. This is frequently uttered while on the wing. The nest is built on the ground, and composed of grass; eggs four to six, white tinged with dull blue, and spotted with blackish. Late in the summer, these birds are seen

Fig. 123.

Bobolink, *D. oryzivorus*, Sw.

in immense flocks around grain-fields. At length they move southward, lingering by the margins of creeks and rivers, where the tops of the reeds are bent with the ripe seeds. Here the Reed-birds become extremely fat, and thousands are shot by the hunters, and sold in the markets.

The Genus *Molothrus* is represented in North America by the Cow-bird, *M. pecoris*, Sw., which is eight inches long, the wing less than four and a half inches; the head, neck, and anterior half of the breast light chocolate-brown; the rest of the body lustrous black. The female is light olivaceous-brown. In spring and summer this bird lingers around the cattle in the pastures. Like the European Cuckoo it makes no nest; but stealthily lays its eggs, only one in a place, in the nests of other birds; especially in those of the Maryland Yellow-throat, several Flycatchers, the Blue-bird, Chipping Sparrow, and Golden-crowned Thrush. The egg is pale grayish-blue, sprinkled with umber-brown dots and short streaks; and it is a remarkable fact, that it hatches before the eggs of the bird in whose nest it is laid. No sooner has the young Cow-bird hatched, than the foster-parents fly off to obtain food for it, and hence their own eggs perish, and are at length thrown from the nest. The young bird is cared for with all tenderness, and fed even long after it has begun to fly about, and after it has become larger than the foster-parents themselves.

The Genus *Agelaius* has the first quill shorter than the second and third, and the outer claw scarcely reaching to the base of the middle one.

The Red-winged Blackbird, *A. phœniceus*, Vieill., of North America, is nine and a half inches long, the wing five inches, the general color lustrous black, the shoulders and lesser wing-coverts bright crimson or vermilion. The female is brown above, the feathers edged or streaked with rufous brown or yellowish, under parts white streaked with brown. The nest is made on low bushes, frequently growing in or hanging over the water, and is composed of coarse grasses and leaves without, and fine materials within; eggs four to six, light blue, sparsely spotted with dusky.

The Red-shouldered Blackbird, *A. gubernator*, Bonap., of the Pacific coast, is nine inches long, the wing five inches, the color lustrous black, the shoulders and lesser coverts rich crimson, the middle coverts brownish yellow at the base, but the exposed portion black. The female is dusky varied with paler.

The Red and White-shouldered Blackbird, *A. tricolor*, Bonap., of California, is over nine inches long, the wing less than five inches; the color lustrous black, the shoulders and lesser wing-coverts brownish red, the median coverts white with sometimes a tinge of brown. The female is dark brown marked with grayish ash.

The Genus *Xanthocephalus* has the bill nearly straight, wings long, first quill longest, claws very long and much curved, and the tail narrow and nearly even.

The Yellow-headed Blackbird, *X. icterocephalus*, Baird, of North America, is ten inches long, the wing over five and a half inches; the general color black; the head and neck all round, and fore part of the breast, yellow, and there is a white patch at the base of the wing. The female is smaller and browner.

The Genus *Trupialis* has the feathers of the crown with the shafts prolonged into stiffened bristles. It is represented by the Red-breasted Lark, *T. militaris*, Bonap., of South America and perhaps California, which is nine and a half inches long, the wing nearly five inches.

The Genus *Sturnella* has the bill slender, elongated, the feathers of the head stiffened and bristly, the hind claw nearly twice as long as the middle one.

The Meadow Lark, *S. magna*, Sw., of North America east of the Central Plains, is over ten and a half inches long, the wing five inches; the upper parts are brown marked with brownish white, and the exposed portions of the wings and tail with transverse dark brown bars; the under parts yellow, with a black pectoral crescent. When this bird first rises from the ground, it flutters like a young bird, then proceeds generally in a straight course, now checking its speed, now resuming it, as if undecided whether to move fast or slow. When pursued by the hunter, it moves more swiftly, alternately sailing and beating with its wings till it is beyond the reach of harm. The nest is built at the foot of a tuft of grass in a cavity scooped out of the ground. It is covered over, except an entrance just large enough to admit one bird at a time; eggs four or five, pure white, sprinkled and blotched with reddish brown. Both birds engage in the work of incubation.

Fig. 124.

Meadow-Lark, *S. magna*, Sw.

The Western Lark, *S. neglecta*, Aud., of Western North America, is ten inches long, and the wing five and a quarter inches, and very closely resembles the preceding.

The Genus *Icterus* has the bill slender, very acute, and as long as the head and a little decurved; the claws short and much curved; prevailing colors yellow, orange, and black.

The Troupial, *I. vulgaris*, Daudin, of South America and the West Indies, and accidental in the United States, is ten inches long, and the wing four and a half inches;

the head and upper neck all round, and beneath from the tail to the upper part of the breast, the interscapular region, wings, and tail, black; the remaining under parts, a collar on the hind part of the neck, rump, and upper tail-coverts, yellowish orange; a band upon the wings and edges of the secondaries, white.

Audubon's Oriole, *I. Audubonii*, Giraud, of the Rio Grande region; Scott's Oriole, *I. parisorum*, Bonap., of Texas; Wagler's Oriole, *I. Wagleri*, Sclat., of the Rio Grande region and Mexico; and the Hooded Oriole, *I. cucullatus*, Sw., of the Lower Rio Grande, are additional species of *Icterus*.

The Orchard Oriole, *I. spurius*, Bonap., of North America east of the Rocky Mountains, is seven and a quarter inches long, the wing three and a quarter inches; the head and neck all round, wings, interscapular region, and tail, black; under parts, lower part of the back, lesser upper wing-coverts, and the lower ones, brownish chestnut; a narrow line across the wing and the outer edges of the quills, white. The female is olivaceous above, greenish yellow beneath, and there are two white bands upon the wings. The nest, generally built in orchards or upon willows, is made of long grasses curiously interwoven and fastened to the smaller twigs; eggs four to six, bluish white sprinkled with dark brown.

The Baltimore Oriole, Golden Robin, or Hangnest, *I. baltimore*, Daud., of North America east of the High Central Plains, is seven and a half inches long, and the wing three and three quarters inches; the color is black, with the rump, upper tail-coverts, lesser wing-coverts, the terminal portion of all but two tail-feathers, and the under parts, orange red; the edges of quills, and a band across the tips of the greater coverts, white. The female is much duller, the black of the head and back replaced by brownish yellow. This well-known and beautiful bird constructs its curious nest on the outer drooping twigs of elms and of other trees. The song consists of few notes, but these are loud, full, and mellow.

Bullock's Oriole, *I. Bullockii*, Bonap., of Western North America, is seven and a half inches long, and the wing three and eight tenths inches. The color is black, with the under parts, the sides of head and neck, forehead and line over the eye, rump and upper tail-coverts, and tail except two central feathers, yellow orange; the outer edges of the quills and a band across the wing, white.

The Genus *Scolecophagus* has the bill shorter than the head, tail even or slightly rounded.

The Rusty Blackbird, *S. ferrugineus*, Sw., of North America east of the Missouri, is nine and a half inches, and the wing four and three quarters inches; the general color black, with purple reflections. The female is dull brown.

Brewer's Blackbird, *S. cyanocephalus*, Cab., of Western North America, is ten inches long, the wing about five and a third inches.

The Genus *Quiscalus* has the bill as long as the head, the tail long, graduated; colors lustrous black.

The Great-tailed Grakle, *Q. macrourus*, Sw., of the Lower Rio Grande, is eighteen inches long, and the wing seven and a half inches.

The Boat-tailed Grakle or Jackdaw, *Q. major*, Vieill., of the Southern Atlantic and Gulf States, is fifteen inches long, and the wing seven inches; the general color lustrous black, head and forward parts glossed with purple.

The Crow Blackbird, *Q. versicolor*, Vieill., of North America east of the Central Plains, is thirteen inches long, the wing six inches; head and neck all round steel-blue, and the rest of the body black, with varied metallic reflections. These beautiful birds are seen in large numbers as soon as the farmers plough their fields, which they visit to search for grubs and worms which the plough uncovers. They also visit newly-sown grain-fields to pick up the exposed kernels, and in the autumn large flocks commit depredations upon the cornfields.

The Florida Blackbird, *Q. baritus*, Vieill., of the West Indies and Florida coast, is over ten and a half inches long, and the wing five inches.

STURNIDÆ, OR STARLING FAMILY.—This Family comprises birds which are nearly related to the preceding one, but which have a rudimentary outer primary, thus making the primaries ten instead of nine, as in Icteridæ.

The Genus *Sturnus* comprises the Common Starling, *S. vulgaris*, Linn., of Europe, which is about the size of a thrush, black with violet and green reflections, and spotted with white or fawn-color. It moves in large flocks, is easily tamed, and may be taught to sing, and even to speak.

CORVIDÆ, OR CROW FAMILY.—This Family comprises the Crows and their allies.

The Genus *Corvus* has the bill thick, culmen much curved, bristly feathers at the base of the bill half as long as the culmen, and the color throughout black.

The American Raven, *C. carnivorus*, Bartram, of North America, but rare east of the Mississippi, is twenty-four inches long, and the wing seventeen inches. It makes its nest on high and rugged cliffs. The eggs are four to six, two inches long, and of a light greenish-blue, and covered with blotches of light purple and yellowish brown.

The Colorado Raven, *C. cacalotl*, Wagler, of California, is twenty-five inches long, and the wing eighteen inches; the color glossy black.

The White-necked Crow, *C. cryptoleucus*, Couch, of the Rio Grande region, is twenty-one inches long, and the wing fourteen inches. The color is glossy black, with the feathers of the neck and breast snow-white at the base.

The Common Crow, *C. americanus*, Aud., of North America, is about twenty inches long, and the wing about thirteen inches. Its stately gait when moving upon the ground, its manner of flight, its shyness and cunning, and its propensity to scratch up and pull up the corn of the farmer, are all well known; but the great benefit which the crow confers upon the farmer, by destroying an almost infinite number of grubs destructive to the crops, is not so well understood.

The Fish Crow, *C. ossifragus*, Wils., of the South Atlantic coast, is fifteen and a half inches long, and the wing ten and a half inches. This species feeds upon small fishes, which it secures with its claws as it passes over the water.

The Western Fish Crow, *C. caurinus*, Baird, of Northwestern North America, is sixteen and a half inches long, and the wing eleven inches.

The Rook, *C. frugilegus*, Linn., of Europe, is nineteen inches long, black, and glossed with purple. It usually builds near human dwellings, and sometimes in large cities.

The Genus *Picicorvus* is represented by Clark's Crow, *P. columbianus*, Bonap., of Western North America, which is twelve inches long, and the wing seven inches; the general color bluish ash, the secondaries and tertials broadly tipped with white, and the tail mainly white.

The Genus *Gymnokitta* has the nostrils naked, and the tail short and even.

Maximilian's Jay, *G. cyanocephala*, Pr. Max., of Western North America, is ten inches long, and the wing five and nine tenths inches; the general color dull blue.

The Genus *Pica* has the bill much curved, tail very long and graduated.

The Magpie, *P. hudsonica*, Bonap., of North America, is nineteen inches long, the wing eight and a half inches, and the tail eleven inches; the general color black; the belly, scapulars, and inner webs of primaries white, and the neck spotted with white. The European Magpie is closely related to the American species, but its voice and habits are said to be different.

Fig. 125.

Magpie, *P. hudsonica*, Bonap.

The Yellow-billed Magpie, *P. Nuttalli*, Aud., of California, is seventeen inches long, and the wing eight inches.

The Genus *Cyanura* has the head crested; wings and tail blue, with transverse bars of black.

The Blue Jay, *C. cristata*, Sw., of North America east of the Missouri, is twelve and a quarter inches long, and the wing less than five and three quarters inches. In beauty of plumage this bird is not surpassed, if equalled, by any other bird in the United States; but its notes are harsh and disagreeable, and its habit of stealing and eating the eggs and young of other birds gives it an unenviable reputation.

Steller's Jay, *C. Stelleri*, Sw., of Western North America, is thirteen inches long, and the wing less than six inches; the head and neck all round, and the fore part of the breast, dark brownish-black; back and lesser wing-coverts blackish brown; under parts, rump, tail-coverts, and wings blue.

The Long-crested Jay, *C. macrolophus*, Baird, of the Rocky Mountains, is twelve and a half inches long, and the wing less than six inches, and is distinguished by its long crest, which is nearly twice the length of the bill.

The Genus *Cyanocitta* has the head without a crest, the wings and tail blue, but not banded.

The California Jay, *C. californica*, Strick., of the Pacific coast, is twelve and a quarter inches long, and the wing five inches; general color above blue without bars; a crescent of blue on the fore part of the breast; under parts before the crescent, white streaked with blue; behind it, dull white.

Woodhouse's Jay, *C. Woodhousii*, Baird, of the Rocky Mountains; the Florida Jay, *C. floridana*, Bonap.; the Mountain Jay, *C. sordida*, Baird, of the Rocky Mountains; and the Ultramarine Jay, *C. ultramarina*, Strick., of the Rio Grande region, are additional species of *Cyanocitta*.

The Genus *Xanthoura* has the bill very stout, head without a crest, the throat black, and the lateral tail-feathers bright yellow.

The Rio Grande Jay, *X. luxuosa*, Bonap., of the Rio Grande, is eleven inches long, and the wing four and three quarters inches; the color above, green; beneath, yellow glossed with green; the inside of the wings and four outer tail-feathers, yellow; the sides of the head, and beneath from the bill to the fore part of the breast, black; the crown and nape brilliant blue, and the sides of the forehead white.

The Genus *Perisoreus* has the bill very short, and notched at the tip; head without a distinct crest.

The Canada Jay, *P. canadensis*, Bonap., of Northern North America, is over ten and a half inches long, and the wing five and three quarters inches; the head and neck and fore part of the breast, white; a plumbeous brown nuchal patch, becoming darker to the back, from which it is separated by a whitish collar; the other upper parts plumbeous; under parts smoky gray.

The Genus *Psilorhinus* is represented by the Brown Jay, *P. morio*, Gray, of the Rio Grande region, which is sixteen inches long, and the wing eight inches.

MENURIDÆ, OR LYRE-BIRD FAMILY. — This Family comprises Australian birds which in some respects are allied to the Thrushes, in others to the Jays, while their large size has induced some authors to place them with the Rasores. With this explanation, they may, for convenience, be mentioned here. They are nearly as large as a pheasant, and are distinguished by the remarkable tail of the male, which is composed of three sorts of feathers; — twelve very long, and with very fine and widely separated barbs; two more, in the middle, only one side of which is furnished with barbs; and two more external, curved into the form of the arms of a lyre, and whose internal barbs, large and thickly set, form a sort of broad ribbon, while the external barbs are very short.

PARADISEIDÆ, OR BIRD OF PARADISE FAMILY. — This Family comprises birds peculiar to New Guinea and adjacent islands, and distinguished for their wonderfully developed and beautiful plumage. The Genus *Paradisæa* is the principal one.

The Emerald Bird of Paradise, *P. apoda*, Linn., is about the size of the American Robin, marroon color, the top of the head and neck yellow, the throat and around the bill emerald. The sides of the tail have a splendid plume of long, loose feathers of a delicate yellow hue, and on either side of these are two slender shafts nearly two feet in length.

BUCERIDÆ, OR HORNBILL FAMILY. — This Family comprises large birds of Africa and India, which have the bill very large, dentated, and generally surmounted with an extraordinary protuberance, which in some cases is as large as the bill itself. By their bill these birds are allied to the Toucans, but their general carriage and habits approximate them to the Crows.

SUB-SECTION IV.

THE ORDER OF RASORES, OR SCRATCHERS.

THE Order of Rasores comprises Birds which have the bill not longer than the head, the terminal portion more or less vaulted, and hard, and with or without a soft skin intervening between it and the head, and the nostril with a fleshy scale extending over its upper edge. With few exceptions, they have a heavy body, short wings, rather stout legs, and large tail. They live mainly upon the ground, are social in their habits, and feed principally upon nuts, berries, buds, tender leaves, and grain.

The Rasores comprise two Sub-Orders, — Columbæ and Gallinæ.

The Sub-Order of Columbæ or Doves comprises those which have the bill shorter than the head, the basal portion covered by a soft skin in which the nostrils are situated, the hind toe on the same level as the others, and the anterior toe without a basal membrane. They live in pairs, lay generally but two eggs for a brood, but breed often, and feed their young, which are hatched in a very feeble condition, with macerated food from their own crops. Bonaparte calls the true Doves or Pigeons Gyrantes, and divides them into four families; — Treronidæ, of the Old World, and especially of the islands of the Pacific, which have the bill robust and tumid, the feathers soft and without lustre, the prevailing color green, and the tail with fourteen feathers; Columbidæ, universally distributed; Calœnidæ, of the East India islands; and Gouridæ, of New Guinea.

Columbidæ, or Dove Family. — This Family comprises those which have the bill horny at the tip, tail-feathers twelve, occasionally fourteen, and the plumage more or less adorned with metallic lustre. About a dozen species are found in North America, and all but two or three belong to the southern and southwestern portions.

The Genus *Columba* has the head large, and the tail short, broad, and rounded. It comprises the Band-tailed Pigeon, *C. fasciata*, Say, of Western North America; the Red-billed Dove, *C. flavirostris*, Wagl., of the Lower Rio Grande; and the White-headed Pigeon, *C. leucocephala*, Linn., of Florida Keys.

The Rock Dove, *C. livia*, Briss., of Europe and Asia, is of a slate-gray color, the neck glossy with greenish hues, the rump white, and a double black band upon the wings. This species is celebrated as the probable stock of most, if not all, of our domestic varieties.

The Carrier Pigeon is a domestic variety which from very ancient times has been employed more or less for the transmission of intelligence. Formerly it was customary to suspend the paper upon which the message was written from the neck; but in later times it has been tied to the upper part of the leg. A message has thus been sent a thousand miles.

The Genus *Ectopistes* has the head very small, bill short and black, and the tail very long and cuneate.

The Wild or Passenger Pigeon, *E. migratoria*, Sw., of North America east of the High Central Plains, is seventeen inches long, and the wing eight and a half inches; the upper parts blue, under parts purplish red passing into whitish behind, and the sides and back of the neck a glossy golden-violet. The female is smaller and much duller in color. This bird is extremely rapid in flight, being able to perform a long journey at an average speed of a mile a minute. The migrations are wholly for the purpose of procuring food, and hence do not take place at any particular season of the year. Pigeons go wherever they can find a supply of grain, rice, or nuts. The

Fig. 126.

Wild Pigeon, *E. migratoria*, Sw.

numbers that sometimes move together are vast beyond conception. Millions associate in a single roost, completely filling a forest for thirty or forty miles in length and several miles in breadth, and literally loading and breaking down large trees. From their roosts they fly off hundreds of miles, in some cases, to feeding-grounds, and return at night. Sometimes, in their migrations, they fill the air like a cloud, and thus continue to pass for a whole day, and even for two or three successive days. The nest is built on high trees, and is composed of a few dry sticks and twigs crossing each other, and supported by the forks of the branches; and more than a hundred nests are sometimes placed on a single tree.

The Genus *Zenaida*, represented by the Zenaida Dove, *Z. amabilis*, Bonap., of the Florida Keys, has the tail short, rounded, and the orbits feathered.

The Genus *Melopelia*, represented by the White-winged Dove, *M. leucoptera*, Bonap., of the Rio Grande and the West Indies, has the tail short, rounded, and the orbits naked.

The Genus *Zenaidura* has the tail excessively lengthened, cuneate, and with fourteen feathers.

The Carolina Dove, *Z. carolinensis*, Bonap., of the United States, is nearly thirteen inches long, the wing five and three quarters inches; the color above, bluish, overlaid with brownish olive; the head, sides of the neck, and under parts generally, light brownish-red, strongly tinged on the breast with purple; the sides of the neck with a patch of metallic purplish-red; the bill black, and feet yellow. The female is smaller, with less red beneath. At night, doves of this species roost upon the ground, and some distance apart.

The Turtle Dove, *C. turtur*, Linn., of the Old World, celebrated for its gentleness and plaintive notes, is eleven and a half inches long; the upper parts tawny slate-color spotted with brown; the breast brownish, and the other under parts white.

The Genus *Scardafella*, represented by the Scaly Dove, *S. squamosa*, Bonap., of the Rio Grande, which is only eight inches long, has the bill lengthened, tail very long and much graduated.

The Genus *Chamæpelia*, comprising the smallest doves known, is represented by the Ground Dove, *C. passerina*, Sw., of the Southern States, which is only six and three tenths inches long, and the wing three and a half inches; color above, grayish olive; under parts light purplish-red.

The Genus *Oreopeleia*, represented by the Key West Pigeon, *O. martinica*, Reich., has the bill lengthened, slender, feet large, and tail suborbicular.

The Genus *Starnoenas* has the bill short, legs stout, tail short and broad. It is represented by the Blue-headed Pigeon, *S. cyanocephala*, Bonap., of Key West and the West Indies, which much resembles the quails.

GOURIDÆ, OR GOURA FAMILY. — This Family comprises Columbæ of large size, with the head conspicuously crested. They belong to the Indian Archipelago.

The Genus *Goura* comprises the Crown Pigeon, *G. coronata*, Temm., which is about the size of a turkey, and the head has a vertical crest of long, slender feathers.

The Sub-Order of Gallinæ comprises the true Rasores, which have the bill short, stout, and the basal portion hard, and generally covered by feathers, and not by soft skin; the legs lengthened, the hind toe in most cases elevated, the toes connected at the base by a membrane. The young are able to run about as soon as hatched. There are five families, — Penelopidæ or Curassow Family, Megapodidæ or Mound-Bird Family, Phasianidæ or Pheasant Family, Tetraonidæ or Grouse Family, and Perdicidæ or Quail Family.

PENELOPIDÆ, OR CURASSOW FAMILY. — This Family comprises birds peculiar to Central and South America. They are mainly of large size, about as large as turkeys, and move in flocks, and build their nests among and often upon the trees. They are known under the names of Curassows, Hoccos, and Guans.

The Genus *Ortalida* is represented by the Chiacalacca, *Ortalida McCalli*, Baird, of New Mexico, which is twenty-three and a half inches long, the wing eight and a half inches; above dark greenish-olive, beneath brownish yellow, head and upper part of the neck plumbeous, tail-feathers lustrous green, and all except the middle one tipped with white.

MEGAPODIDÆ, OR MOUND-BIRD FAMILY. — This Family comprises Indian and Australian birds which are celebrated for building large mounds of vegetation and sand, in which they deposit their eggs.

PHASIANIDÆ OR PHEASANT FAMILY.—This Family comprises birds which have the legs, toes, and nasal fossæ bare, the tarsus in the male with one or more spurs, the hind toe elevated above the others, and the tail-feathers more than twelve. It includes the Turkeys, Peacocks, Guinea Fowls, Jungle Fowls, Domestic Fowls, and Pheasants. All except the Turkeys are indigenous to the Old World, although many of them are now widely distributed over the globe.

The Genus *Meleagris* comprises the Turkeys, of which there are two species, both indigenous to America.

The Wild Turkey, *M. gallopavo*, Linn., of the United States, is about forty-eight inches long, the stretch of wings about sixty inches. The prevailing color is copper-bronze with copper and green reflections, each feather with a black margin. The quills are brown closely barred with white, tail chestnut barred with black, head livid blue, and the legs red. The male averages fifteen to eighteen pounds' weight, and the female about nine pounds, although the former, in some cases, attains thirty to forty pounds. The great size and beauty of this bird, and the fact of its being the origin of all the domestic varieties of turkeys of both hemispheres, together with the well-known delicacy of its flesh as an article of food, render it one of the most interesting of this country. It is rare on the Atlantic coast, but is still common in the regions farther west. The question is now in agitation whether there are not really two or three species of Wild Turkey in the United States.

The Ocellated Turkey, *M. ocellata*, of Central America, is exceedingly beautiful, the plumage exhibiting the most brilliant and varied metallic reflections; and the tail and tail-coverts are ornamented with four series of large ocellated spots.

The Genus *Pavo*—Peacocks—has the head crested,

and the tail of the male excessively elongated. The Common Peacock, *P. cristatus*, Linn., was introduced into Europe by Alexander the Great.

The Genus *Numida*, comprising the Guinea Fowls of tropical Africa, has the frontal bone much developed, producing a vertical crest, the lower jaw with two fleshy lobes, no spur on the tarsi, and the tail very short.

The Common Guinea Fowl, *N. meleagris*, Linn., of the farm-yard, is known by every one. In the wild state, it lives in flocks, and prefers the vicinity of marshes.

The Genus *Gallus* comprises our domestic varieties of the Cock kind, and the Jungle Fowl of India.

The domestic Cock, *G. domesticus*, Linn., is too well known to need description. This bird has been in the possession of man from the earliest times, and the varieties are now almost endless. They have all sprung from the Jungle Fowls of India, of which there are many species, all of which bear great resemblance to our domestic varieties.

The Genus *Phasianus*, or Pheasants, has the tail excessively long, the feathers of which overlap like tiles. There are several species, all natives of Asia.

The Common Pheasant, *P. colchicus*, Linn., of Europe, is thirty-four inches long, the head and neck of metallic lustre, and the rest of the plumage golden fawn-color with markings of green. The female is smaller, and brownish. This bird, now found throughout temperate Europe, is said to have been brought from the banks of the Phasis, a river of Colchis.

The Golden Pheasant, *P. pictus*, Linn., of China, so remarkable for its magnificent plumage, has a golden-colored crest, the neck orange speckled with black, the back green, the rump yellow, the lower parts and wings red, the latter with a blue spot, and the long tail brown spotted with gray.

The Argus Pheasant, *P. argus*, Linn., of India, surpasses in size and splendor all others of this genus. The tail is very long, making the entire length of the bird over sixty inches. The secondaries are excessively elongated and widened, and covered with ocellated spots, so that, when the wings are expanded, the appearance of this bird is splendid beyond description.

TETRAONIDÆ, OR GROUSE FAMILY. — This Family comprises gallinaceous birds which have the nasal fossæ filled and covered with feathers, tarsi densely feathered, toes usually naked and pectinated along their edges.

The Genus *Tetrao* has the tarsus feathered to and between the bases of the toes, and the color mainly black. The Grouse of this genus inhabit wooded regions.

The Dusky Grouse, *T. obscurus,* Say, of Nebraska and to the Cascade Mountains, is twenty and a half inches long, and the wing nine and two fifths inches; the tail has twenty feathers.

The Canada Grouse or Spruce Partridge, *T. canadensis,* Linn., of Northern North America, is over sixteen inches long, the wing nearly seven inches; and the tail with sixteen feathers. The feathers above are banded with plumbeous; beneath, the color is uniform black, with a pectoral white band, and white on the sides of the belly.

Franklin's Grouse, *T. Franklinii,* Douglas, of the Rocky Mountains and westward, is similar to the preceding one.

The Cock of the Woods, or Capercailzie, *T. urogallus,* Linn., of Europe, is larger than the Turkey, being the largest of the Gallinæ.

The Black Cock, *T. tetrix,* Linn., is another European species, about the size of the domestic Cock.

The Genus *Centrocercus* has the lower throat and sides with stiffened spinous feathers, the tail excessively lengthened, and cuneate.

The Sage Cock, or Cock of the Plains, *C. urophasianus,* Sw., of the plains of the northwest portions of America, is twenty-nine inches long, and the wing over eleven inches; the upper parts mottled with black, brown, and brownish yellow; the under parts black and white.

The Genus *Pediocætes* has the central tail-feathers lengthened.

The Sharp-tailed Grouse, *P. phasianellus*, Baird, of the plains of Wisconsin and westward, is eighteen inches long, and the wing eight and a half inches, and distinguished by the tail, which has eighteen feathers, the central pair elongated beyond the rest an inch or more.

The Genus *Cupidonia* has the tail short, the bare space of the neck concealed by a tuft of lanceolate feathers.

The Pinnated Grouse, or Prairie Chicken, *C. cupido*, Baird, of the Western prairies, is sixteen and a half inches long, and the wing nearly nine inches; the colors whitish brown and brownish yellow, the feathers with transverse bars of brown. A tuft of long, pointed feathers on each side of the neck covers a naked, orange-colored air-sack, which is capable of great inflation. These air-sacks enable the males to produce the peculiar booming sounds which are always heard during the pairing season. When the air-receptacles are inflated, the bird lowers his head to the ground, and, opening its bill, utters a succession of sounds, going from loud to low till the air of the sacks is exhausted; then immediately erecting itself, and inflating the sacks, it proceeds as before. These sounds may be heard a mile or more. In autumn and winter, they associate in flocks of hundreds. They are easily tamed. Audubon caught sixty in the early autumn, and, having clipped the tips of their wings, put them in a garden and orchard of four acres; within a week they were not frightened at his approach, and before winter was over they would eat from the hand.

The Genus *Bonasa* has eighteen tail-feathers, the lower half of the tarsi naked, the naked space upon the neck covered by a tuft of broad, soft feathers, and the head with a soft crest.

The Ruffed Grouse, or Partridge, *B. umbellus*, Steph., of the Eastern United States and westward, is eighteen inches long, and the wing over seven inches; the color reddish brown or gray above, the back with spots of lighter; the under parts whitish barred with dull brown; the feathers of the ruff are black; the beautiful tail is tipped with gray, and has a subterminal bar of black. This bird prefers the borders of forests, open woods, thickets of evergreens and birches, and the vicinity of brooks shaded with alder. Nothing can excel the grace with which it moves upon the ground. It walks with a proud step, elevated head, the ruffs more or less raised, and its exquisitely beautiful tail partly spread. It poises itself a second or two on one foot, then on the other, and at almost every movement utters a soft cluck. If disturbed, it lowers its head, spreads its tail wider, and runs rapidly into the thickest bushes. If there be no hiding-place near, it at once takes wing with the loud whirring which all have heard who have had the pleasure of visiting its favorite resorts. According to Audubon, these sounds are never heard when the bird rises of its own accord, but only when flushed by a real or supposed enemy. The flight is straight, rather low, and under ordinary circumstances not more than one or two hundred yards at a time. If, when flushed, it alights upon a tree, as is often the case in regions where it has not been much hunted, it will generally be found, if at all, on the side farthest from the pursuer, and close to the trunk, and standing so still and erect that one can readily mistake it for a stump of a broken limb.

The Oregon Grouse, *B. Sabinii*, Baird, of Western North America, is similar to the preceding one.

The Genus *Lagopus* — Ptarmigans — has the legs

closely feathered to the claws. The members of this genus are snow-white in winter; in summer, more or less marked with black, brown, and yellow.

The Willow Grouse, or White Ptarmigan, *L. albus*, Aud., of Northern America, rare in the United States, is fifteen and a half inches long, the wing about eight inches.

The Rock Ptarmigan, *L. rupestris*, Leach, of Arctic America, is fourteen and a half inches long, and the wing seven and a half inches.

The White-tailed Ptarmigan, *L. leucurus*, Sw., of Western North America, and the American Ptarmigan, *L. Americanus*, Aud., of the Baffin's Bay region, are additional species.

Perdicidæ, or Partridge Family. — This Family comprises birds which differ from the grouse in being much smaller, and in their bare tarsi and naked nasal fossæ. They are very numerous, and widely distributed, and not less than forty species belong to America.

The Genus *Ortyx* has the bill stout, head without a crest, and the tail short.

Fig. 127.

Quail, *O. virginianus*, Bonap.

The Quail, *O. virginianus*, Bonap., of the United States east of the High Central Plains, is ten inches long, the wing nearly four and three quarters inches; prevailing color above, brownish red; the under parts white, tinged with brown before, and marked with obtusely V-shaped spots of black; the head is beautifully marked with pure white and black. The female has the white markings of the head replaced by brownish yellow, and the black wanting. In New England, New York, and westward, this bird is called the Quail; but in Pennsylvania and southward it is called the Partridge. Its clear whistle is composed of three notes, the first and last of equal length, the first being loudest. The nest is built near a tuft of grass; eggs ten to eighteen, pure white.

The Texas Quail, *O. texanus*, Lawr., of Texas, is very similar to the preceding.

The Mountain Quail, *O. pictus*, Baird, of the Mountain ranges of Oregon and California, is ten and a half inches long, and the wing five inches.

Fig. 128.

Mountain Quail, or Plumed Partridge, *Oreortyx pictus*, Baird.

The Genus *Lophortyx* has the head with a crest of lengthened feathers springing from the vertex, the shafts in the same vertical plane.

The California Quail, *L. californicus*, Bonap., of California and Oregon, is nine and a half inches long, the wing over four and a quarter inches; the forward half of the body, and upper parts, plumbeous; the crest black.

Gambel's Partridge, *L. Gambelli*, Nutt., of the Rio Grande to California, is nine and a half inches long, and the wing four and a half inches.

The Genus *Callipepla* has a broad, short, depressed crest of soft, thick feathers. It is represented by the Scaled or Blue Partridge, *C. squamata*, Gray, of the Rio Grande.

The Genus *Cyrtonyx* has the wings long and broad,

and the coverts so much developed as to conceal the quills. It is represented by the Massena Partridge, *C. massena*, Gould, of the Upper Rio Grande.

The Genus *Perdix* comprises the Gray Partridge, *P. cinerea*, of Europe, which is about twelve inches long.

The Genus *Coturnix* includes the Common European Quail, *C. dactylisonans*, which is seven inches long.

SUB-SECTION V.

THE ORDER OF CURSORES, OR RUNNERS.

THE Order of Cursores comprises birds of great size, with the neck and legs very long, and the wings rudimentary.

STRUTHIONIDÆ, OR OSTRICH FAMILY. — This Family comprises the Ostriches, Cassowaries, Apteryx, and the like. They run with great speed, but cannot fly.

The Genus *Struthio* is represented by the Eastern Ostrich, *S. camelus*, Linn., of the deserts of Africa and Asia, which is six to eight feet high. Its feet have but two toes, and the outer one is only half the length of the inner, and destitute of a nail. It is so swift of foot that no animal can overtake it in running. Its eggs weigh about three pounds each, and are laid in the sand, and, in the hot regions, left to hatch, but in cooler regions are brooded with care, and defended with great courage.

The Genus *Rhea* comprises the American Ostrich, of South America, which is much smaller than the preceding, and particularly distinguished by having three toes, all armed with nails; color, gray. It is said that several of them lay their eggs in one and the same nest.

The Genus *Casuarius* has even shorter wings than the Ostriches, and feathers which resemble pendent hairs, and three toes, each furnished with a nail.

The Galeated Cassowary, *C. galeatus*, of the Indian

Archipelago, is about five feet high, and the head surmounted by a horny crest. The wings have some stems without barbs, which seem to serve as weapons of defence.

The Emeu or New Holland Cassowary, *C. novæ hollandiæ*, Lath., of Australia, is five to seven feet high.

The Genus *Apteryx* comprises New Zealand species, which, in addition to their rudimentary wings, are distinguished by their elongated, slender bill, which bears the nostrils at the tip of the upper mandible.

OTIDÆ, OR BUSTARD FAMILY. — This Family comprises birds which have the massive carriage of the Gallinæ, the long neck and legs of the Ostriches, while in the absence of a thumb the smaller species especially approximate the Plovers; and their wings are short — not rudimentary — and seldom used except to assist in running. They belong to the Eastern hemisphere. *Otis* is the only genus.

The Great Bustard, *O. tarda*, Linn., is the largest bird of Europe, — the male attaining the length of four feet, and the weight of thirty pounds or more.

SUB-SECTION VI.

THE ORDER OF GRALLATORES, OR WADERS.

THE Order of Grallatores comprises birds which have the bill, neck, and legs very long, tail short, and the legs bare for some distance above the tarsal joint. They live near the water, upon shores or marshes, or more rarely upon dry plains. The Grallatores may be divided into two Sub-Orders, — Herodiones and Grallæ.

The Sub-Order of Herodiones comprises those which have the bill thick at the base, much longer than the head, the face more or less naked, and the hind toe generally lengthened and nearly on the same level with the anterior ones; and the young are hatched in a weak condition, and reared in a nest. It includes the Gruidæ or Crane Family, Aramidæ or Courlan Family, Ardeidæ or Heron Family, Cancromidæ or Boat-bill Family, Cinconidæ or Stork Family, Tantalidæ or Ibis Family, Plataleidæ or Spoon-bill Family., and Phœnicopteridæ or Flamingo Family.

GRUIDÆ, OR CRANE FAMILY. — This Family comprises very large birds, which have the head more or less bare, the toes connected by a basal membrane, and the hind toe short and much elevated. They inhabit dry plains.

The Genus *Grus* is the only one represented in North America.

The White or Whooping Crane, *G. americanus*, Ord, of Florida and Texas, and occasionally in the Mississippi Valley, is fifty-two inches long, and the wing twenty-four inches.

The Sandhill Crane, or Brown Crane, *G. canadensis*, Temm., of the Mississippi Valley and westward, is forty-eight inches long, and the wing twenty-two inches. It is exceedingly wary, and its sight and hearing are acute. When wounded, it is dangerous to approach it, as a single thrust from its bill may inflict a severe wound.

The Little Crane, *G. fraterculus*, Cass., of New Mexico, is seventeen and a half inches long.

ARAMIDÆ, OR COURLAN FAMILY. — This Family comprises birds which have the head feathered to the bill, toes cleft to the base, and the hind toe long.

The Genus *Aramus* is represented by the Courlan, or Crying Bird, *A. giganteus*, Baird, of Florida and the West Indies, which is twenty-seven and a half inches long.

ARDEIDÆ, OR HERON FAMILY. — This Family comprises waders which have the bill acuminate, compressed, acute, and the edges usually notched at the end; the frontal feathers generally extending beyond the nostrils, the inner toe connected by a basal web to the outer; the claws acute, and the middle one pectinated on its inner edge. It is represented all over the globe.

The Genus *Demigretta* has a full occipital crest of elongated feathers, and the back has free plumes longer than the tail.

Peale's Egret, *D. Pealii*, Baird, of South Florida, is thirty inches long, and the wing thirteen inches. The color is pure white, the terminal half of the bill black.

Reddish Egret, *D. rufa*, Baird, of the Gulf States, is thirty inches long, and the wing twelve and a half inches.

Louisiana Heron, *D. ludoviciana*, Baird, of the Southern States, is twenty-five inches long, and the wing ten and a half inches; slate-blue above; rump, under parts, and the longest occipital feathers, white.

The Genus *Garzetta* has a full occipital crest; middle of the back with long plumes reaching to the tail, and recurved at the tips.

The Snowy Heron, *G. candidissima*, Bonap., of the coast of the Middle and Gulf States, and across to California, is twenty-four inches long, and the wing over ten inches.

The Genus *Herodias* has no crest, the back with plumes longer than the tail, and curving gently downwards.

The White Heron, *H. egretta*, Gray, of the Southern States, and accidental in New England, is thirty-nine inches long, and the wing fifteen and a half inches. The California White Heron is probably a larger variety.

The Genus *Ardea* has the bill very thick, occiput with a few clongated feathers, and no dorsal plumes.

The Great Blue Heron or Crane, *A. herodias*, Linn., of North America and the West Indies, is forty-two inches long, the wing eighteen and a half inches, and the bill five and a half inches. This bird frequents ponds and creeks, where it may be seen standing upon a rock or stump for hours, watching for fish, upon which it feeds.

When a fish comes within reach, it instantly transfixes it with its sharp bill, and afterwards swallows it whole.

Fig. 129.

Great Blue Heron, *A. herodias*, Linn.

It also feeds upon reptiles, mice, and young birds. When wounded, it at once prepares for defence, and the dog or man who comes within reach is sure to receive a severe wound; and the danger is greater, as these birds generally aim at the eye. The nest is placed on a large tree in a dense swamp. It is large and flat, built of sticks, and matted with grass and mosses. The eggs are three, dull bluish-white.

The Florida Heron, *A. Wurdmanii*, Baird, of South Florida, is forty-nine inches long.

The Genus *Audubonia* comprises the Great White Heron, *A. occidentalis*, Bonap., of South Florida and Cuba, which is forty-five inches long.

The Genus *Florida* has the bill convex above, straight below, and very acute.

The Blue Heron, *F. cærulea*, Baird, of the South Atlantic and Gulf States, is twenty-two inches long, the wing eleven inches; the color slate-blue.

The Genus *Ardetta* has claws long and acute, body compressed, the lower neck bare of feathers behind, and the tail with ten feathers. It embraces the smallest known herons.

The Least Bittern, *A. exilis*, Gray, of North America, is thirteen inches long, the wing four and three quarters inches; the head above and back dark glossy green; the upper neck, shoulders, greater coverts, and outer webs of some of the tertials, purplish cinnamon. The female has the green of the head and back replaced by chestnut. The nest is built on low bushes; eggs three to four, dull yellowish-green.

The Genus *Botaurus* has the plumage loose, and the sexes similar.

The Bittern or Stake-driver, *B. lentiginosus*, Steph., of all North America, is twenty-six and a half inches long, the wing eleven inches; the color is brownish yellow finely varied with dark brown and brownish red; and there is a broad, black stripe on each side of the neck. It seldom flies till you are close upon it, and then it moves off very sluggishly.

Fig. 130.

Bittern or Stake-driver, *B. lentiginosus*, Steph.

The Genus *Butorides* has the bill gently curved from the base above, and the tail with twelve feathers. It is represented by the Green Heron or Fly-up-the-Creek, *B. virescens*, Bonap., of the United States generally, which is fifteen inches long, the wing seven and a halt inches.

The Genus *Nyctiardea* has the bill very stout, the end of the upper mandible gently decurved, and an occipital plume of three feathers rolled together.

The Night Heron, *N. Gardeni*, Baird, of the United States generally, is twenty-five inches long, and the wing twelve and a half inches; the head above and middle of the back steel-green; wings and tail ashy blue; forehead, the long occipital feathers, and under parts, white. This species breeds in communities, making nests in trees around stagnant ponds or in swamps; eggs four, light sea-green. These birds perform their migrations in the night, at which times their loud, hoarse note may be heard, which has been represented by the syllable *qua*.

The Genus *Nyctherodius* has the bill very thick.

The Yellow-crowned Night Heron, *N. violaceus*, Reich.,

of the South Atlantic and Gulf States, is twenty-four inches long, and the wing twelve inches.

CANCROMIDÆ, OR BOAT-BILL FAMILY. — This Family comprises waders which resemble herons in all but the bill, which appears like that of a heron very much flattened. Its shape has been aptly compared to a boat with the keel upward.

The Genus *Cancroma* is represented by the Boat-bill, *C. cochlearia*, of South America, which is about the size of the domestic hen. It lives near the water, and feeds upon fish.

CINCONIDÆ, OR STORK FAMILY. — This Family includes the Stork of the Old World, and the Jabirus of both hemispheres. They have the bill thicker than in Ardeidæ, and nearly equal membranes between the bases of the toes.

The Genus *Cinconia* comprises the White Stork, *C. alba*, Cuv., of Europe, which is forty-two inches long, white, with the quills of the wings black, and the feet and bill red. It is held in high estimation on account of its destruction of noxious reptiles. It prefers to build its nest in towers and steeples, and returns to the same spot year after year.

The Pouched Stork, *C. marabou*, Temm., and *C. argala*, Temm., respectively of the tropical regions of Africa and of India, have an appendage under the throat resembling a thick sausage. These birds are six feet high as they ordinarily stand, and seven when the neck is fully erect, and the expanse of wings is fifteen feet. They are black above and white below, and are popularly known as *Adjutants*. The beautiful plumes known as *Marabouts* are obtained from under the wing of these birds.

The Genus *Jabiru* comprises very large birds, which differ from the Storks in having the extremity of the bill curved upward.

TANTALIDÆ, OR IBIS FAMILY. — This Family comprises waders with the bill very long, rounded, much attenuated and decurved, and the toes with a basal web.

The Genus *Tantalus* has the head, in the adult, entirely destitute of feathers.

Fig. 131.

Wood Ibis, *T. loculator*, Linn. — $\frac{1}{16}$ nat. size.

The Wood Ibis, *T. loculator*, Linn., of the Southern States, is forty-five inches long, the wing eighteen and a half inches; the color white, quills and tail a metallic blackish-green. Birds of this species live in flocks, feeding upon fish and aquatic reptiles. Finding shallows that abound in fish, they move about till the water has become muddy, which causes the fish to rise to the surface, when they are struck by the bills of the Ibis, and killed. Soon the surface is covered with dead fishes and reptiles, and the birds swallow them until they have gorged themselves, after which they go to the shore, and arrange themselves in rows, with their breasts turned towards the sun. It is dangerous to approach them when wounded, as they bite severely.

The Genus *Ibis* has the bill very long, moderately thickened at the base, and curves downward towards the tip.

The Red or Scarlet Ibis, or Pink Curlew, *I. rubra*, Vieill., of South America and the West Indies, accidentally in the United States, is twenty-eight inches long, the wing nearly eleven, and the bill nearly seven inches.

The White Ibis, or White Curlew, *I. alba*, Vieill., of the South Atlantic and Gulf States, rarely northward, is twenty-five inches long, the wing eleven and a quarter inches, and the bill seven inches; color white. This species feeds largely upon crawfish, which it often secures by a curious process. The crawfish, in dry weather, burrows to the depth of three or four feet, and in all cases deep enough to reach damp earth or water. The Ibis carefully approaches the hole, drops in pieces of earth, and then retires a step, and silently awaits the result. Soon the crawfish begins to remove the earth thus thrown in, but no sooner does it come to the entrance of its burrow than it is seized.

The Glossy Ibis, *I. Ordii*, Bonap., found sparingly throughout the United States, is twenty and a half inches long, the wing ten inches; color chestnut.

PLATALEIDÆ, OR SPOON-BILL FAMILY. — This Family comprises large birds which have the bill completely depressed and very broad, and widening at the rounded tip. Seven or eight species are known.

The Genus *Platella* comprises our only species, the Rosy Spoonbill, *P. ajaja*, Linn., of the South Atlantic and Gulf States, which is thirty inches long, the wing fifteen inches, and the bill seven; color rose-red.

PHŒNICOPTERIDÆ, OR FLAMINGO FAMILY. — This Family comprises birds with the legs and neck excessively elongated, toes fully webbed, bill bent abruptly in the middle, and the edges lamellated. Some place this family here, and others put it in with Anseres.

The Genus *Phœnicopterus* is represented by the Flamingo, *P. ruber*, Linn., of the warmer parts of America, which is forty-five inches long, the wing sixteen and a half inches; color scarlet.

The Sub-Order of Grallæ comprises waders which have the head feathered to the bill, — the latter, when much longer than the head, slender and contracted at the base, — and whose young run about and pick up food as soon as hatched. Instead of building their nests upon trees or bushes, as in the case of the Herodiones, the members of this group, with some exceptions, lay their eggs in a cavity scooped out in the sand.

The Grallæ comprise the Charadridæ or Plover Family, Hæmatopodidæ or Turnstone Family, Recurvirostridæ or Avoset Family, Phalaropidæ or Phalarope Family, Scolopacidæ or Snipe Family, and Rallidæ or Rail Family.

CHARADRIDÆ, OR PLOVER FAMILY. — This Family comprises waders which have the bill rather cylindrical, as long as the head or shorter, culmen much indented opposite the nostrils, hind toe rarely present and when present only rudimentary, and the outer and middle toes more or less united by a membrane. The wings, when folded back, reach beyond the tail; the head very large, and the neck short and thick.

Fig. 132.

Golden Plover, *C. virginicus*, Borck.

The Genus *Charadrius* is represented by the Golden Plover, *C. virginicus*, Borck., of both hemispheres, which is nine and a half inches long, the wing seven inches; the upper parts brownish black, with numerous spots of golden yellow; under parts black, with a brownish lustre.

The Genus *Ægialitis* has the plumage without spots, the neck and head generally with dark bands.

The Kill-deer, *A. vociferus*, Cass., of North and South America, is nine and a half inches long, and the wing six and a half inches; the head above and the upper parts of the body, light brown; the rump and upper tail-coverts rufous; the front, and lines over and under the eye, white; above the white band in front is one of black; a black band on the breast, and a black ring around the neck; the throat is white, and this color extends upwards around the neck; the under parts, with the exceptions named, are also white. It takes its popular name from its peculiar note.

The Mountain Plover, *A. montanus*, Cass., of Western North America, is about nine inches long, the wing six inches.

Wilson's Plover, *A. Wilsonius*, Cass., of the Atlantic Southern States and South America, is smaller than the preceding.

The King Plover, or Semi-palmated Plover, *A. semipalmatus*, Cab., of all temperate North America, is about seven inches long, and the wing four and three quarters inches; the front, throat, a ring around the neck, and the under parts, white; a band of black across the breast, which extends around the back of the neck below the white ring; a band from the base of the bill under the eye, and a wide frontal band above the white one, black; the upper parts light ashy-brown with a tinge of olive.

The Piping Plover, *A. melodus*, Cab., of the eastern coast of North America, is seven inches long, the wing four and a half inches; the forehead, a ring round the neck, and entire under parts, white; a black band in front above the white one, and a black band encircling the neck before and behind, immediately below the ring of white; the head above and upper parts of the body light brownish-cinereous; the quills dark brown marked with white; the tail tipped with white, and the outer feathers white, and middle ones with a subterminal band of black.

The Western Plover, *A. nivosus*, Cass., of California, is

six and a half inches long, and the wing three and three quarters inches.

The Genus *Squatarola* has a rudimentary hind toe.

The Black-bellied Plover, *S. helvetica*, Cuv., of all North America and the sea-coasts of all countries, is eleven and a half inches long, and the wing seven and a half inches.

The Genus *Aphriza* has the bill shorter than the head, the hind toe distinct, and tail even.

The Surf-bird, *A. virgata*, Gray, of the islands of the Pacific, is ten inches long, and the wing seven inches.

HÆMATOPODIDÆ, OR TURNSTONE FAMILY.—This Family comprises waders which have the bill compressed.

The Genus *Hæmatopus*—Oyster-catchers—has the bill longer than the tarsus, hind toe wanting, tarsus reticulated anteriorly, and the middle and outer toes connected by a basal membrane.

The Oyster-catcher, *H. palliatus*, Temm., of the Atlantic coast, is seventeen and a half inches long, and the wing ten inches; the upper parts light ashy-brown; under parts and upper tail-coverts, and a wide diagonal band across the wings, white.

Backman's Oyster-catcher, *H. niger*, Pallas, of the western coast of the United States, is seventeen inches long, the wing ten and a half inches; the head and neck brownish black; all other parts of the plumage dark brown; the bill bright red, legs pale reddish.

The Dusky Oyster-catcher, *H. ater*, Vieill., of Chili and perhaps of the western coast of North America, is larger than either of the preceding.

The Genus *Strepsilas*—Turnstones—has the bill tapering to rather a blunt point, the tip slightly bent upward, and the hind toe lengthened.

The Turnstone, *S. interpres*, Illig., of North America, and of nearly every country, is nine inches long, and the wing six inches; the upper parts irregularly variegated with black, dark rufous, and white; the abdomen, under wing, and tail-coverts, rump, and back, white; the head and neck above generally white, with spots and stripes of brownish black on the crown and occiput; the throat is white with a black stripe on each side connecting with a black patch on the breast.

Fig. 133.

Turnstone, *S. interpres*, Illig.

The Black Turnstone, *S. melanocephala*, Vig., of Western North America, is darker than the preceding one.

RECURVIROSTRIDÆ, OR AVOSET FAMILY. — This Family comprises birds which are at once distinguished by their excessively elongated legs, long and slender neck, and long and slender bill.

The Genus *Recurvirostra* has the bill extended into a fine point, and recurved at the tip, and toes webbed.

The American Avoset, *R. americana*, Gm., of all North America, is seventeen inches long, the wing about eight and a half inches; the head and neck pale reddish-brown; back, wing-coverts, and quills, black; other parts white.

Fig. 134.

Black-necked Stilt, *H. nigricollis*, Vieill.

The Genus *Himantopus* has the bill nearly straight. It is represented by the Black-necked Stilt, *H. nigricollis*, Vieill., of North America, which is fourteen inches long, the wing about eight and a half inches.

PHALAROPIDÆ, OR PHALAROPE FAMILY. — This Family

comprises waders which have the lateral groove of the bill extending nearly to the tip, toes with a lateral margin, the hinder with a feeble lobe, and the feathers of the breast compact.

The Genus *Phalaropus* — Phalaropes — has the membrane generally more or less scalloped at the joints.

Wilson's Phalarope, *P. Wilsonii*, Sab., of North America, is nine and a half inches long, and the wings five and a half inches; the back, wings, and tail, cinereous; rump and upper tail-coverts, and under parts, white; a wide stripe behind the eye blackish; the neck before, and a stripe running upwards to the back, bright reddish-brown.

The Northern Phalarope, *P. hyperboreus*, Temm., of nearly all countries, is seven inches long, and the wing four and a half inches; the bill short, straight, and pointed; upper parts dark; lower, white; and the neck with a ring of bright ferruginous.

The Red Phalarope, *P. fulicarius*, Bonap., of the temperate regions of North America, Europe, and Asia, is seven and a half inches long, and the wing five and a quarter inches; the under parts deep brownish-red; dark above.

SCOLOPACIDÆ, OR SNIPE FAMILY. — This Family comprises waders which have the bill generally longer than the head, and the hind toe generally present.

The Genus *Philohela* — Woodcocks — has the body very full, head, bill, and eyes very large, wings short and rounded, and the toes cleft to the base.

The American Woodcock, *P. minor*, Gray, of Eastern North America, is eleven inches long, and the wing five and a quarter inches; the bill very long, the upper mandible longer than the under, and fitted to it at the tip; the eyes far from the bill, and the tail short; the occiput with three transverse bands of black alternating with three of pale yellowish-rufous; the upper parts of the

Fig. 135.
American Woodcock, *P. minor*, Gray.

body variegated with ashy reddish and black; the under parts pale rufous, bill brown, and legs pale reddish. The Woodcock is mainly nocturnal in its habits, seldom taking wing in the full light of day unless disturbed. It walks about, however, and feeds by day as well as by night. Its food is mainly earthworms, of which it swallows as many in a day as would equal its own weight; and hence its favorite resorts are where it can obtain these worms in abundance. The moist grounds which these birds frequent are perfectly filled with *bill-holes* which they have made in probing for worms; and these holes become a guide to the hunter, who looks at their frequency and freshness when he would find good shooting. When flushed by the hunter or the dog, the Woodcock ordinarily flies but a short distance, plunging into a clump of bushes or thicket near by, or a thicker part of the swamp. It spends the winter in warm climates, but breeds from the Carolinas to Nova Scotia. The nest, made of dead leaves and grass, is placed under a bush, or beside a fallen trunk. The eggs, which are laid from February to June, according to locality, are usually four, dull yellowish clay-color, irregularly and thickly marked with dark brown. In three or four weeks from the time the young are hatched, they are able to fly; and when six weeks old, they fly almost as well as the old ones.

The Genus *Gallinago* has a more slender body, and longer legs, than *Philohela*.

Wilson's or the English Snipe, *G. Wilsonii*, Bonap., of the temperate regions of North America, is ten and a half inches long, and the wing five inches; upper parts brownish black, marked with light rufous, yellowish

Fig. 136.

Wilson's Snipe, *G. Wilsonii*, Bonap.

brown, or ashy white; under parts white, reddish ashy on the throat. Flushed by the hunter or other enemy, it dashes through the air, in a zigzag course, and when about twenty yards off utters its peculiar *wau-aik*, which Audubon says indicates the best time to fire.

The Genus *Macrorhamphus* has the tarsus longer than the middle toe, and the base of the outer and middle toe connected by a short web.

The Gray Snipe or Red-breasted Snipe, *M. griseus*, Leach, of the temperate regions of North America, is about ten inches long, and the wing five and three quarters inches, and is distinguished from the preceding by the white shaft of the first quill.

The Greater Longbeak, *M. scolopaceus*, Lawr., of all temperate North America, is eleven and a half inches long, the wing five and three quarters inches.

The Genus *Tringa* comprises a large number of small birds which live on the shores of both salt and fresh water in all parts of the world. They are found more or less in flocks, and feed upon small or minute shell-fish and other small aquatic animals. Their wings are long and pointed, tail short, lower portion of tibiæ naked, hind toe very small, and the fore toes with a membranous margin. They are known as Sandpipers.

The Gray-back or Robin Snipe, *T. canutus*, Linn., of Eastern North America and Europe, is ten inches long, and the wing six and a half inches.

Cooper's Sandpiper, *T. Cooperi*, Baird, of Long Island, is rather smaller than the preceding.

The Purple Sandpiper, *T. maritima*, Brünnich, of Eastern North America and Europe, is eight to nine inches long, and the wing five inches.

The Curlew Sandpiper, *T. subarquata*, Temm., of the Eastern hemisphere, and rare on the Atlantic coast of North America, is about nine inches long, the wing five inches; bill slightly curved towards the tip.

The Red-backed Sandpiper, *T. alpina*, var. *americana*, Cass., of temperate North America, is eight and a half inches long, and the wing five inches.

The Jack Snipe, *T. maculata*, Vieill., of all North America, South America, and Europe, is nine inches long, and the wing five and a quarter inches; upper parts brownish black; under parts white and ashy white; breast spotted.

The Least Sandpiper, *T. Wilsonii*, Nutt., of all temperate North America, is five and a half to six inches long, and the wing three and a half inches or more.

Bonaparte's Sandpiper, *T. Bonapartii*, Schlegel, of North America east of the Rocky Mountains, is seven inches long, and the wing four and three quarters inches.

The Genus *Calidris* has the general characters of Tringa, but is destitute of a hind toe.

The Sanderling, *C. arenaria*, Illig., of North America, South America, and Europe, is about eight inches long, and the wing five inches.

The Genus *Ereunetes* has the feet semi-palmated. It is represented by the Semi-palmated Sandpiper, *E. petrificatus*, Illig., of the temperate regions of North and South America, which is six and a half inches long, and the wing three and three quarters inches.

The Genus *Micropalama* has the basal membrane of the toes more deeply emarginate than in the preceding genus. It is represented by the Stilt Sandpiper, *M. himantopus*, Baird, of Eastern North America, which is eight and a half to nine inches long, and the wing five and a quarter inches; legs long and slender.

The Genus *Symphemia* has the bill very thick and recurved.

The Willet, *S. semipalmata*, Hartl., of temperate North and South America, is fifteen inches long, and the wing eight and a quarter inches; upper parts dark ashy; rump, upper tail-coverts, and under parts, white.

The Genus *Glottis* is represented by the Florida Greenshank, *G. floridanus*, Bonap., which is eleven inches long.

The Genus *Gambetta* has the bill much attenuated towards the tip, the outer toe webbed to the first joint, and the inner web very short.

The Tell-tale, or Stone Snipe, *G. melanoleuca*, Bonap., of North America, is fourteen inches long, and the wing about eight inches; upper parts cinereous; under parts white, with longitudinal stripes on the neck, and transverse spots and stripes of dark on the breast and sides.

Fig. 137.

Yellow-Legs, *G. flavipes*, Bonap.

The Yellow-Legs, *G. flavipes*, Bonap., of Eastern North America, is similar to the preceding, but smaller, being ten inches long, and the wing six inches.

The Genus *Rhyacophilus* has the bill curved upward slightly from the middle. It is represented by the Solitary Sandpiper, *R. solitarius*, Bonap., of North America, which is eight and a half inches long, and the wing five inches.

The Genus *Heteroscelus* has the bill stout and compressed. It is represented by the Wandering Tatler, *H. brevipes*, Baird, of Washington Territory, which is about ten and a half inches long, and the wing six and a half inches.

The Genus *Tringoides* has the bill straight, shorter than the head, and the tail much rounded. It is represented by the Spotted Sandpiper, *T. macularius*, Gray, of temperate North America, which is seven and a half to eight inches long, and the wing four and a half inches.

The Genus *Actiturus* is represented by Bartram's Sandpiper or Field Plover, *A. Bartramius*, Bonap., of temperate North America, South America, and Europe, which is about twelve inches long, and the wing six and a half inches; general color above brownish black; under parts yellowish white. It prefers plains and cultivated fields.

The Genus *Philomachus* is represented by the Ruff, *P. pugnax*, Gray, of Northern Europe, and accidental on Long Island, which closely resembles the preceding one.

The Genus *Tryngites* has the wings very long. It is represented by the Buff-breasted Sandpiper, *T. rufescens*, Cab., of North and South America and Europe, which is seven and a half to eight inches long, and the wing five and a half inches. It prefers plains and fields.

Fig. 138.

Marbled Godwit, *L. fedoa*, Ord.

The Genus *Limosa* — Godwits — has the bill lengthened, slender, and curving gently upwards. It is represented by the Marbled Godwit, *L. fedoa*, Ord, of North and South America, which is eighteen inches long, and the wing nine inches; and the Hudsonian Godwit, *L. hudsonicus*, Sw., of Northern North America, which is fifteen inches long.

The Genus *Numenius* — Curlews — has the bill very long, and curved downwards.

The Long-billed Curlew, *N. longirostris*, Wils., of all

temperate North America, is about twenty-five inches long, and the wing ten to eleven inches; the curved bill

Fig. 139.

Long-billed Curlew, *N. longirostris*, Wils.

from five to eight inches long. The upper parts are pale rufous tinged with ashy, every feather marked with brownish black; the under parts pale rufous, with longitudinal lines of black on the neck and sides.

The Short-billed Curlew, *N. hudsonicus*, Lath., of the Atlantic and Pacific coasts of North America, is eighteen inches long, the wing nine inches; and the bill three to four inches long.

The Esquimaux Curlew, *N. borealis*, Lath., of North America, is thirteen and a half inches long, the wing eight and a quarter inches; the bill two and a quarter to two and a half inches long.

Rallidæ, or Rail Family. — This Family comprises waders, with a strong, compressed bill, compressed body, rather short wings, and long toes. They live in marshes, and are but little seen except by hunters and naturalists.

The Genus *Rallus* — Rails — has the bill rather longer than the head, wings and tail very short. It includes about twenty species, inhabiting all countries.

The King Rail, or Marsh Hen, *R. elegans*, Aud., of the Atlantic and Pacific coasts of North America, in the warmer parts, is seventeen inches long, and the wing six and a half inches; upper parts olive brown; under parts rufous chestnut.

The Clapper Rail, or Mud Hen, *R. crepitans*, Gm., of the Atlantic coast, is about fourteen inches long, the wing five and a half inches, and resembles the preceding.

The Virginia Rail, *R. virginianus*, Linn., of temperate North America, is seven and a half inches long, and the wing four inches; upper parts olive brown, with longitudinal stripes of brownish black; throat white; neck before, and breast, bright rufous; abdomen and under tail-coverts with transverse bands of black and white.

The Genus *Porzana* has the bill shorter than the head, and straight. It contains about twenty species.

Fig. 140.

Common Rail, *P. carolina*, Vieill.

The Sora, or Common Rail, *P. carolina*, Vieill., of temperate North America, is eight and a half inches long, and the wing four and a quarter inches; the upper parts greenish brown, with longitudinal bands of black, and many feathers with narrow white stripes at their edges; the sides of the neck and breast bluish ashy, with circular spots and transverse bands of white upon the breast; the bill greenish yellow; legs dark green. The female is similar, but duller.

The Little Black Rail, *P. jamaicensis*, Cass., of the Atlantic coast of North America and the West Indies, five inches long, and the Yellow Rail, *P. noveboracensis*, Baird, of Eastern North America, about six inches long, are additional species.

The Genus *Crex* is represented by the Corn-Crake, *C. pratensis*, Bechst., of Europe, Greenland, and accidental in the United States, which is ten inches long; color dark

The Genus *Fulica* — Coots — has the bill straight, compressed, and extending into the feathers of the forehead, forming a wide and projecting frontal plate; and the toes margined with semicircular lobes.

The Coot, *F. americana*, Gm., of temperate North America, is fourteen inches long, and the wing seven inches.

The Genus *Gallinula* — Gallinules — is distinguished from *Fulica* by the absence of lobes on the toes.

The Florida Gallinule, *G. galeata*, Bonap., of the warmer parts of America, is twelve and a half inches long; the frontal plate large, terminating square on the top of the head; the head, neck, and under parts dark bluish-cinereous; upper parts brownish olive. The frontal plate and bill are bright red tipped with yellow.

Fig. 141.

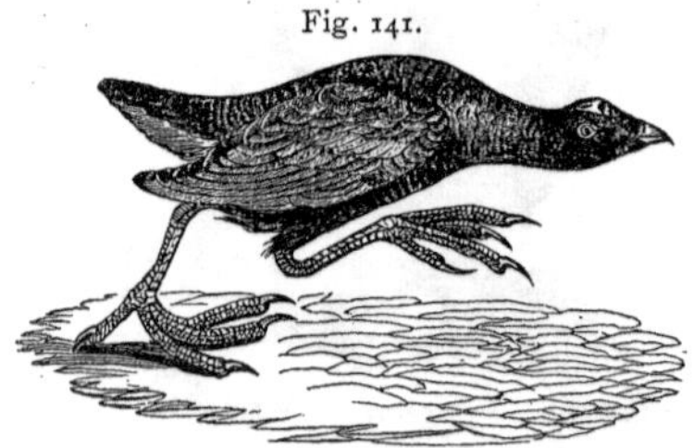

Purple Gallinule, *G. martinica*, Lath.

The Purple Gallinule, *G. martinica*, Lath., of the Southern States and northward, is twelve and a half inches long, and the wing seven inches; head and under parts bluish purple; lower tail-coverts white; upper parts dark olive green; bill bright red tipped with yellow; frontal plate blue, and legs yellow.

SUB-SECTION VII.

THE ORDER OF NATATORES, OR SWIMMERS.

THE Order of Natatores comprises birds which are especially fitted for aquatic life. Their plumage is thick and firm, toes webbed to the claws, tibiæ feathered to near the tarsal joint, the hind toe usually elevated, and rather small. They all swim well, and most of them dive freely. This group may be divided into two Sub-Orders, — Anseres, which have the bill with transverse lamellæ along the edges; and Gaviæ, which have the bill without lamellæ. The first of these is represented by the great group of Anatidæ.

ANATIDÆ, OR DUCK FAMILY. — This family comprises swimming birds whose jaws have transverse lamellæ, the upper mandible ending in an obtuse rounded nail, and a groove running along both jaws to the nail. They are numerous, and found in all parts of the world.

The Genus *Cygnus* — Swans — has the neck very long.

The American Swan, *C. americanus*, Sharp., of North America, is fifty-five inches long, and the wing twenty-two inches; the adult pure white, the bill and legs black; the tail has twenty feathers. The young are brown.

The Trumpeter Swan, *C. buccinator*, Rich., of Western North America, is sixty inches long, and the wing twenty-four inches; the adult pure white, the bill and legs black; the tail has twenty-four feathers. Its notes are more sonorous than those of the preceding.

The Red-billed Swan, *Anas olor*, Gm., and the Black-billed Swan, *A. cygnus*, Gm., belong to Europe. The former is the original of the domestic Swan.

The notion that the Swan sings on the approach of death is erroneous.

The Genus *Anser* has the lamellæ of the upper mandible projecting below the edge as points.

The Snow Goose, *A. hyperboreus*, Pallas, of North America, is thirty inches long, and the wing about sixteen and a half inches; color pure white; bill and legs red.

The White-fronted Goose, *A. Gambelli*, Hartl., of North America, is twenty-eight inches long, and the wing about sixteen and one third inches; color grayish; forehead white, bill and legs red; the tail has sixteen feathers.

The Brown-fronted Goose, *A. frontalis*, Baird, of the interior of North America, closely resembles the preceding.

The Genus *Bernicla* has the teeth of the upper mandible mainly concealed; bill and legs black.

The Canada or Wild Goose, *B. canadensis*, Boie, of North America, is thirty-five inches long, and the wing eighteen inches; the upper parts brownish, the lower parts lighter; the head, neck, bill, and feet black. It spends the winter in the warmer regions, but in spring moves northward in large flocks. Their spring migrations usually take place from the 20th of March to the last of April, but are wholly dependent upon the state of the season. They breed at the North, and linger there till the hard frosts warn them that the lakes and streams will soon be frozen over.

The White-cheeked Goose, *B. leucopareia*, Cass., of the western coast of America, has the general appearance of the preceding, but is darker and somewhat smaller.

Hutchins's Goose, *B. Hutchinsii*, Bonap., of the northern and western portions of North America, closely resembles *B. canadensis*, but is only thirty inches long.

The Brant, *B. brenta*, Steph., of the Atlantic coast of North America and Europe, is twenty-three and a half inches long, the wing twelve and three quarters inches; bill, feet, head, neck, and fore part of the body, primary quills and tail, black. On each side of the neck there is a small white crescent streaked with black.

The Black Brant, *B. nigricans*, Cass., of the Pacific coast of North America, is twenty-nine inches long, and the wing nearly fourteen inches.

The Barnacle Goose, *B. leucopsis*, Linn., of Europe, is twenty-eight inches long, and the wing seventeen inches.

The Genus *Chloephaga* is represented by the Painted Goose, *C. canagica*, Bonap., of the Aleutian Islands.

The Genus *Deudrocygna* has the bill much longer than the head, neck and legs very long, and feet very large.

The Long-legged Duck, *D. autumnalis*, Eyton, of the

Rio Grande, is twenty-four inches long, and the wing ten inches.

The Brown Tree-duck, *D. fulva*, Burm., of California and southward, is twenty inches long, and the wing over nine inches.

Anatinæ, or River Ducks, have the legs shorter than geese, the lobe of the hind toe narrow, and much restricted. Eight genera are found in North America.

The Genus *Anas* has the bill broad, depressed, longer than the head, and the tail pointed.

The Mallard, or Green-Head, *A. boschas*, Linn., of North America and most of the Old World, is twenty-three inches long, and the wing eleven inches; the head and neck bright grass-green, with a violet gloss; a white ring around the middle of the neck, below which and on the fore part and sides of the breast the color is dark brownish-chestnut; speculum purplish violet, terminated with black. This is the original of the common domestic duck.

The Black Duck, *A. obscura*, Gm., of the Atlantic region of North America, is twenty-two inches long, and the wing twelve inches; general color bluish brown.

The Genus *Dafila* has the bill long and narrow, tail pointed. It is represented by the Pintail, or Sprigtail, *D. acuta*, Jenyns, of North America and Europe, which is thirty inches long, and the wing eleven inches.

The Genus *Nettion* has the bill unusually narrow.

The Green-winged Teal, *N. carolinensis*, Baird, of North America, accidental in Europe, is fourteen inches long, and the wing seven and two fifths inches; and distinguished by the broad rich green speculum.

The English Teal, *N. crecca*, Kaup., of Europe, accidental in North America, is similar to the preceding.

The Genus *Querquedula* has the bill narrow, and a little longer than the foot.

The Blue-winged Teal, *Q. discors*, Steph., of North

America east of the Rocky Mountains, is sixteen inches long, and the wing over seven inches.

The Red-breasted Teal, *Q. cyanoptera*, Cass., of Western North America, is nearly eighteen inches long, and the wing seven and a half inches.

The Genus *Spatula* has the bill much longer than the head, and spatulate, widening to the end.

The Shoveller, or Spoonbill, *S. clypeata*, Boie, of North America and Europe, is twenty inches long, the wing nine and a half inches.

The Genus *Chaulelasmus* has the bill as long as the head, and the lamellæ distinctly visible below the edges. It is represented by the Gadwall, or Gray Duck, *C. streperus*, Gray, of North America and Europe, which is twenty-two inches long, the wing ten and a half inches.

The Genus *Mareca* has the bill shorter than the head; tail pointed; upper parts finely waved with black and gray or reddish brown; under parts white.

The Baldpate, or American Widgeon, *M. americana*, Steph., of North America, is nearly twenty-two inches long, and the wing eleven inches; head and neck grayish, spotted and banded with black, a broad green patch around and behind the eye, and top of the head nearly white.

The English Widgeon, *M. penelope*, Bonap., of the Old World, and accidental in North America, is twenty inches long, and the wing over ten and a half inches; head and neck reddish brown, and top of the head cream-color.

The Genus *Aix* has the bill high at the base, the head crested, claws short, curved, and very sharp. It comprises the most beautiful ducks.

The Summer or Wood Duck, *A. sponsa*, Boie, of North America, is nineteen inches long, and the wing nine and a half inches. Its exquisitely beautiful plumage surpasses description. It builds its nest in a hollow tree or limb.

If the nest is over water, the young, the moment they are hatched, drop into the water; but if at a distance from it, they are allowed to fall on the ground, and are then led, or carried, to the water by the parent.

Fig. 142.

Summer or Wood Duck, *A. sponsa*, Boie.

Fuligulinæ, or Sea Ducks, differ from the Anatinæ in having a large lobe or membranous flap attached to the under surface of the hind toe. They are found inland as well as on the sea-coast.

The Genus *Fulix* has the bill as long as the feet; head, neck, body anterior to the shoulders, tail and tail-coverts, rump, and lower back, black; tail rounded, of fourteen feathers; under parts white, finely waved with black behind and on the sides.

The Big Black-Head, or Scaup Duck, *F. marila*, Baird, of North America and Europe, is twenty inches long, and the wing nine inches; speculum white. The female has the head brown.

The Little Black-Head, or Blue-Bill, *F. affinis*, Baird, of North America and Europe, is sixteen and a half inches long, the wing eight inches; similar to the preceding.

The Ring-necked Duck, *F. collaris*, Baird, of North America and accidental in Europe, is eighteen inches long, and the wing eight inches, and is distinguished by a chestnut collar around the middle of the neck.

The Genus *Aythya* comprises two American species.

The Red-Head, *A. americana*, Bonap., of North America, is twenty and a half inches long, and the wing nine and a half inches; the head and neck for more than half its length brownish red.

The Canvas-Back, *A. vallisneria*, Bonap., of all North America, is twenty inches long, and the wing about nine and one third inches; head and neck chestnut; body

Fig. 143.

Canvas-Back, *A. vallisneria*, Bonap.

anterior to the shoulders, the lower back, rump, and tail-coverts, black; under parts white; scapulars and interscapulars white, finely dotted with black in transverse lines.

The Genus *Bucephala* has the bill shorter than the head, and the tail with sixteen feathers.

The Golden-Eye, or Whistle-Wing, *B. americana*, Baird, of North America, is eighteen and three quarters inches long, and the wing eight and a half inches; the head and upper neck glossy green; a white patch before the eye.

Barrow's Golden-Eye, *B. islandica*, Baird, of Northern North America, is twenty-two and a half inches long, and the wing nine and a half inches.

The Butter-Ball, or Dipper, *B. albeola*, Baird, of North America, is fifteen inches long, the wing over six and a half inches; a broad patch on each side of the head, lower neck, and under parts, white.

The Genus *Histrionicus* has the bill very small, and a membranous lobe at its base; the tail pointed, and with fourteen feathers.

The Harlequin Duck, *H. torquatus*, Bonap., of Northern North America, is seventeen and a half inches long, and the wing nearly seven and three quarters inches; general color bluish, under parts dull brownish. There are two white spots on the side of the neck, two on the wings, one on each side of the base of the tail, and the scapulars and tertials are marked with white, and the secondaries have a violet-blue speculum.

The Genus *Harelda* has the bill shorter than the head, and the tail pointed, with fourteen feathers.

The Long-Tail, or Old-Wife, *H. glacialis*, Leach, of the Atlantic and Pacific coasts of North America, and of Europe, is about twenty-one inches long, the wing nearly nine inches; general color blackish; under parts whitish; tail eight inches in length.

The Genus *Camptolæmus* has the bill broad, feathers of the cheek stiffened, and tail rather pointed. It is represented by the Labrador Duck, *C. labradorius*, Gray, of the northeastern coast of North America, which is about twenty-four inches long, and the wing nearly nine inches.

The Genus *Melanetta* is represented by the White-winged Coot, or Velvet Duck, *M. velvetina*, Baird, which is twenty-one and a half inches long, and the wing about eleven and one third inches; color black, with a small white patch round the eye, and a large white speculum upon the wing. The female is brown.

The Genus *Pelionetta* contains two American species.

The Sea Coot, or Surf Duck, *P. perspicillata*, Kaup, is nineteen inches long, and the wing nearly nine and a half inches; color black, a triangular white patch on the top of the head, and one on the nape; bill red. The female is brown, sides and under parts whitish.

The Long-billed Scoter, *P. Trowbridgii*, Baird, of Southern California, is much like the preceding.

The Genus *Oidemia* has the bill much swollen at the base, and the terminal portion much depressed, and very broad. The Scoter, *O. americana*, Sw., of the coast of North America, is nearly twenty-four inches long, and the wing over nine inches; color black. The Huron Scoter, *O. bimaculata*, Baird, of Lake Huron, is eighteen and a half inches long, and the wing ten and a half inches.

The Genus *Somateria* has the bill much compressed, tapering, and terminated by an enormous nail.

The Eider Duck, *S. mollissima*, Leach, of the Atlantic and Arctic coasts, is twenty-six inches long, the wing about eleven and a quarter inches; prevailing color white; under parts, rump, tail, quills, and stripe above the eye, black. It is an expert diver, often going down in search of food eight or ten fathoms. The celebrated eider-down is obtained from the nest of this species, the birds having plucked it from their breasts to place around their eggs.

The King Eider, *S. spectabilis*, Leach, of the Arctic regions, is twenty-one and a half inches long, and the wing nearly ten and three quarters inches; colors black and white. The females of both these species are mainly brown.

The Pacific Eider, *S. V-nigra*, Gray, has a V-shaped mark on the chin.

The Genus *Erismatura* has the bill much depressed and bent upwards; tail of eighteen very stiff feathers.

The Ruddy Duck, *E. rubida*, Bonap., of North America, is sixteen inches long, and the wing about six inches; chestnut-red above, grayish white below.

The Black Masked Duck, *E. dominica*, Eyton, of Lake Champlain, is an additional species.

The Genus *Mergus* has the bill very narrow, slender, longer than the head, and conspicuously serrated.

The Sheldrake, Goosander, or Fish Duck, *M. americanus*, Cass., of North America, is twenty-six and a half inches long, and the wing eleven inches; the male is without a conspicuous crest, the head and neck green, fore part of the back black; lower parts salmon-color. The female has a depressed occipital crest; head and neck chestnut, upper parts ashy, lower like the male.

The Red-breasted Merganser, *M. serrator*, Linn., of North America and Europe, is twenty-three and a quarter inches long, the wing eight and three fifths inches; head and upper part of neck all round, dark green; under parts reddish white; head with a conspicuous, pointed occipital crest. The female has the head chestnut-brown, body ashy above, and reddish white beneath.

The Genus *Lophodytes* has the bill shorter than the head, the serrations short, and the head with a very conspicuous, compressed, circular, erect crest.

The Hooded Merganser, *L. cucullatus*, Reich., of all North America, is seventeen and a half inches long, and the wing nearly eight inches; the head, neck, and back, black; the under parts, and centre of the crest, white. The female has the crest shorter and more pointed, the head and neck reddish brown.

The Sub-Order of Gaviæ includes three great tribes;—Totipalmi, comprising Pelicanidæ or Pelican Family, Sulidæ or Gannet Family, Tachypetidæ, Phalacrocoracidæ or Cormorant Family, Plotidæ, and Phætonidæ;—Longipennes, comprising Procellaridæ or Petrel Family, and Laridæ or Gull Family;—and Brachypteri, comprising Colymbidæ or Diver Family, and Alcidæ or Auk Family.

PELICANIDÆ, OR PELICAN FAMILY.—This Family comprises swimming birds which have the bill long, hooked at the end, nostrils hardly perceptible, wings long, pointed, and tail rather short. They have a pouch under the lower mandible and opening into the throat, which is capable of great distention. Their flight is heavy.

The Genus *Pelicanus* is the only one found in the United States, where it is represented by two species.

The Rough-billed Pelican, *P. erythrorhynchus*, Gm., is seventy inches long, and the wing twenty-four and a half inches; the prevailing color white.

The Brown Pelican, *P. fuscus*, Linn., of California and Texas, is fifty-six inches long, and the wing twenty-two inches; bill thirteen and a half inches; color dark.

SULIDÆ OR GANNET FAMILY. — This Family comprises swimming birds which have the bill rather long, straight, strong, compressed, and tapering to the point, which is a little decurved, and the nostrils hardly perceptible. The wings are very long, tail long and cuneate.

The Genus *Sula* is the only one in North America.

The Common Gannet, *S. bassana*, Briss., of Labrador to the Gulf of Mexico, is thirty-eight inches long, and the wing nineteen and a half inches; general color white. It feeds upon fish, and obtains them by plunging from a height, often remaining under water for a minute or more. This species breeds in immense numbers on the rocky islands near the coast of Labrador.

The Booby Gannet, *S. fiber*, Linn., of the Southern coast of the United States, is thirty-one inches long, and the wing sixteen and a half inches; brown above; white below; throat yellow.

TACHYPETIDÆ, OR MAN-OF-WAR BIRD FAMILY. — This Family is characterized by a very long, strong, acute, hooked bill, and hardly perceptible nostrils. The wings are exceedingly long, and the tail long and much forked.

The Genus *Tachypetes* is represented by the Man-of-War Bird, or Frigate Pelican, *T. aquila*, Vieill., of Florida to California, which is forty-one inches in total length, the wing twenty-five, and the tail eighteen inches long; the prevailing color brownish black. In swiftness and power of flight this bird is not surpassed by any other.

PHALACROCORACIDÆ, OR CORMORANT FAMILY. — This Family comprises swimming birds which have the tip of the bill much hooked, acute, and the nostrils not perceptible. They are abundant on the coasts of all countries, and breed on rugged cliffs and on trees, and feed upon fish, which they obtain with great expertness.

The Genus *Graculus* comprises eight species.

The Common Cormorant, *G. carbo*, Gray, of Labrador, is thirty-seven inches long, and the wing fourteen inches; the color bluish black, gular sac yellow, with a broad white band at the base.

The Double-crested Cormorant, *G. dilophus*, Gray, of the Atlantic and Pacific coasts; the Florida Cormorant, *G. floridanus*, Bonap., of the Southern States and of the Mississippi; the Mexican Cormorant, *G. mexicanus*, Bonap.; Brandt's Cormorant, *G. penicillatus*, Bonap.; and the Violet-Green Cormorant, *G. violaceus*, Gray, of the Pacific coast, are additional species.

PLOTIDÆ, OR DARTER FAMILY. — This Family is characterized by a long, straight bill, long wings and tail, and short tarsi.

The Genus *Plotus* comprises four species, one in America and three in the Old World. They inhabit the warm regions, and are found in flocks.

The Snake Bird, Darter, or Water Turkey, *P. anhinga*, Linn., of the Southern coast of the United States, is thirty-five inches long, and the wing fourteen inches; the general color greenish black.

PHÆTONIDÆ, OR TROPIC-BIRD FAMILY. — This Family is characterized by a long bill, long wings, tail with central feathers extremely elongated, tarsi short.

The Genus *Phæton* is represented by the Yellow-billed

Tropic-Bird, *P. flavirostris*, Brandt, of Florida, which is thirty inches long, and the wing eleven inches; the general color white, the wings banded with black.

PROCELLARIDÆ, OR PETREL FAMILY.—This Family comprises swimming birds which have the bill more or less lengthened, compressed, deeply grooved, and appearing as if formed of several distinct parts; and nostrils opening from distinct tubes. They are all oceanic.

The Genus *Diomedia*—Albatrosses—has the bill powerful, much curved, acute, and the upper mandible grooved

Fig. 144.

Sooty Albatross, *D. fuliginosa*, Gm.

its whole length, and great extent of wing. The Wandering Albatross, *D. exulans*, Linn., of the Pacific, is forty-four inches long, and the wing twenty-four inches; color white, with narrow transverse lines of black above. The Short-tailed Albatross, *D. brachyura*, Temm., of the North Pacific, is about thirty-six inches long, and the wing twenty inches; general color white. The Yellow-nosed Albatross, *D. chlororhynchus*, Gm., of the Pacific, is thirty-six inches long, and the wing twenty-two inches.

The Sooty Albatross, *D. fuliginosa*, Gm., of the Pacific coast of Oregon and California, is thirty-four inches long, and the wing twenty-one inches; color sooty brown.

The Genus *Procellaria* — Fulmar Petrels — has the bill rather stout, and the lower mandible with a lateral groove.

The Gigantic Fulmar, *P. gigantea*, Gm., of the Pacific, is thirty-six inches long, and the wing twenty inches.

The Fulmar Petrel, *P. glacialis*, Linn., of the North Atlantic, is twenty inches long, and the wing thirteen inches; the back and wings bluish, under parts white.

The Pacific Fulmar, *P. pacifica*, Aud., of the Pacific, closely resembles the preceding one.

The Slender-billed Fulmar, *P. tenuirostris*, Aud., of the Pacific, is eighteen and a half inches long, and the wing thirteen inches.

The Tropical Fulmar, *P. meridionalis*, Lawr., of the Atlantic, is sixteen inches long, and the wing twelve inches.

The Genus *Daption* is represented by the Cape Pigeon, *D. capensis*, Steph., of the coast of California, which is fifteen inches long.

The Genus *Thalassidroma* — Stormy Petrels — has the bill short and slender. It comprises the smallest of web-footed birds; but they are able to contend with the most terrific storms. While flying close to the water, they project their feet, and thus give the appearance of walking upon its surface.

The Fork-tailed Petrel, *T. furcata*, Gould, of the Pacific, is eight inches long, and the wing six inches. Hornby's Petrel, *T. Hornbyi*, Gray, of the Pacific, is eight and a quarter inches long. Leach's Petrel, *T. Leachii*, Temm., of the North Atlantic, is eight inches long, and the wing six and a half inches; color sooty brown; rump white. The Black Stormy Petrel, *T. melania*, Bonap., of the coast of California, much resembles *T. Leachii*. Wilson's Stormy Petrel, *T. Wilsoni*, Bonap., of the Atlantic, is

Fig. 145.

Leach's Petrel, *T. Leachii*, Temm.

seven and a quarter inches long, the wing six inches; color dark sooty-brown, rump white; tail slightly emarginate. The Stormy Petrel, or Mother Carey's Chicken, *T. pelagica*, Bonaparte, of the Atlantic, is five and three quarters inches long, and the wing five inches; color grayish black above, sooty brown below, rump white.

The Genus *Fregetta* is represented by Lawrence's Black and White Stormy Petrel, *F. Lawrencii*, Bonap., of the Florida coast, which is eight inches long.

The Genus *Puffinus* has the bill compressed near the end, and a straight spur in place of the hind toe.

The Greater Shearwater, *P. major*, Faber, of the Atlantic, is twenty inches long, the wing thirteen and a quarter inches; brownish ash above, grayish white below.

The Sooty Shearwater, *P. fuliginosus*, Strick., of the Atlantic; Mank's Shearwater, *P. anglorum*, Temm., of the coast from New Jersey to Labrador; the Dusky Shearwater, *P. obscurus*, Lath., of the Southern coast of the United States; and the Cinereous Petrel, *P. cinereus*, Gm., of the Pacific, are additional species.

LARIDÆ, OR GULL FAMILY. — This Family comprises swimming birds which have the bill generally shorter than the head, straight at the base, more or less curved at the tip, nostrils linear, and wings long and pointed. They frequent the shores of all countries, and also wander far inland. They swim with facility, but do not dive. They feed upon fish, shell-fish, and other aquatic animals.

The Genus *Stercorarius* comprises the Skua Gulls, or Jagers, hardy birds, about twenty inches long, of the northern regions of both hemispheres. They are piratical, chasing other species and robbing them of their prey.

The Genus *Larus* — Gulls — has the bill strong, tail nearly even, colors light, and head white.

The Glaucous Gull, or Burgomaster, *L. glaucus*, Brünn, of the Arctic regions and southward, is thirty inches long, with an extent of wing of sixty to sixty-five inches. The Glaucous-winged Gull, *L. glaucescens*, Licht., of the northwest coast of North America, is twenty-seven and three quarters inches long, and the wing sixteen and a half inches. The White-winged Gull, *L. leucopterus*, Faber, of Labrador to the Arctic regions, is twenty-six inches long, and the wing seventeen and a half inches. The Great Black-backed Gull, *L. marinus*, Linn., of the Atlantic, is thirty inches long, and the wing nine inches. The Herring or Silvery Gull, *L. argentatus*, Brünn, of the Atlantic coast and the interior of the United States, is twenty-three inches long, the wing eighteen inches; the head, neck, under parts, rump, and tail, pure white; back and wings light pearl-blue. The Western Gull, *L. occidentalis*, Aud., of the Pacific, is twenty-five inches long, and the wing seventeen inches. The California Gull, *L. californicus*, Lawr., of the Pacific, is twenty-two inches long, and the wing sixteen and a half inches. The Ring-billed Gull, *L. delawarensis*, Ord, of North America, is twenty inches long, and the wing fifteen inches; bill yellow, with a dark brown band. Suckley's Gull, *L. Suckleyi*, Lawr., of the Pacific, is seventeen and a half inches long, and the wing thirteen and three quarters inches.

The Genus *Blasipus* has the bill long and rather slender. It is represented by the White-headed Gull, *B. Heermanni*, Bonap., of California, which is seventeen and a half inches long, and the wing thirteen and a half inches.

The Genus *Chroicocephalus* has the bill rather slender and much compressed, wings long, narrow, and tail usually even. The colors are mainly dark above and light below, and these are in beautiful contrast. In spring and summer the head is dark, but in winter white. Five species belong to North America. The Laughing Gull, *C. atricilla*, Linn., of New England to Texas, is seventeen inches long, the wing thirteen inches. Bonaparte's Gull, *C. philadelphia*, Lawr., of all North America, is fourteen and a half inches long, the wing ten and a half inches.

The Genus *Rissa* comprises the Kittewakes, of the northern regions, which are fourteen to seventeen inches long.

The Genus *Pagophila* — Ivory Gulls — has the webs of the feet indented. The gulls of this genus are Arctic, found far from land, are about eighteen or nineteen inches long, and mostly pure white.

The Genus *Rhodostethia* contains the Wedge-tailed Gull, *R. rosea*, Jard., of the Arctic regions, which is fourteen inches long, and the wing ten and a half inches.

The Genus *Creagrus* contains the Swallow-tailed Gull, *C. furcatus*, Bon., of California; and *Xema* the Fork-tailed Gull, *X. Sabinii*, Bon., of the North Atlantic, which is thirteen and a half inches long, and the wing eleven inches.

Fig. 146.

Roseate Tern,
S. paradisea, Brünn.

The Genus *Sterna* comprises the Terns, which have the bill slender, wings long and pointed, tail long and forked. They are generally white or light below; black, and bluish or bluish gray, above. They feed upon small marine animals, which they dart down upon. Fourteen or more species are found on the coasts of North America. The Marsh Tern, *S. aranea*, Wils., of the Atlantic coast as far north as Connecticut, is nearly fourteen inches long. The Caspian Tern, *S. caspia*, Pallas, of New Jersey and northward, is twenty-one and a half inches long. The Royal Tern, *S. regia*, Gamb., of the Middle and Southern States, is twenty-one inches long. Trudeau's Tern, *S. Trudeauii*, Aud., of New Jersey, is fifteen inches long. Wilson's Tern, *S. Wilsoni*, Bonap., from Texas to Labrador, is nearly fifteen inches long. The Arctic Tern, *S. macroura*, Naum., of the coast of New England and northward, is fourteen and a half inches long. The Roseate Tern, *S. paradisea*, Brünn, from New York to Florida, is sixteen inches long. The Least Tern, *S. frenata*, Gamb., of North America, is eight and three quarters inches long. The Short-tailed Tern, *S. plumbea*, Wils., of North America, is nine and a half inches long.

The Genus *Rhynchops* has the lower mandible longer than the upper. The Black Skimmer, *R. nigra*, Linn., of the Atlantic, is nineteen inches long, wing fourteen and a half inches. It skims its food from the surface of the water.

COLYMBIDÆ, OR DIVER FAMILY.—This Family comprises birds which are remarkable for their power of swimming and diving, and which move with difficulty upon the ground.

Fig. 147.

Great Northern Diver, or Loon, *C. torquatus*, Brünn.

The Genus *Colymbus*—Divers proper—has the bill

compressed and acute, tail short and rounded. Birds of this genus excel all others in diving and in making progress beneath the surface of the water. They are solitary, keen-sighted, and wary.

The Great Northern Diver, or Loon, *C. torquatus*, Brünn, of North America, is thirty-one inches long, and the wing fourteen inches. The Black-throated Diver, *C. arcticus*, Linn., of the northern regions, is about twenty-eight inches long, and the wing twelve and a half inches. The Red-throated Diver, *C. septentrionalis*, Linn., of the northern regions, is twenty-seven inches long, and the wing eleven and a half inches.

Fig. 148.

Crested Grebe, *P. cristatus*, Lath.

The Genus *Podiceps* — Grebes — has the bill long, slender, and pointed; the head in the spring ornamented with tufts. These birds frequent lakes, rivers, and the sea-coast. When alarmed, they remain beneath the surface of the water, exposing only the bill. Nine species belong to North America, varying from thirteen to twenty-nine inches in length.

The Crested Grebe, *P. cristatus*, Lath., of North America, is twenty-three and a half inches long, the wing seven and three quarters inches; umber brown above, silvery white below.

The Genus *Podylimbus* has the bill short, and head without ruffs. It contains the Pied-bill Grebe, *P. podiceps*, Lawr., of North America, which is fourteen inches long, and the wing over five inches.

Alcidæ, or Auk Family. — This Family comprises swimming birds which have the bill compressed and pointed, hind toe usually wanting, wings very short, and the legs placed far back. They are all marine, and con-

Fig. 149.

Penguin, *A. patagonica*, Linn.

fined to cold climates. The Genus *Alca*—Auks proper—has the bill rather long, flattened, slightly resembling a knife-blade in form, grooved upon the sides; the color dark above, white below.

The Great Auk,* *A. impennis*, Linn., of the Arctic regions, is about thirty inches long, and the wing five and a half inches; it is incapable of flight. The Razor-billed Auk, *A. torda*, Linn., of the Arctic regions, and southward to New Jersey in winter, is seventeen inches long, the wing eight and a half inches.

The Genus *Aptenodytes* comprises the Penguins of the cold regions of the Southern hemisphere. They have the wings very small, are incapable of flight, and go on shore only to lay their eggs. The Great Penguin, *A. patagonica*, Linn., is as large as a goose, slate-color above, white below, with a large black patch in front, surrounded by a yellow band.

* Professor Steenstrup reports that this bird has become extinct.

Fig. 150.

Puffin, *M. arctica*, Illig.

The Genus *Mormon* — Puffins — has the bill short, compressed, very high, and the sides obliquely grooved. Four species or more belong to the northern portions of America.

The Arctic Puffin, *M. arctica*, Illig., is eleven and a half inches long, and the wing six and a half inches. Puffins make their nests in burrows, which they dig to the depth of four to five feet in some cases. Each lays but a single egg in a season.

The Genus *Uria* — Guillemots — has the bill rather long, straight and pointed, wings short, and claws curved. Six species belong to North America.

The Black Guillemot, *U. grylle*, Lath., of the North Atlantic, is thirteen inches long, and the wing six and a half inches; color black; a white patch on the wing.

The Genus *Brachyrhamphus* comprises the Short-billed Guillemots. Six species inhabit the North Pacific.

The Genus *Mergullus* has the bill short, thick, and slightly lobed on its edges. It contains the Little Auk, Sea-Dove, or Dovekie, *M. alle*, Vieill., of the North Atlantic, which is seven and a half inches long, and the wing four and a half inches; breast and upper parts brownish black; under parts white.

Evidences of extinct species of birds exist in the rocks of both this and other countries, and some species have become extinct in comparatively recent times.* Tracks of birds which lived in the Triassic Period are common in the Sandstone of the Connecticut valley.

* The *Dodo* was a large bird, weighing about fifty pounds, and with rudimentary wings, which in the seventeenth century inhabited Mauritius and adjacent islands; but of which there is now not even one perfect specimen

SECTION III.

THE CLASS OF REPTILES.

THE Class of Reptiles comprises cool-blooded, oviparous vertebrates, which are covered with scales, and which lay their eggs upon the land, and whose young closely resemble the parents from the time they leave the shell. They breathe by lungs, have a heart with two auricles and one ventricle, and their digestion is slow. The reptilian heart, at each of its contractions, transmits to the lungs only a portion of the blood which comes to it from the various parts of the system, and the remainder goes into the circulation again without having been subjected to respiration. Thus the blood of these animals is acted upon by oxygen far less than in Mammals and Birds, and their temperature is correspondingly lower, and their habits more sluggish. The smallness of the pulmonary vessels enables Reptiles to suspend the process of respiration without arresting the progress of the blood; and hence they are able to remain much longer beneath the surface of the water than any of the animals described in the previous pages. The cells of the lungs are less numerous, and larger, than in the higher animals. The brain of Reptiles is comparatively small, and their sensations blunt. They continue to live and exhibit voluntary motions long after losing the brain, and even after the

in existence. It is allied to the Columbæ. The *Solitaire* is another large extinct bird of the same region.

The *Moa*, or *Dinornis*, a bird whose remains are found in New Zealand, was twelve or fourteen feet high, and the tibial bone thirty-two inches long; and its egg, which has been found fossil, fills a man's hat!

The *Æpyornis* of Madagascar was, as its bones show, twelve feet high, and its fossil egg is thirteen and a half inches long! This and the Moa, and some other extinct birds whose remains are found in the same regions, are allied to the Struthionidæ.

head is cut off. The muscles preserve their irritability for a considerable time after being severed from the body, and even the heart pulsates for hours after it is removed; nor does its loss prevent the animal from moving about. The vertebræ of Reptiles are convex at one end and concave at the other; and their teeth, when set in sockets, never have more than one prong.

This Class comprises three Orders,—Testudinata or Turtles, Sauria or Lizards, and Ophidia or Serpents. Some writers also include the Batrachians in this group.

SUB-SECTION I.

THE ORDER OF TESTUDINATA, OR TURTLES.

THE Order of Testudinata comprises scaly reptiles which have a continuous shield upon the back, called *carapace*, which is connected by bridges to another shield below, called *plastron*, the whole forming a hard covering for the soft organs of the body. This hard covering is formed of the greatly expanded ribs and sternum, together with ossified skin. The head, neck, and tail are the only movable parts of the spinal column. The jaws are covered with a horny substance, and destitute of true teeth; tongue short, thick, and covered with fleshy filaments; nostrils anterior, and near together; eyes with three lids; and lungs large, and

Fig. 151.

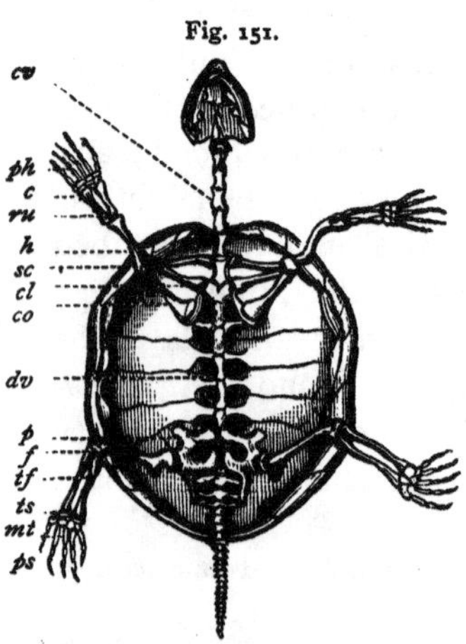

Skeleton of a Turtle, plastron removed.

cv, cervical vertebræ; *ph*, phalanges; *c*, carpus; *ru*, radius and ulna; *h*, humerus; *sc*, scapula; *cl*, clavicle; *co*, coracoid bone; *dv*, dorsal vertebræ; *p*, pelvis; *f*, femur; *tf*, tibia and fibula; *ts*, tarsus; *mt*, metatarsus; *ps*, phalanges.

placed in the same cavity with the other viscera. The Testudinata are divided into two Sub-Orders, — Chelonii or Sea Turtles, and Amydæ or Fresh-water and Land Turtles. All lay their eggs in holes which they dig in dry ground, and, covering them with earth, leave them to hatch.

The Sub-Order of Amydæ comprises Testudinina or Land-Tortoise Family, Emydoidæ or Terrapin Family, Cinosternoidæ or Mud-Turtle Family, Chelydroidæ or Snapping-Turtle Family, Hydraspidæ, Chelyoidæ, and Trionychidæ or Trionyx Family. The two families preceding the last belong to South America; and the last but one contains the curious turtle called Matamata.

TESTUDININA, OR LAND-TORTOISE FAMILY. — This Family comprises turtles which have the shell high and arched, sternum broad and flat, and the legs and feet so arranged that the body is raised free from the ground.

The Genus *Xerobates* contains the Gopher, *X. carolinus*, Ag., *Testudo polyphemus*, Daudin, of the Southern States, which has the shell fourteen to eighteen inches long, and which burrows in the ground, digging holes four or five feet deep. It will be observed that its popular name is the same as that given to certain members of the Sciuridæ.

The Genus *Testudo* contains the European Land-Tortoise, *T. græca*, Linn., six to ten inches long; and the Galapago or Indian Tortoise, *T. indica*, Linn., which is three feet long, and is the largest land tortoise known.

EMYDOIDÆ, OR TERRAPIN FAMILY. — This Family comprises turtles which have the shell highest in the middle, and the sternum flat, broad, and long. It is the largest of all the turtle families, and its representatives present so wide a

Fig. 152.

Wood Tortoise, *G. insculpta*, Ag.

range of differences, that Agassiz suggests its subdivision into about five sub-families. Most of its members inhabit bogs, marshes, still streams, and ponds. Some, however, live upon the land; nearly all are perfectly harmless. Their food is both vegetable and animal. Their eggs are more or less elongated, and covered with a shell which is in most cases flexible. About a dozen genera and about twenty species belong to North America.

The Genus *Pseudemys, Ptychemys*, Ag., contains the Red-bellied Terrapin, *Ps. serrata*, Gray, *Pt. rugosa*, Ag., *Emys rubriventris*, Lec., of the Middle States, which has the shell about eleven inches long; the Mobile Turtle, *Pt. mobilensis*, Ag., which has the shell fifteen inches long; *Ps. concinna*, Gr., *Pt. concinna*, Ag., of the Southern States, which has the shell about eight inches long, dusky, and marked with yellow lines; and *Ps. hieroglyphica*, Gr., *Pt. hieroglyphica*, Ag., of the Middle, Western, and Southern States, which has the shell over eight inches long, olive brown, reticulated with brownish orange.

The Genus *Trachemys* contains three species, common in the Southern and Western States, — *T. scabra*, Ag., *T. Troostii*, Ag., and *T. elegans*, Ag.

The Genus *Graptemys* contains the Geographical or Map Turtles, of the Middle and Western States.

The Genus *Malacoclemmys* contains the Salt-water Terrapin, *M. palustris*, Ag., *E. terrapin*, Holbr., of the salt-water marshes from New York to South America, which has the shell about seven and a half inches long.

The Genus *Chrysemys* contains the Painted Turtle, *C. picta*, Gray, from New Brunswick to Florida, and westward, which has the shell about six and a half inches long, nearly black, with yellow lines, and the marginal plates generally marked with bright red.

The Genus *Deirochelys* contains the Reticulated Turtle, *D. reticulata*, Ag., of the Southern States.

The Genus *Emys* contains Blanding's Tortoise, *E. meleagris*, Ag., *Cistuda Blandingii*, Holbr., from New England to Wisconsin, which has the shell about eight inches long, color black, with numerous yellow spots.

The Genus *Nanemys* contains the Speckled Tortoise, *N. guttata*, Ag., of North America east of the Rocky Mountains, which has the shell four and a half inches long, black, dotted with orange; and *Calemys*, Mühlenberg's Tortoise, *C. Mühlenbergii*, Ag., of the Middle States, which has the shell about four inches long, dark, and an orange patch on each side of the neck.

The Genus *Glyptemys* contains the Wood Tortoise (Fig. 152), *G. insculpta*, Ag., *E. insculpta*, Lec., of the Northern States, which has the shell about eight inches long, and presenting a beautifully carved appearance. It is found in woods and fields.

The Genus *Actinemys* contains *A. marmorata*, the only Emydoid known from the western slope of North America.

Fig. 153.

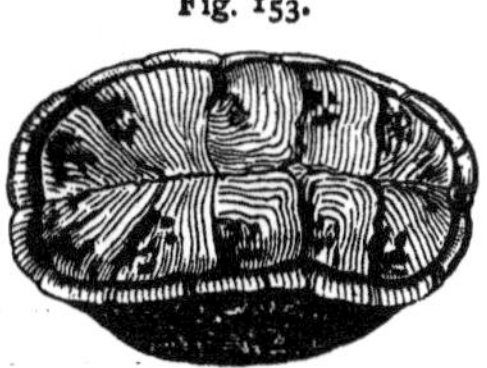

Box Turtle, *C. virginea*, Ag.

The Genus *Cistudo* contains the Box Turtles, which have the plastron composed of two parts that are movable upon one axis, and which can be brought into close contact with the carapace, and thus completely conceal all the extremities of the animal. They are found in dry woods.

The Common Box Turtle, of the United States, *C. virginea*, Ag., is about six and a half inches long.

Cinosternoidæ, or Mud-Turtle Family. — This Family comprises turtles which are long and narrow, and whose average size is less than that of any other family of Testudinata. They are aquatic, but come out of the water to bask in the sunshine, yet remain so near as to drop in on the slightest alarm. They lay three to five hard-shelled eggs, with a glazed surface. Three genera and about half a dozen species are found in North America.

The Genus *Aromochelys, Ozotheca*, Ag., contains the Musk Tortoise, *A. odoratum*, Gr., *O. odorata*, Ag., of the United States, which has the shell three and a half inches long, and emits a strong odor of musk.

The Genus *Thyrosternum* contains the Mud Tortoise, *T. pennsylvanicum*, Ag., *Cinosternum penn.* of authors, from Pennsylvania southward and westward, which is three and a half inches long, and emits a slight odor of musk.

CHELYDROIDÆ, OR SNAPPING-TURTLE FAMILY. — This Family comprises turtles which have the body high in front, low behind, head large, neck large and long, both jaws strongly hooked, the tail long and powerful, and the sternum small. They are aquatic, but are very frequently found upon the land near the water. They are exceedingly powerful and voracious, devouring smaller reptiles, fishes, young ducks, and other animals. When molested, they take the defensive, raise themselves upon their legs and tail, open wide the mouth, and, forcibly throwing the body forward, snap the jaws upon the assailant with fearful power. They are fully a match for anything which they are likely to meet with except man. The eggs are numerous and spherical. Three genera are known, each with a single species, and two of these are American.

Fig. 154.

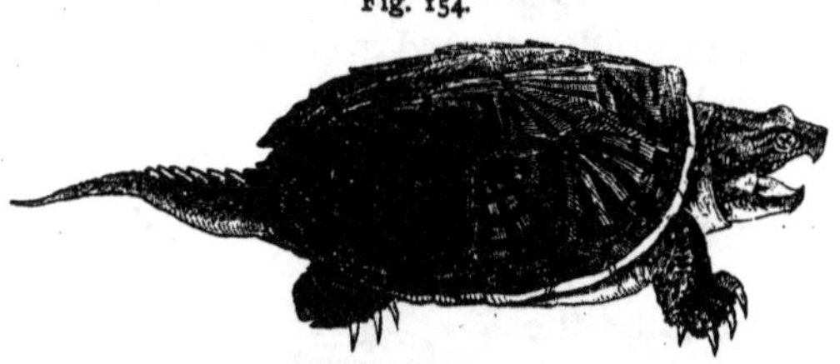

Snapping Turtle, *C. serpentina*, Schw.

The Genus *Macroclemys, Gypochelys*, Ag., contains the Alligator Turtle, *M. lacertina*, Ag., of the Southern States, which sometimes attains the weight of two hundred pounds. It seizes fish, and, holding them down with its feet, devours them somewhat as a hawk devours its prey.

The Genus *Chelydra* contains the Common Snapping-Turtle, *C. serpentina*, Schw., from Canada to Flórida and westward to the Missouri, which is found of all sizes, from a few inches to a total length of four feet. It has two long, flexible warts on the chin, and a crest of wedge-shaped tubercles on the tail. This turtle exhibits its ferocious disposition even before it is hatched.

TRIONYCHIDÆ, OR TRIONYX FAMILY.—This Family comprises turtles which have the body flat, circular, slightly elongated, the shield more or less flexible, or not completely ossified, neck and head long, the latter terminated by a long leathery snout, feet broad and webbed, five-toed, but with only three nails to each foot. They inhabit principally the muddy bottoms of shallow waters, often burying themselves in the soft mud, leaving only the head exposed. They take breath from time to time by carrying the snout above the water without even moving the body. They feed upon small animals; and lay from twelve to twenty or more spherical eggs, with a thick and very brittle shell. Three genera and about half a dozen species belong to North America.

The Genus *Platypeltis* contains the Soft-shelled Turtle, *P. ferox*, Fitz., *Trionyx ferox*, Sch., of the Southern States, which attains sixteen inches in length; color umber-brown above, with dusky blotches; below white, marked by the red bloodvessels.

The Genus *Amyda* is represented by *A. mutica*, Fitz., of New York and southward and westward; and *Aspidonectes*, by *A. spinifer*, Ag., from Lake Champlain to the Rocky Mountains.

The Sub-Order of Chelonii, or Sea-Turtles, comprises Testudinata which are perfectly adapted for swimming, or flying through the water, and which never leave the sea except to lay their eggs, which are placed near the shore in the sand. It contains Chelonioidæ or Loggerhead Family, and Sphargididæ, or Sphargis Family.

CHELONIOIDÆ, OR LOGGERHEAD FAMILY. — This Family comprises turtles whose general form is something like a heart flattened on one side, and furnished in front with a pair of large flat, wing-like, scaly flippers, and behind with a pair of scaly rudders.

The Genus *Chelonia* contains the Green Turtles. The Green Turtle, *C. mydas*, Schw., of the warm parts of the Atlantic coast of America, attains a weight of two hundred to three hundred pounds, or more. It is highly prized for food, and great numbers are caught on shore at night, as they come there to deposit their eggs.

The Genus *Eretmochelys* contains the Tortoise-shell Turtles. The Hawk-bill, or Tortoise-shell Turtle, *E. imbricata*, Fitz., of the warm parts of the Atlantic, approaches

Fig. 155.

Hawk-bill Turtle, or Tortoise-shell Turtle, *E. imbricata*, Fitz.

the Green Turtles in size, and the plates of its shell furnish the well-known and highly prized *tortoise-shell.*

The Genus *Thalassochelys* contains true Loggerheads.

The Loggerhead Turtle, *T. Caouana*, Fitz., of the Atlantic and Mediterranean, is the largest of all the turtles, except Sphargis, to be hereafter noticed. Holbrook gives the dimensions of one specimen as follows: head twelve and a half inches long and ten wide, shell forty-two inches long and thirty-four wide, and sternum twenty-eight inches long. It is frequently seen in mid-ocean floating upon the waves, apparently asleep.

SPHARGIDIDÆ, OR SPHARGIS FAMILY. — This Family comprises turtles whose general form is something like that of a flattened pyramid, and the body is covered with a thick coriaceous skin, instead of a hard shell. *Sphargis* is the only genus, and is represented by *S. coriacea*, Gray, of the Atlantic and Mediterranean, which is the largest of all turtles, attaining the weight of twelve hundred to two thousand pounds, in some cases. One caught in Chesapeake Bay had a total length of almost eight feet.

Fossil turtles are found in both continents.

SUB-SECTION II.

THE ORDER OF SAURIA, SAURIANS, OR LIZARDS.

THE Order of Sauria comprises scaly reptiles whose body is destitute of a shell, much elongated, and the tail generally long, and whose mouth is large, and armed with teeth. With few exceptions, they have four feet, which are generally furnished with nails. The ribs are movable, partially connected to the sternum, and are raised and depressed in respiration. The lungs generally extend far into the abdominal cavity. It comprises the Crocodilidæ or Crocodile Family,* Lacertinidæ or true Lizard Family, Iguanidæ or Iguana Family, Geckotidæ or Gecko Family, Chameleonidæ or Chameleon Family, Scincoidæ or Skink Family, Chalcidæ or Glass-snake Family, and many fossil reptiles.†

* Regarded as an Order by many authors.

† The Triassic, Jurassic, and Cretaceous rocks, in some localities, abound with the remains of extinct reptiles, and many of them, in the two latter groups especially, are those of Saurians of the most wonderful, and, in many cases, gigantic forms. The *Megalosaurus* was a terrestrial Saurian, thirty feet long. The *Iguanodon*, herbivorous in its habits, was thirty feet

CROCODILIDÆ, OR CROCODILE FAMILY. — This Family comprises the Crocodile of the Nile and the Gavial of the Ganges, — bulky reptiles, that attain the length of thirty feet, — and the Alligators of America.

The Genus *Alligator* — Alligators — has the muzzle broad, obtuse, and greatly resembling that of the pike. The teeth are unequal, the fourth lower ones largest, and entering into holes in the upper jaw, and the toes semi-palmate. Alligators are found from the Carolinas to Paraguay, and within these limits there are several species.

The Common Alligator, *A. mississippiensis*, Gray, of the Southern States, attains the length of fifteen feet or

Fig. 156.

Alligator, *A. mississippiensis*, Gray.

more. It is exceedingly voracious, devouring all kinds of animal substances, and is particularly attracted by fish, muskrats, dogs, ducks, or other animals in motion.

long, the femur thirty-three inches in length. The *Hylæosaurus* was twenty feet long; and the *Mosasaurus*, twenty-five feet. The *Ichthyosauri* were ten to forty feet long, with paddles adapting them for swimming; and the jaws were in some cases six feet long, and the cavity for the eye fourteen inches in diameter. The *Plesiosauri* were also swimmers, twenty-five to forty feet long, with the neck long, snake-like, and containing twenty to forty vertebræ. The *Pterodactyli* were flying Saurians, a foot long, with a spread of wing of about three feet. These strange forms, and others, represented by many species, are found in England and on the continent of Europe.

LACERTIDÆ, OR LIZARD FAMILY. — This Family comprises scaly reptiles which have the head in the form of a quadrangular pyramid with the apex in front; tongue thin, more or less extensible, and with its base in some cases lodged in a sheath, and its apex always bifid; body and tail much elongated, the latter in some cases several times the length of the former; feet four, five-toed, and armed with nails; the scales beneath the body and around the tail arranged in transverse parallel bands, and the abdominal scales always larger than those upon the back. They are the most agile, beautiful, and innocent of all the Sauria. The larger members live upon the ground, but many of the smaller ones upon trees. The species are numerous.

The Genus *Ameiva* has the tail round, and the scales of the belly broader than long. It contains the Six-lined or Striped Lizard, *A. sex-lineata*, Cuv., of the Southern States, which is nine to ten inches or more long, dark brown above, and with six yellow, longitudinal lines; abdomen bluish-silvery white. It is timid, harmless, quick in its movements, runs swiftly, and feeds upon insects, which it seeks at the close of the day.

Fig. 157.

Six-lined Lizard, *A. sex-lineata*, Cuv.

The Genus *Lacerta* contains the Monitor, *L. nilotica*, Linn., of the Nile, which is five or six feet long, and is said to destroy the eggs of the crocodile. It is sculptured on the ancient monuments.

IGUANIDÆ, OR IGUANA FAMILY. — This Family comprises scaly reptiles which are lizard-like in general appearance, but which have their tongue thick, fleshy, nonextensible, and only emarginated at the tip.

The Genus *Iguana* has the body and tail covered with small imbricated scales, the back furnished with a range of spines, and throat with a pendent, compressed dewlap.

The Common Iguana, *Ig. tuberculata*, of South America, is four to five feet long, greenish, the tail banded with brown. It lives upon trees. Its flesh is used for food.

The Genus *Draco* comprises the Dragons of the East Indies. They have their first six false ribs extending outwards, and supporting a fold of skin, and thus forming a sort of wing which acts like a parachute in sustaining them as they leap from one tree or branch to another; but which does not enable them to truly fly.

The Genus *Anolius* has the skin of the four external toes developed beneath to form an oval disk; the tail is cylindrical and very long.

The Green Lizard, *A. carolinensis*, Cuv., of the Southern States, is six and three quarters inches long, golden green above, and the abdomen greenish white. It is very active, running up trees and moving from branch to branch with swiftness. It is common about the garden and buildings, frequently entering the houses, and sometimes moves over the tables and other furniture in search of flies. It is able to walk upon the walls and ceilings, and even upon the window-panes.

The Genus *Tropidolepis* is represented by *T. undulatus*, of the United States south of the forty-third parallel. It is over seven inches long, grayish above with transverse, undulating black bands, which have their posterior margins marked with white. Below, on each side of the abdomen, is a band of green, surrounded with black; and a light central band forms a cross with a similarly colored transverse band between the anterior extremities. It inhabits pine-forests, runs to the tops of the tallest trees, and feeds upon insects.

The Genus *Phrynosoma* — Horned Toads — has the

head short, rounded in front, and bordered at the sides and behind with spines; the body short, oval, much depressed, with a dentated margin at the flank, and covered above with three-sided tubercles arising from small imbricated scales. Several species are found in the central and western parts of North America.

Fig. 158.

Horned Toad, *P. cornuta*, Gray.

The Horned Toad, *P. cornuta*, Gray, from Missouri to Texas, is over four and a half inches long, ash-color, marked with yellowish and dark; abdomen and thorax silvery white with dusky spots. It moves with rapidity upon the ground, but never climbs. It is sluggish in confinement.

Geckotidæ, or Gecko Family. — This Family comprises Saurians which have not the elongated, graceful form of the preceding lizards, but are more or less flattened, and their gait is a heavy kind of crawling. The tongue is fleshy, and not extensible; their jaws are furnished with a range of very small teeth; and their toes have a flattened disk, which enables them to move even on walls and ceilings. Many genera and species are known in the warmer parts of both continents.

Chameleonidæ, or Chameleon Family. — This Family comprises lizards which have the body compressed, skin roughened, tail round and prehensile, and feet five-toed. The tongue is cylindrical, fleshy, and extremely extensible; teeth trilobate, and eyes large, but covered with skin except a small hole opposite the pupil, and possessing the faculty of moving independently of each other. Their lungs are so enormous, that, when inflated, their body seems to be transparent, — a circumstance which led the ancients to believe that these animals fed on air. They live upon trees, are excessively slow in their movements, and often remain motionless upon a branch for hours. The great size of their lungs is probably the source of the power of the Chameleons to change their color, which takes place according to their feelings, and not in conformity with the hues of the bodies on which they rest, or near which they pass. The very extensible tongue has the extremity covered with a viscid secretion; and when the animal has marked an insect, it darts forth this organ, and quicker than a glance of the eye secures the prize for food. Eight or ten species or more inhabit the warmer parts of the Old World.

SCINCOIDÆ, OR SKINK FAMILY. — This Family comprises lizards which have the body cylindrical, and covered with smooth scales, variable in form and size, and disposed in the form of a quincunx. The head is covered with large, thin, angular plates; the neck is of the same size as the thorax; tongue free, flat, and notched, and not retractile into a sheath; and the jaws are furnished with closely set teeth. The body and tail seem to be one continued and uniform piece. The genera and species are quite numerous.

The Genus *Plestiodon* has the head very large and broad behind, but contracted in front of the eyes.

The Scorpion, or Red-headed Skink, *P. erythrocephalus*, Holbrook, of the Southern Atlantic and Gulf States, is twelve or thirteen inches long, olive brown above, the throat and abdomen yellowish white, and the head above bright red. It lives in hollow trees, and seldom comes to the ground except for food and water. It is generally timid, but when captured is very fierce, and bites severely; but its bite is not venomous, as is generally believed.

The Genus *Scincus* contains the Five-lined Skink, *S. quinquelineatus*, Daud., of the Southern States, which is about ten inches long, the head pale red, with six obscure white lines, and the body above olive brown, with five pale white, longitudinal lines, and a black lateral band; and the Blue-tailed Lizard, *S. fasciatus*, Holbr., of the United States, which is eight and a half inches long; the head bluish black, with six straw-colored lines; body bluish black, with five longitudinal, straw-colored lines; tail ultramarine blue, the throat and abdomen white.

The Genus *Lygosoma* contains the Ground Lizard, *L. lateralis*, D. & B., of the southern and western portions of the United States, which is four and three quarters inches long, with a short head and very long tail; the body, tail, and legs of a bronze or chestnut color, throat silver-white, abdomen yellow, and there is a broad, lateral black band from the head nearly to the extremity of the tail. It may be seen by thousands in the thick forests of oak and hickory in Carolina and Georgia, after sunset, when they emerge from their hiding-places to hunt for worms and insects.

The Genus *Seps* comprises reptiles which have a longer body and smaller feet than the Skinks; and *Bipes*, those which differ from Seps in the absence of forefeet. Found in South America and in the Eastern hemisphere.

CHALCIDÆ, OR GLASS-SNAKE FAMILY. — This Family comprises lizards whose body is elongated, serpent-like, without feet, or with those but slightly developed, and there is generally a deep groove along the flanks. The genera placed under this family really represent several families, and from their general resemblance to snakes have been regarded by some authors as belonging to the next order.

The Genus *Ophisaurus* is represented by the Glass-snake, *O. ventralis*, Daud., of the Southern and Western States, which is twenty to forty inches long, the body and tail above yellowish green spotted with black, and the under surface yellow. This snake-shaped lizard inhabits dry places, and spends much of the time in the ground. The vertebræ of the tail are so easily separated, that it is broken by a very slight blow; and this fragility of the tail has gained for this animal its popular name.

The Genus *Anguis* contains the Blind-Worms.

The Common Blind-Worm, *A. fragilis,* of Europe, is over a foot long, silvery yellow above, blackish beneath.

The Genus *Amphisbæna,* comprising the Double Walkers, has the head and body nearly of uniform size,—a form which enables them to move backwards or forwards with equal facility. They inhabit the hot regions of South America.

The Genus *Typhlops* comprises reptiles that externally resemble earth-worms. They are found in the hot regions of both continents. The eye is a mere point, and scarcely visible through the skin.

SUB-SECTION III.

THE ORDER OF OPHIDIA, OR SERPENTS.

The Order of Ophidia comprises scaly reptiles which are exceedingly long, and without feet, and which move by the alternate folds of their long and slender body. The ribs and vertebræ make up most of the skeleton, the sternum being wholly wanting. The vertebræ join each other as a ball and socket joint, thus giving great freedom of motion. The bones of the jaws and mouth, which in the higher animals are more or less firmly united, are connected in animals of this order only by extensible ligaments,—an arrangement by which the mouth may be distended so as to receive an object of much greater diameter than the serpent itself. Both the jaws and palatine arches are almost always armed with teeth. These are solid, and situated on the margins of the maxillary bones. As serpents do not masticate their food, their teeth are adapted simply to seizing, killing, and retaining prey, and are accordingly pointed, smooth, and arched towards the throat. The tongue is long, slender, bifid, extensible, and retractile within a sheath placed at the root; the eyes are without movable lids; trachea very long; lung single, and extending nearly the whole length of the body. Serpents cast their skins at least once a year. They lay eggs with a calcareous, flexible shell; some of them, however, are ovoviviparous; that is, the eggs are hatched while still in the body of the parent. This is especially true of venomous snakes. The Order comprises a thousand or more known species. Baird and Girard enumerate over thirty genera and a hundred and nineteen species belonging to North America.

BOIDÆ, OR BOA FAMILY.—This Family comprises serpents which have both jaws armed with teeth, and rudiments of hind limbs, or spur-like appendages. Some of them are the largest of all serpents, attaining the length of thirty to forty feet, and are able to swallow dogs, deer, and even oxen, after having crushed them in their powerful folds. Such are the Boas and Anacondas of South America, and the Pythons of Africa and the East Indies. *Wenona*, with one small species found in Oregon, is our only genus of this family.

COLUBRIDÆ, OR COLUBER FAMILY.—This Family comprises serpents which have both jaws fully provided with teeth, but have no rudiments of hind limbs. It includes a large proportion of all snakes, and is represented in all countries where snakes are found.

The Genus *Eutænia*—Striped Snakes—has the body moderately stout, or slender, scales carinated, the skin very extensible. The general color is three light stripes on a darker ground, the intervals with alternating spots. It is represented by several species in the United States. Length twelve to thirty-six inches.

The Genus *Nerodia* — Water Snakes — has the body generally stout, tail one fourth or fifth the total length, and the scales carinated. The general color is three series of dark blotches on a lighter ground; sometimes almost uniform brown or blackish. The abdomen is uniform or spotted. There are four or five species in the United States. Length twelve to fifty inches.

The Genus *Regina* comprises aquatic snakes with a slender body, tail very much tapering, and forming one third or one fourth the total length, and scales carinated. The general color is five or more longitudinal dark bands on a lighter ground, and the abdomen uniform, or also provided with similar bands. Four species are found in the Middle, Southern, and Western States. Length nine to twenty-four inches.

The Genus *Heterodon* — Hog-nose Snakes — has a short stout body, short tail, and the head, neck, and body capable of excessive dilatation. Two or three species are found in the Southern and Southwestern States, and one extends into Massachusetts. Length twelve to thirty-six inches.

The Genus *Pituophis* contains the Pine Snake, or Bull Snake, *P. melanoleucus*, Holbr., of the pine forests of New Jersey, and southward and westward, which attains the length of nine feet or more; whitish above, with a series of very large dark blotches. Other species are found from Texas to California.

The Genus *Scotophis* has the body cylindrical, very long, color brown or black, in quadrate blotches on the back and sides. Several species are found in North America, from two to five feet or more in length.

The Genus *Ophibolus* has the body thick, tail and head short, and eyes very small; color black, brown, or red, crossed by lighter. Baird and Girard mention nine species found in the United States. *O. Boylii*, B. & G., of California, is black, with more than thirty broad, ivory-white transverse bands; length about thirty inches. The King Snake, *O. Sayi*, B. & G., of the Gulf States, is black above, each scale with a large yellow spot in the centre; length about forty inches.

The Thunder Snake, or Chain Snake, *O. getulus*, B. & G., from New York to Mississippi, is black, crossed above by about thirty narrow lines bifurcating on the flanks; length thirty to forty inches or more.

The House, Milk, or Chicken Snake, *O. eximius*, B. & G., of the Northern and Middle States, is grayish ash, with a dorsal series of upwards of fifty transverse elliptical, chocolate blotches, and with two other alternating series on each side; length twenty-five to forty inches.

The Genus *Georgia* is represented by the Indigo or Gopher Snake, *G. Couperi*, B. & G., of Georgia, which is sixty inches long, black above, slate beneath.

The Genus *Bascanion* contains the Black Snakes, three species of which are found in the United States.

Fig. 159.

Black Snake, *B. constrictor*, B. & G.

The Common Black Snake, *B. constrictor*, B. & G., of North America, is three to five feet long or more, lustrous black above, greenish black beneath; chin and throat white. It climbs trees and bushes, and devours the young of birds, but is perfectly harmless to man.

The Genus *Masti-*

cophis — Coach-whip Snakes — has the general features of the preceding one, but is more slender; length four to five feet or more. Six species inhabit the Southern and Southwestern States.

The Genus *Leptophis* — comprising the Green Snakes of the Southern and Southwestern States — has a conical head, very small neck, and very long tail; length twenty to thirty inches.

The Genus *Chlorosoma* is represented by the Green Snake, *C. vernalis*, B. & G., of the Northern States and southward, which is uniform green, darker above and lighter beneath; length twelve to twenty inches.

The Genus *Contia* is represented by *C. mitis*, B. & G., of the Pacific Slope, which is brown, with two longitudinal light bands, below which is a series of black dots.

The Genus *Diadophis* — Ring-necked Snakes — has the head distinct, body slender, color uniform, with a light ring on the occipital region; length ten to twenty inches. Three or four species are found in the United States.

The Genus *Rhinostoma* is represented by the Scarlet Snake, *R. coccinea*, Holbr., of the Southern States, which is crimson, crossed by pairs of black rings enclosing a yellow one; length twelve to eighteen inches.

The Genus *Rhinocheilus* contains *R. Lecontii*, B. & G., of California, which is crossed by about thirty-three quadrate black blotches; length about twenty inches.

The Genus *Haldea* contains the Brown Snake, *H. striatula*, B. & G., of the Southern States, which is grayish above, salmon beneath; length eight to nine inches.

The Genus *Farancia* contains the Red-bellied, or Horn Snake, *F. abacurus*, B. & G., of the Southern States, which is bluish black, with subquadrate red spots on the flanks; under parts red, with bluish black irregular spots; length about thirty inches.

The Genus *Abastor* is represented by *A. erythrogram-*

mus, Gray, of the Southern States, which is bluish black above, with three longitudinal lines of red; length about fifteen inches.

The Genus *Virginia* is represented by *V. Valeriæ*, B. & G., of the Southern States, which is yellowish above, with minute black dots; length eight to ten inches.

The Genus *Celuta* is represented by the Ground or Worm Snake, *C. amœna*, B. & G., of the Connecticut valley and southwestward, which is uniform chestnut-brown, bright salmon beneath; length seven to twelve inches.

The Genus *Tantilla* is represented by *T. coronata*, B. & G., of Mississippi, which has the body reddish brown, with a black band across the neck above; and *T. gracilis*, B. & G., of Texas, which is greenish brown above, lighter beneath; length about eight inches.

The Genus *Osceola* is represented by *O. elapsoidea*, B. & G., of the Southern States, which is red, crossed by pairs of black rings, each pair enclosing a white one.

The Genus *Storeria* has the head subelliptical, and distinct from the body.

S. Dekayi, B. & G., of the United States, is gray above, with a clay-colored dorsal band, on each side of which is a double series of black dots. The lower parts are generally straw-color, each plate with four minute black dots; length nine to thirteen inches.

S. occipito-maculata, B. & G., of the United States, is gray salmon-color below, and there are three light-colored spots behind the head; length nine to eleven inches.

CROTALIDÆ, OR RATTLESNAKE FAMILY. — This Family comprises serpents whose upper jaw contains but few teeth, but is armed with sharp-pointed, perforated or grooved, movable poison-fangs. These fangs are concealed in a fold of the gum, or raised, at the will of the animal. They connect with a gland situated near the eye, which furnishes the fluid poison. When the snake bites, the fangs are raised, and the pressure of the temporal muscles upon the gland forces the poison along the fang into the wound. These animals have a deep pit between the eye and the nostril, and the rattlesnakes proper have the tail furnished with a rattle, with which they make a peculiar noise when they apprehend danger.

The Genus *Crotalus* comprises the Rattlesnakes proper. Several species are found in the United States.

The Common Rattlesnake, *C. durissus*, Linn., is three to four feet long, sulphur brown above, with two rows of confluent, lozenge-shaped brown spots; tail black. It is generally sluggish, and never attacks animals unless disturbed or hungry. But the slightest noise will arouse it, when it immediately coils, moves the rattles violently, and strikes at whatever comes within reach. It never pursues the object of its anger, but strikes on the spot, and, recoiling, repeats the blow as often as it can. This snake feeds upon young hares, squirrels, and birds, which it secures by lying in wait for them. Its reported charming power is probably a mere notion which has no foundation in fact. It was formerly supposed that the number of rattles indicates the age, — one rattle being added each year; but this is not so. In some cases there are more than one added in a year, and in others, one or more is lost.

The Diamond Rattlesnake, *C. adamanteus*, Beauv., of Carolina and southward, attains the length of eight feet.

The Genus *Crotalophorus* has the tail with a rattle, and top of the head covered with plates, as in Coluber. Five species are stated to be found in the United States.

The Prairie Rattlesnake, or Massasauga, *C. tergeminus*, Holbr., of the Prairies, is twenty to thirty inches long.

The Genus *Ancistrodon* has the tail without a rattle, and loral plates present.

The Copperhead, *A. contortrix*, B. & G., of the Southern

States and northward, is about twenty-four inches long, light chestnut, with darker transverse bars. It chooses dark, shady places, or meadows of high grass.

The Genus *Toxicophis* has the tail without a rattle, and loral plates absent. The species are aquatic.

The Water Moccasin, or Cotton Mouth, *T. piscivorus*, B. & G., of the Southern States, is about twenty-four inches long, dark chestnut-brown, with transverse bars of black. In summer, this snake is seen on the low branches which hang over the water, into which it falls on the slightest alarm. It is more to be dreaded than even the rattlesnake, as it attacks everything that comes in its way, and without warning.

VIPERIDÆ, OR VIPER FAMILY. — This Family comprises venomous snakes which are without pits behind the nostrils. They belong to the Old World.

ELAPIDÆ. — This Family comprises venomous snakes which have fixed and permanently erect fangs.

The Genus *Elaps* — Harlequin Snakes — has a slender body, never exceeding three or four feet long, and the upper jaw with an erect fang on each side.

The Harlequin Snake, *E. fulvius*, Cuv., of the southern and western portions of the United States, is red, annulated with black, margined with yellow; length eighteen to thirty inches.

The Genus *Naia* comprises vipers which can raise up and draw forward the anterior ribs, so as to dilate the forward part of the body into a more or less broad disk.

The Cobra, or Spectacled Viper, *N. tripudians*, Schleg., of India, is distinguished by a black line, resembling the figure of a pair of spectacles, traced on the widened portion of its disk. Its bite is deadly. The jugglers of India extract its fangs, and then teach it to dance.

The Haje, *N. haje*, Geoff., of Africa, has the neck less wide than that of the preceding. The ancient Egyptians made it the emblem of the protecting divinity of the world, and sculptured it on the sides of a globe upon the gates of their temples. By pressing this snake on the nape, the jugglers of Egypt throw it into a stiff and immovable condition, which they call turning it into a rod. It is probably the Asp of Egypt, and Asp of Cleopatra.

HYDROPHIDÆ, OR SEA-SNAKE FAMILY. — This Family comprises serpents which have the posterior parts of the body and tail much compressed, and raised vertically. They inhabit the warm parts of the Indian and Pacific Oceans, and the streams of the East Indies, and are very venomous. They are mainly of small size.

SECTION IV.

THE CLASS OF BATRACHIA, BATRACHIANS, OR AMPHIBIANS.

THE Class of Batrachians comprises cool-blooded, oviparous vertebrates which are destitute of scales, and which in most cases lay their numerous eggs in the water or in damp places,* and whose young hatch in an immature condition, and afterwards undergo various changes before they acquire the form of the parents. The young breathe by gills similar to those of fishes, and in most cases live in the water; but in the adult state, these animals, with few exceptions, breathe by lungs, like scaly reptiles. The

* Some Tree-toads lay their eggs on trees and places overhanging water, and the young, as soon as hatched, drop into the water.

Pipa, or Surinam Toad, of South America, lays its eggs in the water, after which they are collected by the male, and placed on the back of the female, the skin enlarging in such a manner as to enclose the eggs in cells; here the development goes on till the young come forth as perfectly formed toads.

A small frog of Venezuela has a pouch upon the back in which the eggs are carried and hatched.

lungs are two, equal, and the heart is composed of only one auricle and one ventricle. In the higher forms the vertebræ are convex at one end and concave at the other; but in some cases the vertebræ are concave at both ends.

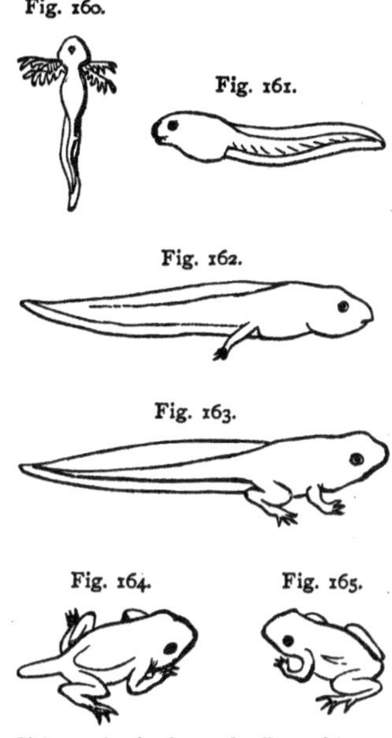

Changes in the form of a Batrachian — the Frog — from the time of hatching.

Batrachians comprise three orders, — Anoura or Tailless Batrachians, Urodela or Tailed Batrachians, and Apoda or Cæcilians. Besides these, there is a group of extinct batrachians called Labyrinthodonts, from the peculiar internal structure of their teeth. They have scales, and are first found in the Carboniferous, and culminate and end in the Triassic rocks. Some of the species were very large.

SUB-SECTION I.

THE ORDER OF ANOURA, OR TAILLESS BATRACHIANS.

THE Order of Anoura comprises batrachians which have the body short, thick, and covered with a skin which does not adhere to the muscles, but covers them loosely like a sack. The tongue is long, and fixed to the front of the jaw, and its tip turned backwards in the mouth, whence it can be darted forth with almost lightning swiftness; and it is in this way that these animals secure the living insects which constitute their food. The young are tadpoles (Figs. 160–163), which have a large head, short, thick body, and a long compressed tail, and feed upon vegetable food. As they grow older, the extremities appear, the tail is gradually absorbed, the gills are superseded by lungs, and the animal becomes air-breathing and carnivorous. This Order comprises three families, — Ranidæ or Frog Family, Hyloidæ or Tree-Toad Family, and Bufonidæ or Toad Family.

RANIDÆ, OR FROG FAMILY. — This Family comprises tailless batrachians which have the fingers and toes free, and never dilated into a disk, tympanum visible, upper jaw and palate provided with teeth, and the throat of the males with vocal vesicles, which communicate internally with the mouth. The genera and species are numerous.

The Genus *Rana* comprises the Frogs proper.

The Bull-frog, *R. Catesbiana*, Shaw, *R. pipiens*, Latr., of the United States, attains a total length of twenty-one inches, in some instances. It is mainly solitary, and is the most aquatic of all the frogs. The deep croakings of the male may be heard a mile. *R. horiconensis*, Holbr., of Lake George, is a closely related species.

The Green Frog, *R. clamitans*, Bosc., *R. fontanalis*, LeC., of the United States, is three and a half inches long, green above, with dusky spots behind, tinged with yellow below. There is a ridge from the orbit to the posterior extremities.

Fig. 166.

Leopard Frog, *R. halecina*, Kalm.

The Leopard Frog, *R. halecina*, Kalm, of North America, is over three inches long, green above, with spots of dark brown margined with yellow; beneath yellowish white. This beautiful frog is very active, leaping sometimes eight or ten feet at a single bound.

The Pickerel-frog, *R. palustris*, LeC., of the Northern States and southward, is two and three quarters inches long, slender, pale brown above, marked with dark brown square spots; under parts yellowish white; posterior half of the thighs bright yellow mottled with black. Its flesh is delicate, and often used as pickerel bait. Its call is a singular prolonged grating sound, which it utters while floating on the surface of the water.

The Wood-frog, *R. sylvatica*, LeC., of the Northern States and southward, is about two inches long, pale reddish brown above, yellowish white beneath, head with a dark brown stripe on the side. It is found in woods, and frequents water only early in spring, when it lays its eggs.

The Genus *Cystignathus* is represented by *C. ornatus*, Holbr., of South Carolina, one and a quarter inches long; and *C. nigritus*, Holbr., of Georgia and Carolina, one and a half inches long.

HYLOIDÆ, OR TREE-TOAD FAMILY. — This Family comprises frogs which have the extremities of the toes and fingers enlarged into a disk or viscous pellet, by means of which they sustain themselves on the sides of trees, branches, leaves, and all kinds of smooth surfaces. They inhabit trees, shrubs, or plants, except in the breeding season, when they resort to the water.

The Genus *Hyla* comprises Tree-frogs or Tree-toads.

The Tree-toad, *H. versicolor*, LeC., of the Northern and Middle States, is two inches long, flattened, warty above, color varying from palest ash to dark brown, with several large irregular blotches of brown. The under surface is mainly white, granulated. It is very noisy towards evening and in cloudy weather, or before a rain. In the latter part of spring or early summer it resorts to the pools to lay its eggs.

The Green Tree-frog, *H. viridis*, Laurenti, of the Southern States, is one and three quarters inches long. Several additional species, from an inch and a quarter to an inch and a half in length, are found in the Southern States.

The Genus *Hylodes* comprises the Cricket Frogs.

The Savannah Cricket, *H. gryllus*, Holbr., of the Atlantic and Gulf coast, is one and a half inches long, cinereous above, vertebral line green or red, and the sides with three oblong black spots, edged with white; under parts silver white. It is found on the leaves of aquatic plants, is very agile, and makes long leaps to secure insects, which constitute its food. It is constantly chirping like a cricket, is easily domesticated, and sings merrily even in confinement.

Fig. 167.

Pickering's Hylodes, *H. Pickeringii*, Holbr.

Pickering's Hylodes, *H. Pickeringii*, Holbr., of New England and the Middle States, is less than one inch long, body yellowish brown, with small, dusky rhomboidal spots, and lines of the same color, sometimes arranged in the form of a cross; abdomen pale flesh-color, throat tinged with yellow. It is found in woods upon the ground, or upon plants growing near water. In the spring-time, its shrill, piping note is heard throughout the night.

Bufonidæ, or Toad Family.—This Family comprises batrachians which have a short, thick body, warty above and granulated beneath, the upper jaw and palate in most cases destitute of teeth. They are mainly nocturnal. They confer a great benefit upon the farmer and gardener by destroying insects.

The Genus *Bufo* contains the Toads proper, about half a dozen species of which are found in the United States.

The Carolina Toad, *B. lentiginosus*, Shaw, of the Southern States, is two and a half inches long, with a large head.

The Common American Toad, *B. americanus*, LeC., of the United States, is two and a half inches long, with large, brilliant eyes. In spring, when it resorts to the pools to lay its eggs, its prolonged trill is heard day and night.

B. obstetricans, Laur., of Europe, is remarkable for the habit the male has of fastening the eggs, when produced by the female, to his thighs, by means of some glutinous threads, and carrying them about until near the time of hatching, when he seeks a pool and the young escape from the eggs.

The Genus *Scaphiophus* contains the Toad-frog, *S. solitarius*, Holbr., from New England to Georgia, which is two and a quarter inches long, olive above, with two yellow lines extending from the eyes to the hind extremity; yellowish white below. The eyes are very large and prominent, pupil black, iris golden, and subdivided into four parts by two black lines. This singular animal combines to some extent the characteristics of both frog and toad.

SUB-SECTION II.

THE ORDER OF URODELA, OR TAILED BATRACHIANS.

THE Order of Urodela comprises batrachians which have a tail at all periods of life, and generally four feet. The body is long, round, and covered with skin adherent to the muscles. It comprises Salamandridæ or Salamander Family, Amphiumidæ or Amphiuma Family, and Sirenidæ or Siren Family.

SALAMANDRIDÆ, OR SALAMANDER FAMILY.—This Family comprises Salamanders proper and the Tritons. Its representatives are numerous. More than twenty species are found in the United States, from two and a half to twelve inches long. They have no sternum, ribs rudimentary, legs four, fingers four, and toes five in most genera; and, contrary to what is seen in frogs and toads, the fore feet are developed before the hind ones. In their adult state, most Salamanders proper live upon the land, approaching the water only at the season in which they lay their eggs. Some are terrestrial throughout life, laying their eggs under stones and old logs in damp places. The Tritons have the tail compressed, and are entirely aquatic; yet, as they respire by means of lungs, they come to the surface of the water from time to time for atmospheric air. They have the most wonderful power to reproduce mutilated or lost parts. The limbs may be removed, and in less than a year they will grow again; and the new-formed limbs may in turn be amputated, and will in turn be replaced by others. Even the eye, when destroyed, is said in time to be reproduced.

The Genus *Pseudotriton* contains the Red Salamander, *P. ruber*, Tsch., of the Atlantic States, which is less than five inches long, red, with numerous small, black points above; and *Spelerpes*, the Long-tailed Salamander, *S. longicauda*, Baird, of the Northern United States, which is less than six inches long, lemon-colored, marked with small spots of black. The tail is more than twice the length of the body, and marked with transverse black bands.

The Genus *Amblystoma* contains *A. punctatum*, Baird, of the Northern States and southward, which is five and three quarters inches long, bluish black above, with a row of yellow spots on each side; below bluish black.

Fig. 168.

Salamander, *A. punctatum*, Baird.

The Genus *Plethodon* contains the Red-backed Salamander, *P. erythronotus*, Baird, of the Northeastern and Middle States, which is less than three inches long, with a broad vertebral band of reddish-brown. It inhabits decayed trees and other damp places, and lays its eggs, in bunches of about a dozen, in moist, decaying wood; and though the young never go into the water, they are still true to the character of their family, and have gills when first hatched, although they soon disappear.

The Genus *Diemictylus* contains the American Water-Newt, *D. viridescens*, Rafin., of the Atlantic States, which is about three and three quarters inches long; color above olive brown tinged with green, and on each side of the vertebral line there is a row of bright vermilion circular spots; below, orange, with numerous small black spots. This species is almost entirely aquatic, and is thus a Triton. Several other species of Triton are found in the United States, from four and a half to twelve inches long.

Fig. 169.

Triton, or Water-Newt, *D. viridescens*, Rafin.

AMPHIUMIDÆ, OR AMPHIUMA FAMILY. — This Family comprises batrachians which are said to be destitute of gills at all periods of their existence, and which breathe by means of exposed spiracles or branchial orifices at the sides of the neck. According to Holbrook, they undergo no metamorphosis after they are hatched, but at once appear in the forms which they are permanently to retain.

The Genus *Amphiuma* has an eel-shaped body, four imperfectly developed legs, and a single spiracle on each side of the neck.

The Congo Snake, *A. means*, Linnæus, of the Southern States, is about twenty-eight inches long, bluish black. It lives in muddy waters, or in mud, sometimes penetrating the latter to the depth of three feet.

The Genus *Menopoma* has a large, flat head, short, stout body, large tail, a single spiracle on each side; legs short and thick; fingers four, toes five, and all palmated.

Fig. 170.

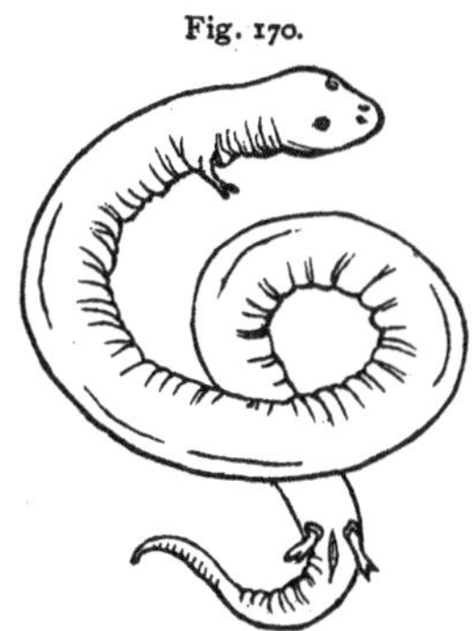

Congo Snake, *A. means*, Linn.

The Hellbender, *M. alleghaniensis*, Latr., of the Alleghany River and westward, is fifteen inches long, pale cinereous, with dusky blotches. It lives entirely in water, and is very voracious, feeding on fish, shell-fish, and other small animals.

SIRENIDÆ, OR SIREN FAMILY.—This Family comprises batrachians which have permanent external branchiæ that occur in tufts, covering the branchial orifices. They also have lungs like others of their class, and are thus true amphibians. It will be observed that, even in their adult state, these animals represent the embryonic forms of the higher batrachians.

The Genus *Siren* has an eel-shaped body, with anterior legs, three spiracles, and three tufts; no posterior legs.

The Siren, or Mud-Eel, *S. lacertina*, Linn., of South Carolina, attains the length of twenty-four inches; color above nearly black, with numerous light spots; abdomen purplish. It lives in mud, and is common in the ditches of rice-fields. Two more species are found in the South.

The Genus *Menobranchus* has the head and mouth large, two spiracles on each side of the neck, and these covered with three branchial tufts; tail compressed; feet four, and four-toed.

The Menobranchus, or Mud-Puppy, *M. maculatus*, Barnes, of Lakes Erie and Champlain, is twelve inches long, dusky cinereous-gray, with darker spots; under parts nearly white. It generally remains at the bottom of the waters it inhabits, where it swims or creeps slowly, with a serpentine motion. It is often taken with the hook, and fishermen regard it as poisonous. *M. late-*

Fig. 171.

Menobranchus, *M. lateralis*, Say.

ralis, Say, of the tributaries of the Mississippi, on the eastern side, much resembles the preceding.

The Genus *Siredon* contains the celebrated Axolotl of Mexico, and species very closely related from other parts of Western North America. Siredons are from six to ten inches long, and every way similar in form to young aquatic salamanders. They live mainly in the water. The Axolotl is eaten by the Mexicans. *Proteus*, represented by a species a foot long in the waters of Adelsberg Cave, Carniola, is a related genus.

Fig. 172.

Siredon, Western North America.

SUB-SECTION III.

THE ORDER OF APODA, OR CÆCILIANS.

THE Order of Apoda comprises snake-shaped batrachians. They are destitute of limbs, and move like serpents. They are found in the marshes of tropical regions, and are named Cæcilians on account of their exceedingly minute eyes; and these members are also apparently wanting in some cases. Length, one to three feet.

Some authors recognize and place here a group called Lepidota. Its representatives inhabit the fresh waters of the hot parts of South America and Africa, and are known under the name of *Lepidosiren*. They are one to three feet long, fish-like in form, scaly, and with simple styliform legs. During the dry season they are said to bury themselves in the mud.

SECTION V.

THE CLASS OF FISHES.

THE Class of Fishes comprises cool-blooded vertebrates which live exclusively in water, and whose respiration is effected in that medium by means of gills. Though living in a liquid of nearly the same specific gravity as themselves, they are adapted, by their whole structure, for easy motion, and for the most rapid progression. The limbs corresponding to the locomotive members of higher vertebrates are comparatively little developed, and are called fins. The parts corresponding to arms and legs are extremely short, or entirely concealed. The parts corresponding to fingers and toes are called rays, and serve as supports to the membranous parts of the fins. Those fins which correspond to the anterior locomotive members of higher vertebrates are called *pectorals*, and those which correspond to the posterior, *ventrals*. Besides the pectorals and ventrals, there are other fins which are vertical. Those upon the back are called *dorsal*,

Fig. 173.

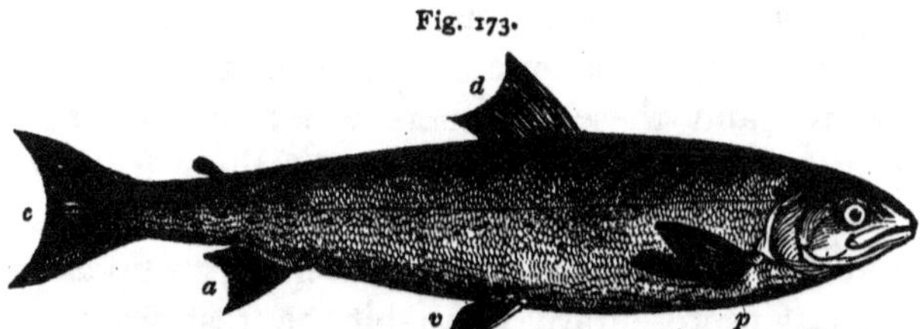

p, pectorals; *v*, ventrals; *d*, dorsal; *a*, anal; *c*, caudal.

those beneath the tail *anal*, and the fin at the end of the tail *caudal*. Most of this class are scaly, but some are naked, others spinous, and others still are mailed. The head of Fishes varies much in form, and consists of a greater number of bones than are found in other oviparous vertebrates, if we consider every piece a distinct bone. The jaws are armed with teeth, and in many cases these are placed in all parts of the mouth, and even in the gullet. The spinal column is made up of vertebræ which are concave at each end, and the cavities which thus occur between the vertebræ are filled by a soft membranous and gelatinous substance, which extends from one cavity to another, through a hole with which each vertebra is pierced. The spinal column bends with perfect freedom laterally, but not in a vertical direction; and it is chiefly by the lateral motions of the tail and body that Fishes are propelled; although some swim principally by the undulation of the dorsal fin. The fins are employed mainly in balancing and directing. The muscular system is highly developed. Their flesh, except certain muscles which are deep red, is paler than that of Birds or Mammals, and in some cases is pure white. One large and complicated muscle on each side, and filling up the space from the head to the tail, furnishes the principal motive power. The brain is exceedingly small, and seldom fills the cavity in which it is situated. The senses of smell, sight, and hearing are conferred on Fishes by organs analogous to those of other vertebrates, and are arranged in nearly a similar manner. The nostrils are simple cavities at the end of the muzzle, almost always perforated by two holes. The position, direction, and size of the eye in Fishes are almost endlessly varied. With few exceptions, the eye has no motion; the iris neither contracts nor dilates, and the pupil is never altered, whatever be the quantity of light. The ear of Fishes is enclosed on every side in the bones of the head, and consists merely of a sac, representing the vestibule, and of three membranous semicircular canals. In the former are suspended small bodies, generally of stony hardness. All the vibrations which reach the ear of

Fishes come through the hard covering of the head; hence they hear scarcely more than the loudest sounds.* The taste is generally regarded as feeble; † nor are they highly endowed with the sense of touch. The vegetative functions of Fishes follow the same order as those of higher vertebrates. Respiration, as stated above, is performed by means of branchiæ or gills, an apparatus on each side of the neck consisting of fringes suspended on arches attached to the hyoid bone, and traversed by innumerable bloodvessels. In most species, the great opening of the gills is closed by means of the branchial membrane, which is supported by rays attached to the hyoid bone, together with a sort of lid composed of three pieces, plainly or obscurely indicated, and called the *operculum*, *suboperculum*, and *interoperculum*. This lid or gill-cover is articulated with the tympanal bone, and plays on one called the *preoperculum*. The general relative positions of these parts is shown in Fig. 174. In some groups, however, the gill-covers are wanting. By motion of their jaws and the opercular and hyoidean apparatus, Fishes keep currents of water flowing through their gills, where the blood which is continually sent from the heart is purified. The blood, having undergone respiration, is poured into an arterial trunk situated under the spine, whence it is distributed to every part of the body, and in due time returns by the veins to the heart. The heart of Fishes contains but two cavities,

Fig. 174.

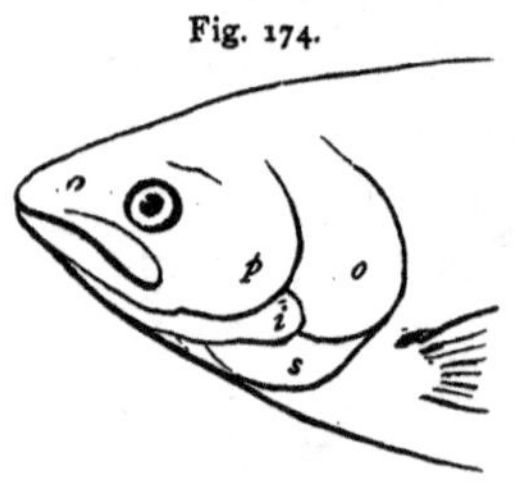

p, preoperculum; *o*, operculum; *s*, suboperculum; *i*, interoperculum.

* Some naturalists believe that the sense of hearing is much less obtuse than it is here represented to be.

† It is well known that some fishes show a decided preference for one kind of bait over another, not only in taking it, but in clinging to it.

corresponding to the right auricle and ventricle of the warm-blooded vertebrates. The process of respiration now described is as indispensable to Fishes as the respiration of air to other animals; and they exhibit the same symptoms of distress when it is stopped, and rapidly perish. Still, it is not by itself, nor by the oxygen which enters into its chemical composition, that the water acts upon the blood. It is only the small quantity of air that is mixed with the water which serves for the respiration of these animals, and if they be put in water which has been deprived of air, they immediately die. It is sometimes necessary for them to come to the surface for atmospheric air, when the supply of that substance has become deficient in the water in which they live.

Most fishes are furnished with a membranous bag filled with air, and called the swimming-bladder. This organ is probably a rudimentary lung, but its true function is not known; although some believe that by it Fishes have the power of varying their specific gravity, and thus more easily rising and descending. Others believe it aids in hearing, as there is a connection between it and the chamber of the ear.

Fishes in general are characterized by great voracity. They feed mainly upon smaller members of their class, and other small animals; although some are vegetable feeders. Most of them swallow their prey whole. Some, which feed on shell-fish, crush their food by means of the powerful crushing and grinding teeth in the gullet.

Most fishes are oviparous in their manner of reproduction; but some species bring forth living young. They produce a far greater number of eggs than any other vertebrates.* Some species prepare a place for their

* A Salmon sometimes contains as many as 20,000 eggs; a Perch, 28,320; a Herring, 36,960; a Mackerel, 546,000; a Flounder, 1,357,400; a Sturgeon, 7,635,200; a Cod, 9,344,000; and a species of Upeneus, 13,000,000!

eggs, and defend them with great spirit; but most abandon them as soon as laid. With few exceptions, Fishes have no care of their young, but devour them as readily as they do any other food. In the Sygnathi, or Pipe-Fishes, the eggs are conveyed into a pouch under the abdomen or at the base of the tail of the *male*, where they are hatched.

Although the lowest class of the Vertebrates, their varied forms, and colors which often rival those of precious stones and burnished gold, the wonderful power and velocity of some, the wholesome food furnished by many, and the exciting sport of their capture, combine to render Fishes subjects of great interest to the casual observer, as well as to the amateur and the professional naturalist.

The number of known species of Fishes is about ten thousand. According to the earlier writings of Agassiz, they are divided into four orders, the scales being taken as the basis of classification, as follows: — Ctenoids, embracing fishes which have the scales toothed on the edge, as Perch, Breams, Bass, and the like; Cycloids, comprising fishes whose scales are rounded and smooth on the edge, as Salmon and Cod; Ganoids, embracing fishes with enamelled scales, as the Gar-pike; and Placoids, fishes with fine point-like or stellate scales, as Sharks and Skates.

Cuvier divides fishes into two great groups, — Bony and Cartilaginous Fishes, — which together include nine orders; and since so many of the books upon natural history to which the student may have occasion to refer follow him more or less closely, and as there is no generally accepted classification of this important group of animals, I have thought it best to adhere mainly to his arrangement.

Bony Fishes, or Fishes proper, are those with a true bony skeleton, and include six orders, as follows: —

Acanthopterygians, or Spine-finned Fishes, as Perch, Breams, Bass, Mackerel, and the like ;

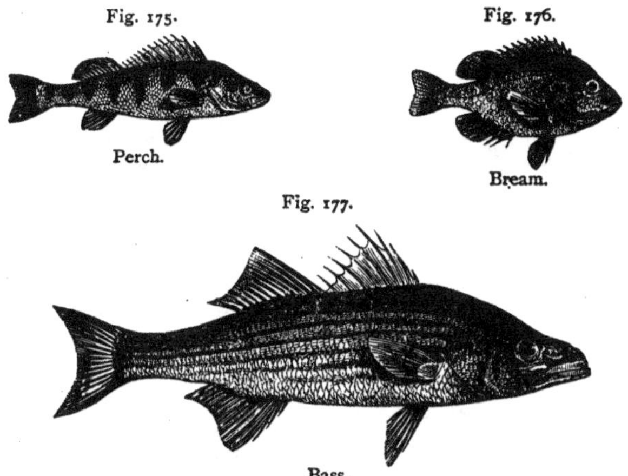

Fig. 175. Perch.

Fig. 176. Bream.

Fig. 177. Bass.

Abdominal Malacopterygians, or Soft-finned Fishes with the ventrals behind the pectorals, as Salmon, Pike and Pickerel, Trout, Carp, Herring, and their allies ;

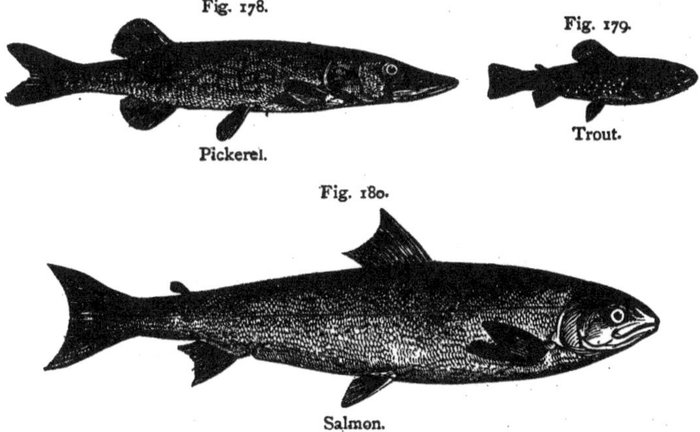

Fig. 178. Pickerel.

Fig. 179. Trout.

Fig. 180. Salmon.

Sub-brachian Malacopterygians, or Soft-finned Fishes

with the ventrals under or forward of the pectorals, as Cod, Haddock, and their allies ;

Fig. 181.

Cod.

Apodal Malacopterygians, or Soft-finned Fishes without ventrals, as Eels ;

Fig. 182. Fig. 183.

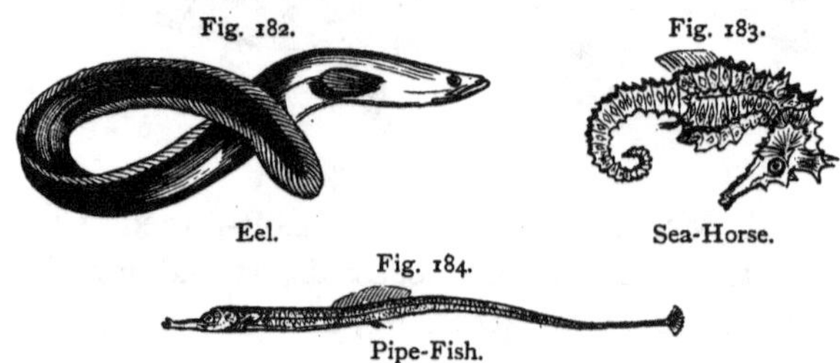

Eel. Sea-Horse.

Fig. 184.

Pipe-Fish.

Lophobranchiates, or Fishes with the gills in tufts, as Pipe-Fishes and Sea-Horses ; and

Plectognathes, or Fishes which have the maxillary bones united to the intermaxillary, as the Puffers, Sun-Fishes, File-Fishes, and Trunk-Fishes.

Fig. 185. Fig. 186.

Trunk-Fish. Puffer.

Cartilaginous Fishes have their skeleton essentially cartilaginous, calcareous matter being present only in small portions. They have no sutures in the cranium, and the gelatinous substance which in other Fishes fills the spaces between the vertebræ, and only extends from one space to another by means of a small aperture, forms in part of this group a long cord which traverses nearly all the vertebræ without materially varying in its diameter. This group includes three orders: —

Sturiones, or chondropterygians with free gills, as Sturgeons;

Fig. 187.

Sturgeon.

Selachians, or chondropterygians with fixed gills, as Sharks and Skates; and

Fig. 188.

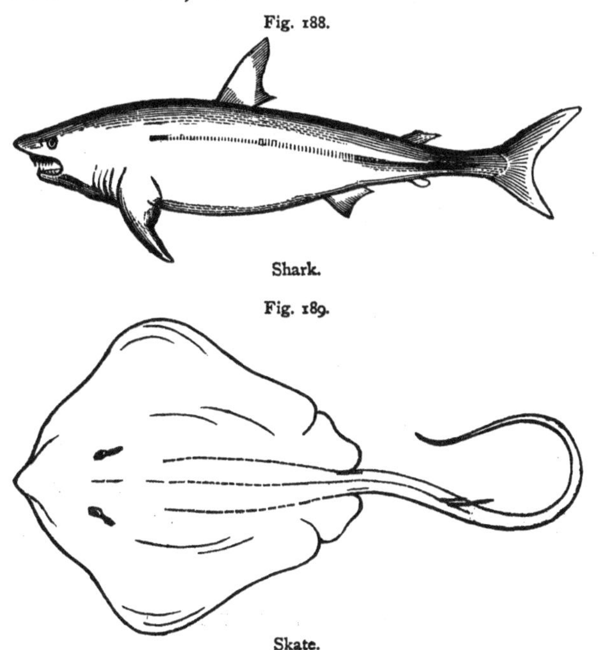

Shark.

Fig. 189.

Skate.

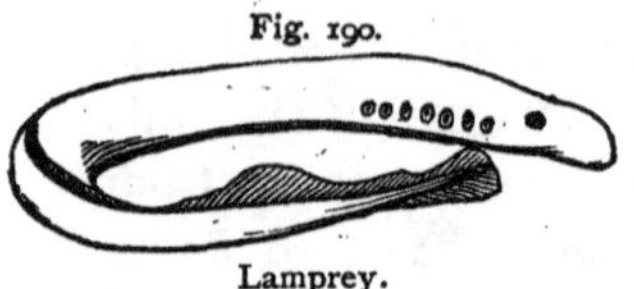
Fig. 190.
Lamprey.

Cyclostomes, or Suckers, chondropterygians with skeleton very slightly developed, and the body terminated before with a circular or semicircular lip. Such are the Lampreys.

SUB-SECTION I.

THE ORDER OF ACANTHOPTERYGIANS, OR SPINE-FINNED FISHES.

THIS Order, the most extensive of the Class of Fishes, is at once recognized by the spines which occupy the place of the first rays of the dorsal, or which alone support the first fin of the back whenever there are two dorsals. In some cases, instead of a first dorsal, there are only a few free spines. The first rays of the anal fin are also spines, and there is generally one spine to each ventral.

PERCIDÆ, OR PERCH FAMILY.—This Family comprises the Perch and its immediate allies, embracing about six hundred species, distributed among fifty or sixty genera. About one fifth of these inhabit fresh water, and the others are marine. Some are remarkably beautiful. The flesh of all is wholesome, and that of many is highly prized, and much used for food.

The Genus *Perca* — Perches proper — has a dentated preoperculum, bony operculum terminated by two or three sharp points, and a smooth tongue.

The American Yellow Perch, *P. flavescens*, Cuv., is from six to twelve inches long, greenish and golden above, the sides golden yellow, with six to eight dark vertical bands, which extend over the back; the pectorals, ventrals, and anal orange. This is one of the most common and best known of the fresh-water fishes of the United States.

Fig. 191.
Yellow Perch, *P. flavescens*, Cuv.

The Genus *Labrax* — Bass — is distinguished from Perca by scaly opercula terminating in two spines, and by a rough tongue.

The Striped Bass, *L. lineatus*, Cuv., is from one to four feet long, brown above, silvery beneath, and with from seven to nine blackish longitudinal stripes on each side of

Fig. 192.

Striped Bass, *L. lineatus*, Cuv.

the body. This is a salt-water fish, which keeps near the land, ascending fresh-water streams in the spring to breed. It is very common on the coast of New England. It readily bites the hook, and is taken in large quantities with the seine. Some specimens weigh seventy-five pounds each.

The White Perch, or Ruddy Bass, *L. rufus*, Dekay, Eastern North America, is twelve to fifteen inches long, and is highly prized for food. It is most common in waters which are in easy communication with the sea.

The Genus *Lucioperca* adds to the characteristics of the Perch those of the Pike.

The Yellow Pike-Perch, *L. americana*, Dekay, of the North American Great Lakes and adjacent regions, is from twelve to eighteen inches long, cylindrical and tapering, and is popularly known as the Common Pike, Glass-Eye, and Yellow Pike.

The Genus *Centropristes* has the teeth small and crowded, preoperculum dentated, and the operculum spinous.

The Black Sea-Bass, or Black-Fish, *C. nigricans*, C. & V.,

of the Atlantic coast of the United States, is from six to twelve inches long, bluish black, dorsal fin mottled with white, and the caudal, when perfect, divided into three lobes. Its flesh is highly prized for food. It is abundant in the markets from May to July.

The Genus *Pomotis* — Breams — is characterized by an oval, much-compressed body, and a membranous prolongation at the angle of the operculum. There are several species, all of which inhabit fresh water, and all are American. They make circular cavities in the sand for nests, often two feet in diameter, and six inches deep, where they lay their eggs and courageously defend them.

The Common Pond-fish, Bream, or Pumpkin-seed, *P. vulgaris*, authors, is from five to eight inches long, greenish olive, and with numerous reddish spots scattered over the body, and with the appendix of the operculum black, bordered behind with scarlet.

Fig. 193.

Bream, *P. vulgaris*, Cuv.

The Genus *Uranoscopus* — Star-gazers — has the eyes so placed that they appear constantly looking at the heavens, the mouth cleft vertically, a stout spine at each shoulder, and in the mouth and before the tongue a long filament which can be protruded at will, and which serves, it is said, to attract small fish, while its owner remains concealed in the mud. Star-gazers inhabit the Atlantic and Mediterranean. One species, *U. anoplos*, Cuv. & Val., two inches or more in length, is found on the southeastern coast of the United States.

Fig. 194.

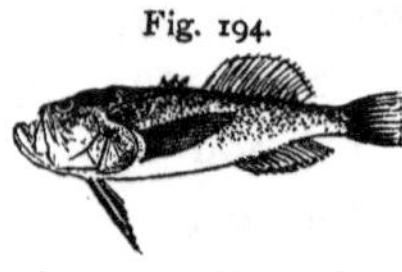

Star-gazer, *U. anoplos*, Cuv. & Val.

TRIGLIDÆ, OR MAILED CHEEK FAMILY. — This Family embraces spine-rayed fishes, which have the head variously mailed. About thirty genera have been described.

The Genus *Trigla* — Gurnards — is characterized by

nearly a square head, and dense teeth in the jaws and before the vomer. The species are numerous and marine.

The Red Gurnard, *T. cuculus*, Linn., of the Atlantic, is from eight to ten inches long, and of a red color.

The Genus *Prionotus* has very large pectorals, and a belt of dense teeth on the palatines.

The Sea Robin, or Grunter, *P. lineatus*, Dekay, is from twelve to eighteen inches long, color dark-brown above, sides and abdomen cream-colored; and there is a broad, reddish stripe below the lateral line.

Fig. 195.

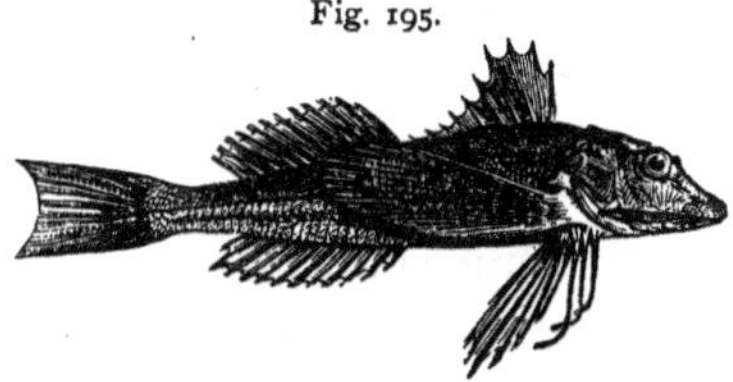

Sea Robin, *P. lineatus*, Dekay.

The Web-fingered Gurnard, *P. palmipes*, Storer, of the Atlantic, is from fifteen to eighteen inches long, with the pectoral processes dilated at their extremities.

The Genus *Dactylopterus* is characterized by pectoral fins, which are excessively developed, and composed of two parts, forming a large fin, which, like a wing, supports the fish in the air for a short time.

The Sea-Swallow, *D. volitans*, Cuv., of the Atlantic coast of America, is six to fourteen inches long.

The Genus *Uranidea* has the head much depressed, second dorsal higher than the first, and only one small spine at the angle of the preoperculum. The species inhabit fresh water.

The River Bull-head, or Miller's Thumb, *U. gracilis*, Putnam, of the Northern States, is only two or three inches long.

The Genus *Cottus* has spines upon each of the opercular bones, and the head armed with spines. The species are marine.

The Greenland Sculpin, *C. grœnlandicus*, C. & V., of

Fig. 196.

Greenland Sculpin, *C. grœnlandicus*, Cuv. & Val.

the North Atlantic, is about twelve inches long, dark brown above, with clay-colored blotches; abdomen with yellowish and white spots. Four tubercles on the top of the head enclose a quadrangular area.

The Common Sculpin, *C. octodecimspinosus*, Mitch., of the North Atlantic, is about twelve inches long, light brown above, with darker blotches; white below. When first taken from the water, its head is excessively expanded, its spines and fins erect, and it makes a croaking noise. The head is slenderer than in *C. grœnlandicus.*

The Genus *Boleosoma* has the body in the form of a dart, and the head very short and rounded.

Fig. 197.

Darter, *B. Olmstedi*, Ag.

The Tessellated Darter, *B. Olmstedi*, Ag., of the small streams of the Northeastern States, is about three inches long.

The Genus *Hemitripterus* has the first dorsal deeply emarginate, thus making the two dorsals look like three; head bristly and spinous, and with cutaneous appendages.

The Sea-Raven, or Deep-water Sculpin, *H. acadianus*, Storer, of the North Atlantic, is from twelve to twenty-four inches long, with the color exceedingly variable; often yellow, or blood-red varied with brown.

Fig. 198.

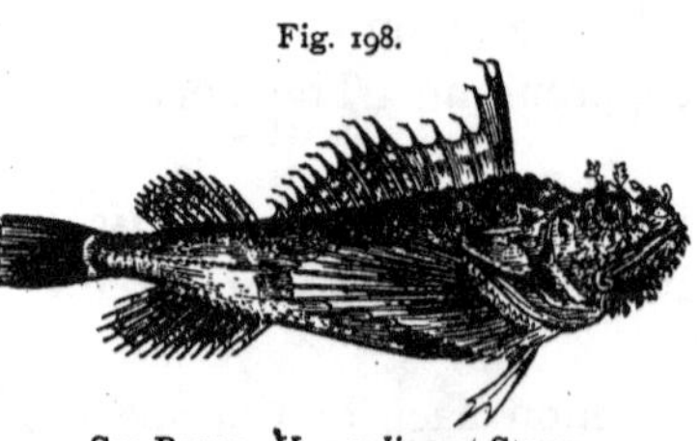

Sea-Raven, *H. acadianus*, Storer.

The Genus *Scorpæna* has the head mailed, roughened, compressed, scaly, and a single dorsal; head and body with fleshy appendages.

The Sea-Scorpion, *S. porcus*, Linn., of the Atlantic and Mediterranean, is from eight to ten inches long, color brown, rosaceous beneath. Wounds from its spines are considered dangerous.

The Spotted Sea-Scorpion, or Sea-Toad, *S. bufo*, Rich., of the Atlantic, is about seven inches long, brown, marbled with rosaceous and violet.

The Genus *Sebastes* differs from the preceding one in the more compressed form, the absence of cutaneous appendages, and in the smoother head.

The Norway Haddock, *S. norvegicus*, Cuv., of the North Atlantic, is from twelve to twenty-four inches long, red above and silvery below.

The Genus *Gasterosteus* — Sticklebacks — has mailed cheeks, one dorsal fin, with free spines before it, and the bones of the pelvis united, forming a shield pointed behind; and their ventrals are reduced to a single spine. The species are quite numerous, very small, from one to two inches and a half long. They inhabit the ocean and fresh-water streams, and are very active, pugnacious, and voracious. A single individual has been known to devour seventy-five young fish in five hours. They construct nests, which are protected by the female fish.

Fig. 199.

Many-spined Stickleback, *G. Dekayi*, Ag.

Scienidæ, or Scienoid Family. — This Family comprises fishes which are closely related to the Percidæ; the bones of the cranium and face are generally cavernous, muzzle more or less gibbous, and vertical fins somewhat scaly. The species are very numerous, inhabiting mainly the tropical seas. Cuvier enumerates over thirty genera, and the species known are more than two hundred and fifty.

The Genus *Otolithus* is characterized by feeble or obsolete anal spines, two dorsals, and the air-bladder bifid in

Fig. 200.

Weak-Fish, *O. regalis*, Cuv. & Val.

front. The Squeteague, or Weak-Fish, *O. regalis*, Cuv. & Val., of the Atlantic, is from twelve to twenty-four inches long, bluish above, varied with dusky; ventrals and anals orange.

The Genus *Amblodon* has the lower pharyngeals soldered together, and covered with thick, heavy, pavement-like crushing-teeth, arranged in regular rows, and opposed by the same kind of teeth in the upper pharyngeals.

The Lake Sheepshead, *A. grunniens*, Rafin., of the Northern and Western lakes, is from twelve to thirty-six inches long, silvery above, grayish-white below.

The Genus *Umbrina* is characterized by a single cirrus on the point of the lower jaw. The species are marine.

The King-Fish, *U. nebulosa*, Storer, of the Atlantic coast of the United States, is from twelve to eighteen inches long, the color dark gray, with transverse dusky bars and bands.

SPARIDÆ, OR PORGEE FAMILY. — This Family comprises spine-rayed fishes which have no teeth in the palate, no spines or teeth on the opercular bones, the muzzle not gibbous, and the bones of the head not cavernous. The genera are thirteen, species nearly two hundred.

The Genus *Sargus* has trenchant incisors in front, similar to those of man.

The Sheepshead, *S. ovis*, Cuv., of the Atlantic coast of the United States, is from twelve to thirty-six inches long, with the tail abruptly diminished from the body; color dull silvery on the sides, brassy on the back, with dark transverse bands. Its flesh is very highly prized.

The Genus *Pagrus* has from four to six stout conical teeth in the front of each jaw, and two series of round teeth on the sides.

The Scupaug, Scup, or Big Porgee, *P. argyrops*, Cuv., of the Atlantic coast of the United States, is from eight to twelve inches long, with a short recumbent spine in front of the dorsal fin, and the sides with brilliant metallic reflections.

Fig. 201.

Scupaug, *P. argyrops*, Cuv.

CHÆTODONTIDÆ, OR CHÆTODON FAMILY.—This Family comprises spine-rayed fishes which have the body compressed and scaly, and the dorsal and anal fins scaly. The Chætodons are so named from their teeth, which resemble bristles collected in rows like those of a brush. Their mouth is small, and colors brilliant. They abound in the hot seas. Eighteen genera and one hundred and fifty species are enumerated.

The Genus *Chætodon* contains *C. rostratus*, Bl., of Java, which has the faculty of spirting drops of water so as to hit insects on the plants near by, and bring them down so that it can secure them for food.

The Genus *Ephippus* contains those known as Horsemen. They have a dorsal deeply emarginate between the spinous and soft rays; and the spinous part can be folded into a groove formed by the scales of the back.

The Banded Ephippus, *E. faber*, Cuv. & Val., of the Atlantic, is from five to eighteen inches long, brownish, with six broad, vertical dusky-bluish bands.

The Genus *Pimelepterus* has the fins much thickened by the scales which cover them. It contains the Razor-Fish, *P. Boscii*, Cuv. & Val., of the Atlantic, near the Southern coast. It is six inches long.

The Genus *Toxotes* contains the Archer, *T. jaculator*, Cuv., of Java, celebrated for the same faculty which distinguishes *C. rostratus*, that of spirting drops of water so as to bring down insects from the plants above it, forcing water three or four feet, and rarely missing its aim.

SCOMBRIDÆ, OR MACKEREL FAMILY. — This Family comprises fishes with a smooth body and small scales, and whose tail and caudal fin are extremely powerful. Over fifty genera and more than four hundred species are known, many of which are of the highest utility to man.

The Genus *Scomber* — Mackerels proper — is characterized by a fusiform body, two dorsals widely separated, finlets behind the dorsal and anal fins, and two cutaneous crests on the sides of the tail.

The Mackerel, *S. vernalis*, Mitch., of the Atlantic, is from sixteen to eighteen inches long, dark steel-blue above, becoming lighter on the sides, and with twenty-four to thirty vertical deep-blue half bands; beneath silvery, with metallic reflections. This species appears on the coast of New England in the spring and summer, sometimes in the most astonishing numbers. Dr. Storer states that in 1837 Massachusetts fishermen caught over two hundred thousand barrels of this Mackerel.

Fig. 202.

Mackerel, *S. vernalis*, Mitch.

The Spanish Mackerel, *S. Dekayi*, Storer, of the Atlantic, is from twelve to twenty-four inches long, light-green above, with numerous undulating darker green lines; lower, dull bluish, with grayish-brown spots on the sides; and the abdomen light, with metallic reflections.

The Genus *Thynnus* has a corselet around the thorax, formed of larger scales than those of the rest of the body, a long and elevated crest on each side of the tail, and the front dorsal reaching nearly to the hind one.

The American Tunny, or Horse-Mackerel, *T. secundo-dorsalis*, Storer, is from nine to twelve feet long, and attains a weight of a thousand pounds.

The Genus *Pelamys* has separate, stout, and acute

teeth. It contains the Skip-Jack, or Striped Bonito, *P. sarda*, Cuv., of the Atlantic, which is from twelve to twenty inches long, dark plumbeous above, abdomen silvery white, with six or more parallel, longitudinal, somewhat oblique dark stripes on the body and sides.

The Genus *Xiphias* — Sword-Fishes — has a very long beak, or sword-like upper jaw. The Common Sword-

Fig. 203.

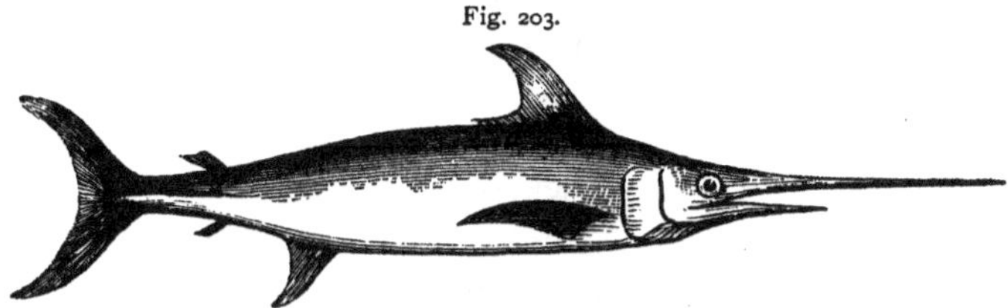

Sword-Fish, *X. gladius*, Linn.

Fish, *X. gladius*, Linn., of the Atlantic and Mediterranean, is from ten to sixteen feet long. It attacks the largest animals of the sea, and swims with astonishing swiftness.

The Genus *Naucrates* — Pilot-Fishes — has a fusiform body, free dorsal spines, a crest on the sides of the tail, and two free spines before the anal fin; species marine.

Fig. 204.

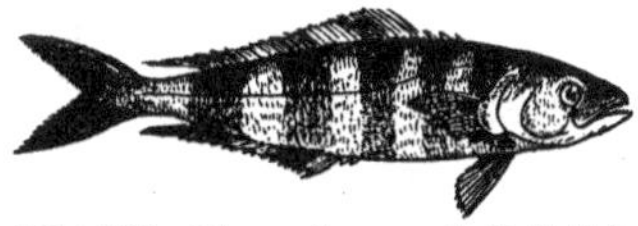

Pilot-Fish, *N. noveboracensis*, C. & Val.

The Genus *Temnodon* has the first dorsal in a furrow, teeth on the outer row separate, flat, and lancet-shaped; inner series crowded, and the teeth dense upon the vomer, palatines, and tongue. The Blue-Fish, *T. saltator*, Cuv., of almost all seas, is about eighteen inches long, bluish above, lighter below. It is prized for food.

Fig. 205.

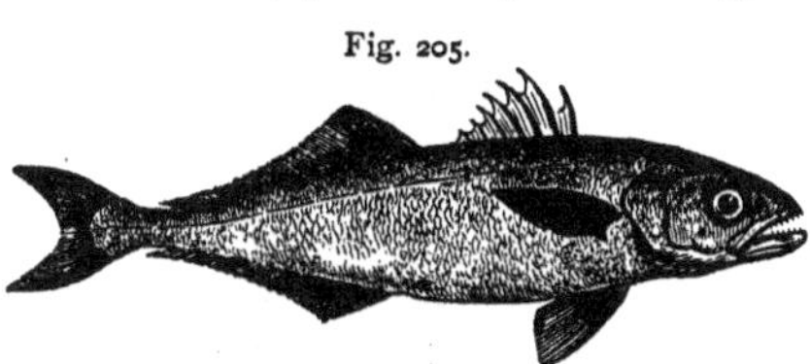

Blue-Fish, *T. saltator*, Cuv.

The Genus *Caranx* has a lateral line more or less mailed with scaly plates, carinated, and frequently spinous. It contains the Yellow Mackerel, *C. chrysos*, Cuv., which is from six to eight inches long, and found on the Atlantic coast of the United States.

The Genus *Vomer* has the body deep, much compressed, and the profile nearly vertical. The Blunt-nosed Shiner, *V. Brownii*, Cuv. & Val., of the tropics, and northward to New York, is about eight inches long; color lustrous silvery.

Fig. 206.

Blunt-Nosed Shiner, *V. Brownii*, Cuv. & Val.

The Genus *Coryphæna* contains the Dolphins, which are large, and beautiful in their colors, and celebrated for the war which they wage against the flying-fishes, and for the brilliant hues which they exhibit when dying. They inhabit the Atlantic and Mediterranean.

Fig. 207.

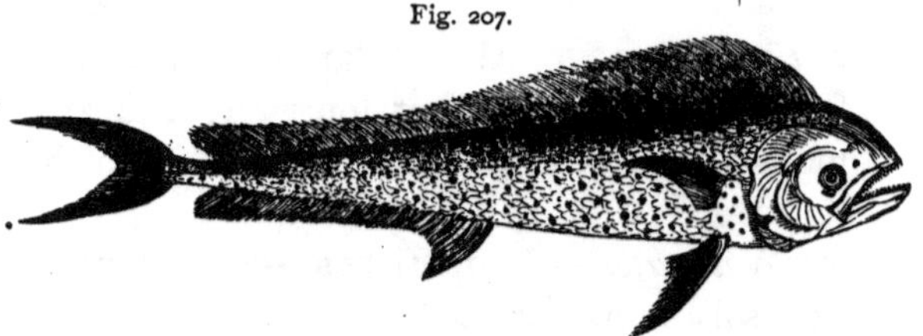

Dolphin, *C. doradon*, Cuv. & Val.

TEUTHIDÆ, OR LANCET-FISH FAMILY.—This Family comprises spine-rayed herbivorous fishes which inhabit the warm seas. About a hundred species have been described.

Fig. 208.

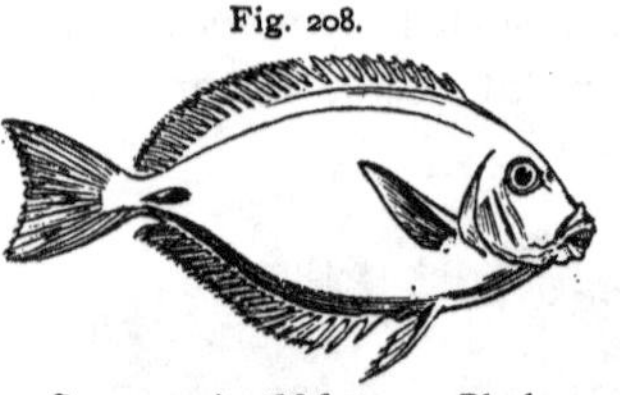

Surgeon, *A. phlebotomus*, Bloch.

The Genus *Acanthurus* contains the Surgeon, *A. phlebotomus*, Bloch, from six to ten inches long, which has a movable spine at the base of the tail.

LABYRINTHICI, OR CLIMBING PERCH FAMILY.—This

Family includes spine-rayed fishes which have an apparatus by which they retain water for the purpose of moistening their gills while they are upon shore; for at times they quit the water and crawl for a considerable distance upon dry land. They inhabit the fresh waters of the East Indies and South Africa.

MUGILIDÆ, OR MULLET FAMILY. — This Family comprises spine-rayed fishes with a nearly cylindrical body, large scales, two distinct dorsals, head somewhat depressed and covered with large scales or plates, and the muzzle short. They inhabit the fresh waters and coasts of temperate and tropical regions, and about eighty species are known.

The Genus *Mugil* is represented by several species on the Atlantic coast of the United States, all of which are small, varying from six to eight or nine inches in length, and of a silvery color.

Fig. 209.

Striped Mullet, *M. lineatus*, Mitch.

The Genus *Atherina* contains the Silversides. The Dotted Silverside, *A. notata*, Mitch., of the Atlantic coast of the United States, is three to four inches long, with a broad silvery band from the branchial aperture to the tail, and dark points on each scale.

GOBIDÆ, OR GOBY FAMILY. — This Family comprises spine-rayed fishes with a more or less elongated body, small scales or none, slender and flexible dorsal spines, and small branchial apertures. About thirty genera and about three hundred species are described.

The Genus *Blennius* — Blennies — has a single dorsal, smooth skin, and ventrals under the throat. The species are found in small communities among the rocks near the shore, and are capable of living without water for some time. They are all small, some of them only one or two inches long, and covered with a slimy mucus.

The Genus *Gunnellus* has a much-compressed body, spinous dorsal rays, and ventrals often reduced to a single spine.

The American Butter-Fish, *G. mucronatus*, Cuv., of the Atlantic, is from four to twelve inches long, grayish, with a series of dusky oval rings along the sides.

The Genus *Zoarces* has an elongated body, dorsal, anal, and caudal united, and no spinous rays in the dorsal except in its posterior part. The Eel-shaped Blenny, or

Fig. 210.

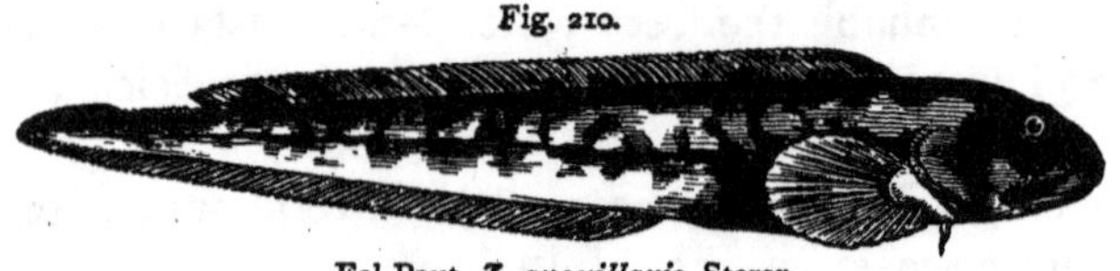

Eel-Pout, *Z. anguillaris*, Storer.

Eel-Pout, *Z. anguillaris*, Storer, is from twenty-four to thirty-six inches long, dark olive-brown, varied with dusky blotches. It is caught in fishing for Cod.

The Genus *Anarrhicas* — Wolf-Fish — has a smooth, elongated, and slimy body, globular head, dorsal and anal distinct from the caudal fin, and teeth of two kinds, one kind long, curved, and sharp, the other truncated or abruptly rounded. Their dentition furnishes fishes of this genus with powerful weapons, which, added to their great size and ferocity, make them very dangerous antagonists.

The Sea-Wolf, or Sea-Cat, *A. vomerinus*, Ag., of the Atlantic, is from three to five feet or more in length, of a gray color.

The Genus *Gobius* — Gobies — has the thoracic ventrals united, either along their whole length, or at least at the base, forming a hollow disk. The species are small, some of them only two or three inches long, and live among the rocks near the shore Some of them are viviparous. Over a hundred species are known.

PEDICULATI, OR ANGLER FAMILY. — This Family embraces fishes that are usually without scales, or these are replaced by bony plates, or grains bearing spines, and whose carpel bones are elongated, forming a sort of arm to support the pectorals. Eight genera and about forty species have been described.

The Genus *Lophius* has the head and mouth enormously large, two dorsals, the anterior rays distant, and forming long filaments bearing fleshy slips.

The American Angler, Fishing-Frog, or Goose-Fish,

Fig. 211.

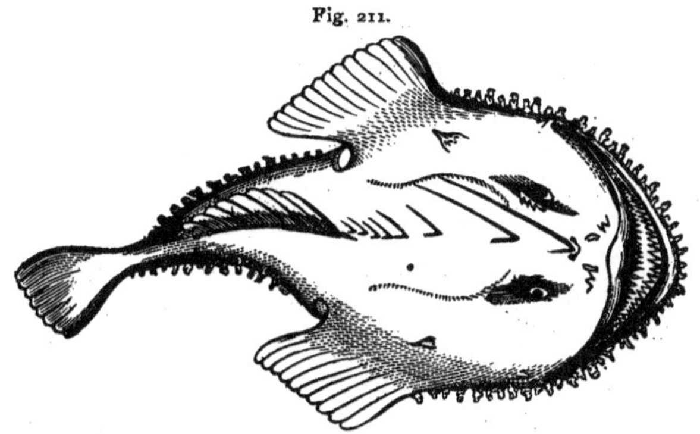

Angler, or Goose-Fish, *L. americanus*, Cuv.

L. americanus, Cuv., of the Atlantic, is from two to three feet long, and attains a weight of seventy pounds in some cases. It is exceedingly voracious, and its enormous mouth enables it to swallow fishes about as large as itself. Large sea-birds, as gulls, are frequently found whole in its stomach.

The Genus *Chironectes* — Hand-Fishes — has a compressed head and body, vertically cleft mouth, and fins suited to creeping. The species belong mostly to the warm seas. The smooth Chironectes, or Mouse-Fish, *C. lævigatus*, Cuv., of the Atlantic coast of the United States, is from two to four inches long.

The Genus *Batrachus* contains the Toad-Fishes. The Common Toad-Fish, *B. tau*, Linn., of the Atlantic coast of the United States, is from six to twelve inches long, olive green above, mottled with darker; light below. It is often found in cavities under stones, and seems to show a care for its young, which are found in such situations.

Fig. 212.

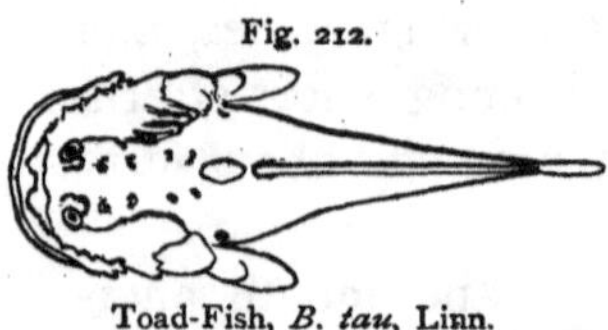

Toad-Fish, *B. tau*, Linn.

Labridæ, or Tautog Family. — This Family comprises spine-rayed fishes which have the body oblong and scaly, and a single dorsal supported in front by spines. A large number of genera and several hundred species are known.

The Genus *Ctenolabrus* has a denticulated preoperculum, and three spinous rays to the anal fin. The Conner, Blue-Perch, or Burgall, *C. Burgall*, C. & V., of the North Atlantic, is from six to twelve inches long, color exceedingly variable, but generally bluish, passing into reddish or bronze. From June to September it is extremely abundant on the coast, and is taken in immense numbers with hook and net. It is considered an excellent fish for the table, when fried.

Fig 213.

Conner, *C. Burgall*, C. & V.

The Genus *Tautoga* has the operculum and preoperculum without spines or denticulations. The Tautog, *T. americana*, Dekay, of the coast of New England and southward, is from six to eighteen inches long, the color generally bluish black, with irregular darker blotches and bands. It bites freely from early spring till late in autumn, and is highly prized for food. It is called Black-Fish at New York and on the Southern coast. It averages only one or two pounds in weight; but individuals have been taken which weighed sixteen pounds.

SUB-SECTION II.

THE ORDER OF ABDOMINAL MALACOPTERYGIANS.

THIS Order comprises fishes in which the ventrals are suspended to the under part of the abdomen, and behind the pectorals, without being attached to the bones of the shoulder. It contains a large majority of all fresh-water fishes.

CYPRINIDÆ, OR CARP FAMILY. — This Family comprises scaly fishes which have a slightly cleft mouth, and weak jaws without teeth. Of all fishes, they are the least carnivorous. About two hundred and seventy species have been described.

The Genus *Cyprinus* contains the Carps proper.

The Common Carp, *C. carpio*, Linn., of Europe, attains the length of four feet, is olive green, yellowish beneath. It is bred in fish-ponds, and esteemed for food. In 1831, Henry Robinson, Esq., of Newburgh, N. Y., introduced this species into his fish-pond, where it increased and grew so rapidly, that he had more than enough to supply his table. He afterwards put many of them into the Hudson River, and these have so increased, that they are often taken by the fishermen.

The Golden Carp, or Gold-Fish, *C. auratus*, Linn., common in aquaria and vases, is indigenous to China. It was introduced into Europe early in the seventeenth century. It breeds in ponds in various parts of the United States.

The Genus *Leuciscus*, as formerly defined, includes the Dace and Shiners of the United States; but writers now refer these fishes to several genera. The Brook-Minnow, or Black-nosed Dace, *L. atronasus*, Cuv. & Val., *Rhinichthys atronasus*, Ag., is three inches long, tail forked, color greenish above, abdomen silvery, and a dark band from the nose to the tail. The Common Shiner, *Plargyrus*

Fig. 214.

Shiner, *P. americanus*, Putnam.

americanus, Putn., *L. americanus*, Storer, is from three to six inches long, head small, tail forked, general color golden, dark above. The Red-Fin, *Hypsolepis cornutus*, Ag., *L. cornutus*, Storer, is from three to six inches long, fins and opercles margined with crimson, the male with numerous tubercles on the head. It is often found with trout.

The Beautiful Leuciscus, Dace, or Chivin, *Semotilus argenteus*, Putn., *L. pulchellus*, Storer, of the Eastern States and New York, is from twelve to fourteen inches long, back slightly arched, and the color brown. The Black-headed Dace, *L. atromaculatus*, of New York, is six to twelve inches long. It is known as the Lake Chub.

CATOSTOMI, OR SUCKER FAMILY. — This Family contains soft-finned fishes which have a single dorsal, the mouth beneath the snout, lips plaited, lobed, or carunculated, and suitable for sucking. It contains the well-known Suckers, of which there are many species, and the Chub-Suckers, of the ponds and streams of the United States. Large numbers move together, and some of the former attain the weight of ten pounds.

CYPRINODONTIDÆ. — This Family contains fishes whose mouth is constructed as in the Cyprinidæ, but with teeth upon the jaws; and the dorsal is opposite the anal fin. The Genera *Fundulus* and *Hydrargyra* are closely allied; but the latter has a more flattened head than the former, and six branchial rays instead of five. The former includes the Ornamented Minnow, Mummachog, or Cobler, *F. pisculentus*, Cuv. & Val., of the brackish waters of the coast of New England, and several fresh-water species. The Cobler is extensively used for bait.

ESOCIDÆ, OR PIKE FAMILY. — This Family comprises soft-finned fishes which have the body long, one dorsal generally opposite the anal, and a very large mouth extensively armed with very sharp teeth. Twenty-five or more species are known, all inhabiting fresh waters.

The Genus *Esox* is characterized by an oblong, broad, and depressed snout. The species are very voracious. The Muskallunge, or Pike, *E. estor*, LeS., of the North American lakes, is twelve to forty-eight inches long, and sometimes attains the weight of thirty pounds. The Common Pickerel, *E. reticulatus*, LeS., of the Eastern States, is from twelve to thirty-six inches long.

Fig. 215.

Pickerel, *E. reticulatus*, LeSueur.

The Short-nosed Pickerel, *E. fasciatus*, Dekay, is smaller than the preceding, with a short snout, and is common in the brooks and rivers of New England.

Scomberesocidæ, or Bill-Fish Family. — This Family is allied to the preceding one, with which it was formerly united. All the representatives are marine.

The Genus *Belone* — Gar-Fishes — has the head and body greatly elongated, the jaws narrow, pointed, and armed with numerous small teeth. The bones are remarkable for their green color. The Gar-Fish, *B. truncata*, LeS., of the Atlantic, is from twelve to twenty-four inches long, green above and silvery beneath, with a dark green longitudinal band upon the sides.

Fig. 216.

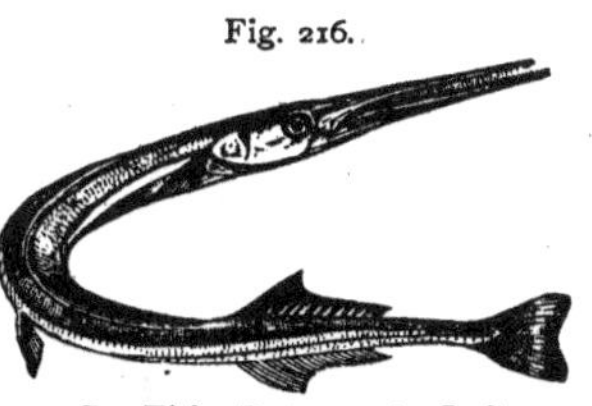

Gar-Fish, *B. truncata*, LeS.

The Genus *Scomberesox* has the last rays of the dorsal and anal detached. The Bill-Fish, *S. Storeri*, Dekay, of the Atlantic coast of the United States, is from ten to twelve inches long, dark green above, silvery below.

Fistularidæ, or Flute-Mouth Family. — This Family embraces fishes which have a long tube in front of the cranium, at the extremity of which is the mouth. They inhabit the warm seas, and are sometimes called Tobacco-Pipe Fishes. The Genus *Fistularia* has a very long filament extending from between the two lobes of the tail. The Tobacco-Pipe Fish, *F. serrata*, Bloch, of the southern coast of Massachusetts and southward, is nineteen inches long without the filament, or twenty-eight including it.

Exocœtidæ, or Flying-Fish Family.—This Family is characterized by the excessive development of the pectorals, which are about the length of the body, and enable the possessors to support themselves in the air for a few moments. Fishes of this family are found in all warm and temperate seas, and there are many species from three to twelve inches in length.

Fig. 217.

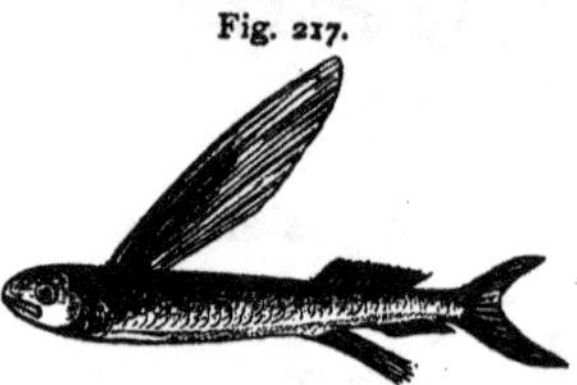

Flying-Fish, *Exocœtus*.

Hypsæidæ, or Blind-Fish Family.—This Family contains the Blind-Fish, *Amblyopsis spelæus*, Dekay, of the Mammoth Cave, Kentucky. This celebrated fish is about three inches long, with the vent before the base of the pectorals, and the eyes concealed under the skin so as to make the fish perfectly blind, and thus adapted to the dark waters of the cave.

Fig. 218.

Blind-Fish, *A. spelæus*, Dek.

The Siluridæ, or Cat-Fish Family.—This Family is readily distinguished from all other abdominal malacopterygians by the absence of scales, the skin being either naked or covered with large bony plates. The head in most cases is large, depressed, and with several fleshy filaments. In a majority of cases, the first ray of the dorsal and pectoral has a strong spine, which is so articulated that the fish can bring it close to the body, or immovably extend it, thus constituting it a dangerous weapon. There are about thirty-three genera and three hundred species, and they abound in nearly all fresh waters, especially those that are sluggish, or with muddy bottoms. Some, however, are marine.

The Genus *Silurus* contains the Silurus, *S. glanis*, of the rivers of Germany and Hungary, which sometimes exceeds six feet in length, and weighs three hundred pounds.

The Genus *Pimelodus* contains the Cat-Fishes of the United States, of which there are about thirty species.

The Cat-Fish of the Great Lakes, *P. nigricans*, LeS., is from two to four feet long, and attains the weight of thirty pounds. The Common Horned Pout, *P. atrarius*, Dekay, is from six to ten inches long, and abounds in ponds and slow streams.

Fig. 219.

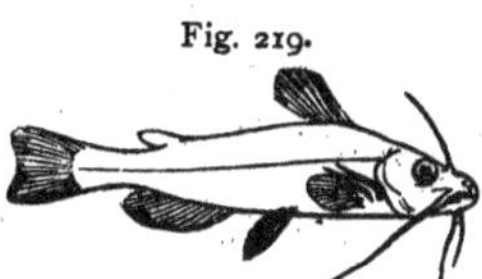

Horned Pout, *P. atrarius*, Dek.

SALMONIDÆ, OR SALMON FAMILY.—This Family comprises abdominal malacopterygians which have the body more or less scaly, a first dorsal with soft rays, followed by a second small one, which is fatty, and unsupported by rays. They inhabit both salt and fresh water; are very voracious, and highly prized for food. The Genus *Salmo* is the principal one.

Fig. 220.

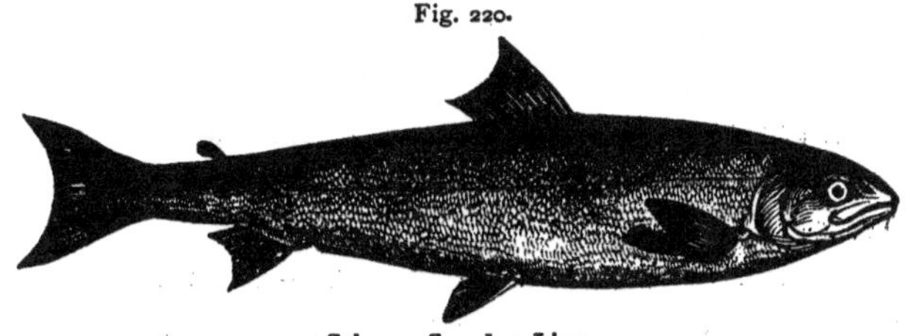

Salmon, *S. salar*, Linn.

The Salmon, *S. salar*, Linn., of the Arctic seas, whence it visits the rivers of both continents, is one of the largest of the genus, and is celebrated for its delicious flesh. It is from twenty-four to thirty-six inches long, and attains a weight of thirty pounds or more.

The Great Trout of the Lakes, *S. amethystus*, Mitch., of the Northern lakes of North America, is from twenty-four to sixty inches long, dark gray, with numerous lighter spots on the back and sides; under parts light ashy-gray or cream-color. It sometimes attains the weight of one hundred and twenty pounds, and is often called Mackinaw Trout. It is also known as the Longe.

The Speckled Trout, or Brook Trout, *S. fontinalis*, Mitch., of the clear streams of Northern North America, is from six to twenty inches long, horn-color above with irregular darker markings, sides bluish mixed with silvery white, and ornamented with yellow spots and vermilion dots. There are many varieties of trout, and probably some of the so-called varieties are distinct species. All are highly prized on account of the delicacy of their flesh.

Fig. 221.

Trout, *S. fontinalis*, Mitch.

The Genus *Osmerus* contains the American Smelt, *O. viridescens*, LeS., which is six to twelve inches long, greenish above, silvery beneath, with an obscure satin-like longitudinal band. It abounds on the coast, and ascends rivers, from New York northward.

The Genus *Coregonus* contains the White-Fish, *C. albus*, LeS., which is from eighteen to twenty inches long, bluish-gray above and white below, and inhabits the Great Lakes; and the Common Shad Salmon, *C. clupeiformis*, Dekay, of Lakes Erie and Ontario.

Clupeidæ, or Herring Family.—This Family embraces abdominal malacopterygians which have the body compressed, very scaly, and the inferior portion of the body forming an edge more or less serrated. The Genus *Clupea* comprises the Herrings proper. The Common

Herring, *C. elongata*, LeS., of the Atlantic coast of North America, is about twelve inches long, deep blue, tinged with yellow above, and silvery beneath. Three thousand barrels of this herring have been taken at Martha's Vineyard in a single year. Herrings live in the Arctic seas, and come southward in spring to deposit their eggs.

Fig. 222.

Herring, *C. elongata*, LeS.

The Pilchard, *C. pilchardus*, Bl., of the coast of England, is about the size of the herring. The Sardine, *C. sardina*, Cuv., is taken in the Mediterranean, and is celebrated for its extreme delicacy of flavor.

The Genus *Alausa* is distinguished from the Herrings by a deep notch in the middle of the upper jaw, and by the roof of the mouth and the tongue, which are destitute of teeth. It contains the Shad, Alewive, Menhaden, Autumnal Herring, and allied species.

The American Shad, *A. præstabilis*, Dekay, is about twenty inches long. It appears upon the coast, and ascends the rivers of South Carolina in January and February, of the Middle States in March or the first of April, and of Massachusetts in May.

The Genus *Elops* has a cylindrical body, and a flat spine on the upper and under edges of the caudal fin. The Saury, *E. saurus*, Linn., of the Atlantic, is from eleven to twenty-two inches long, body silvery, with a greenish tinge above.

SAURIDÆ, OR GAR-PIKE FAMILY. — This Family comprises elongated fishes covered with scales of stony hardness, which are extended into imbricated spines upon the first rays of all the fins. About fifteen species, all American and West Indian, are known.

The Genus *Lepidosteus* is characterized by elongated slightly unequal jaws, which are furnished over their

Fig. 223.

Gar-Pike, *Lepidosteus.*

whole inner surface with rasp-like teeth, and a row of long, pointed teeth along their edges. It contains the Gar-Pikes of the Northern lakes and the Western and Southern rivers, one species in Central America, and another in Cuba.

SUB-SECTION III.

THE ORDER OF SUB-BRACHIAN MALACOPTERYGIANS.

This Order embraces soft-rayed fishes which have the ventrals inserted under the pectorals, and the pelvis directly attached to the bones of the shoulder.

Gadidæ, or Cod Family. — This Family includes the Cod and its allies, about sixty species of which are known, most of them inhabiting cold or temperate seas.

The Genus *Morrhua* — Cods — has three dorsals, two anals, and a barbel at the point of the lower jaw.

The American Cod, *M. americana*, Storer, is from twenty-four to thirty-six inches long, olive-green above, dusky-

Fig. 224.

American Cod, *M. americana*, Storer.

white beneath, and the back and sides marked with yellowish spots. There are several varieties, differing in their color and markings. This species attains the weight of a hundred pounds in some instances. In 1840, the tonnage of vessels engaged in our cod-fisheries was 75,000, and the number of fishermen more than 18,000. The Tom-Cod, or Frost-Fish, *M. pruinosa*, Dekay, of the North Atlantic, is from four to twelve inches long, olive-green above, and silvery below. It is unusually abundant in the mouths of the rivers after the first frosts of autumn; hence one of its popular names. The Haddock, *M. æglefinus*, Linn., of the North Atlantic, is twelve to twenty-four inches long, blackish-brown above, silvery gray below, the lateral line jet black.

The Genus *Merlangus* has three dorsal and two anal fins, and no barbels on the chin. The Pollack, *M. purpureus*, Storer, is from eighteen to thirty-six inches long, the caudal deeply concave.

The Genus *Merlucius* has the head flattened, body elongated, only two dorsal fins, barbels wanting.

The American Hake, or Whiting, *M. albidus*, Dekay, is from twelve to thirty-six inches long, reddish-brown above, soiled-white below.

The Genus *Lota* has two dorsal fins, one anal, and barbels on the chin. The Spotted Burbot, *L. maculosa*, LeS., of our Northern lakes and rivers, is twenty-four inches long.

Fig. 225.

Burbot, *L. maculosa*, LeS.

The Genus *Brosmius* has a single dorsal extending the whole length of the back. The Cusk, *B. flavescens*, LeS., of the Atlantic, is twenty-four to thirty-six inches long.

The Genus *Phycis* has two dorsals, the first short, the second long, the ventrals with a single ray, and a single barbel at the chin.

The Hake, *P. americanus*, Storer, of the North Atlantic, is from twelve to thirty-six inches long, the color reddish-brown. It is usually taken with the hook at night, on muddy bottoms.

PLEURONECTIDÆ, OR FLOUNDER FAMILY. — This Family comprises fishes which have the body flat, being compressed vertically, both eyes on the same side of the head, sides of the mouth unequal, and a dorsal extending the whole length of the back. The side upon which the eyes are placed is always uppermost when the animal is swimming, and is deeply colored; while that on which the eyes are wanting is always whitish. They have no natatory bladder, and seldom quit the bottom. The want of symmetry between the two sides of the fishes of this family is seen in no other vertebrates. About one hundred and thirty species are known, all of which are marine.

Fig. 226.

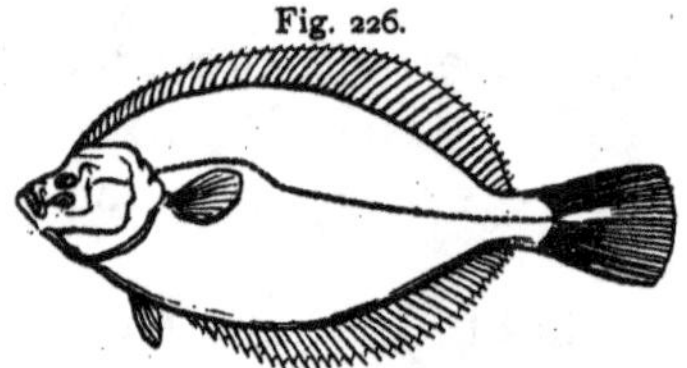

Flounder, *B. plana*, Mitch.

The Genus *Platessa* contains the Flounders proper, which are from six to twenty-five inches long. Turbots and Soles are representatives of genera closely allied to the preceding one.

The Genus *Hippoglossus* embraces the Halibuts, which have the eyes and color on the right side.

The Halibut, *H. vulgaris*, Cuv., of the North Atlantic, is from two to eight feet long, ashen-gray on the right side, white on the other. It attains the weight of six hundred pounds in some cases.

CYCLOPTERIDÆ, OR LUMP-FISH FAMILY. — This Family comprises fishes whose ventrals are united into a disk or cup-shaped form. By means of the disk, these fishes are able to attach themselves to the surface of the rocks with great firmness. The Genus *Cyclopterus* contains the Lump-Sucker, *C. lumpus*, Linn.,

Fig. 227.

Lump-Fish, *C. lumpus*, Linn.

which is from ten to twenty inches long, and inhabits the North Atlantic. Pennant states that, upon putting one into a pail of water, it adhered so firmly, that he lifted the whole pailful, several gallons, by taking hold of the fish by the tail.

ECHENEIDÆ, OR REMORA FAMILY. — This Family embraces fishes which have a flattened disk upon the head, composed of a number of transverse cartilaginous laminæ directed obliquely backwards, serrated or spiny on the hind edge, and movable, so that by creating a vacuum between them, or by hooking on to various bodies by means of the serrated edges, they are enabled to attach themselves very firmly. The Genus *Echeneis* is the prin-

Fig. 228.

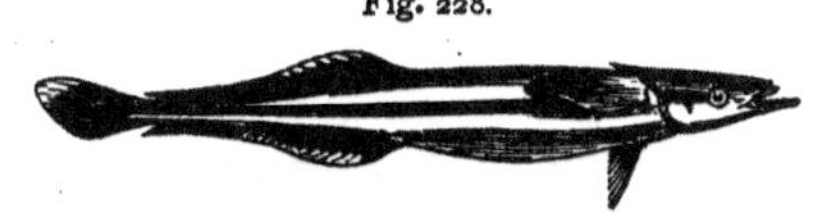

White-tailed Remora, *E. albicauda*, Mitch.

Fig. 229.

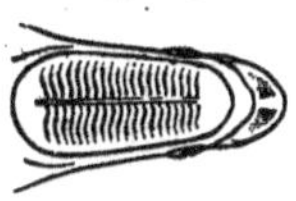

Disk of Remora.

cipal one. The species are few, and chiefly tropical; but some are taken on the coast as far north as Labrador. They are from twelve to thirty inches long, and often found attached to other marine animals.

SUB-SECTION IV.

APODAL MALACOPTERYGIANS.

THIS Order is represented by one great family, containing about one hundred known species.

ANGUILLIDÆ, OR EEL FAMILY. — This Family has the body much elongated, cylindrical, and covered with a thick, soft skin. The scales, when present, as in the common Eel, are scattered, and deeply imbedded in the skin.

The Genus *Anguilla* — Eels proper — has the dorsal

Fig. 230.

Eel, *A. bostoniensis*, LeS.

and anal continued around the end of the tail, forming by their union a pointed caudal. The common fresh- and salt-water Eel, *A. bostoniensis*, LeS., of the United States, is twelve to sixty inches in length.

The Genus *Muræna* has no vestige of pectorals. The Roman Muræna, *M. helena*, Linn., of the Mediterranean, attains the length of thirty-six inches or more, and is mottled with brown and yellowish. This fish was highly prized by the ancients, who fed it in ponds constructed expressly for it. Vedius Pollio caused his transgressing slaves to be flung alive into these ponds as food for the murænæ. The Romans domesticated these eels so that they would approach at call.

The Genus *Gymnotus* has the anal fin beneath a greater part of the body, and generally as far as the end of the tail, but no dorsal. The Gymnotus, or Electrical Eel, *G. electricus*, Linn., of the warm regions of South America, is five or six feet long, and is celebrated for its ability to communicate such electrical shocks that men and animals are struck down by them.

The Genus *Ammodytes* has the dorsal fin extending nearly the whole length of the back, the anal long, and both separated from the caudal, which is forked. The species are marine, and live in the sand, and are known as Sand-Eels, or Sand-Launces.

SUB-SECTION V.

THE ORDER OF LOPHOBRANCHIATES, OR TUFT-GILLED FISHES.

THIS Order comprises fishes which have the gills in small round tufts arranged in pairs along the branchial arches, instead of resembling, as in other fishes, the teeth of a comb.

SYNGNATHIDÆ, OR PIPE-FISH FAMILY. — This Family is usually made to include all the Lophobranchiates.

The Genus *Syngnathus* — Pipe-Fishes — has the body exceedingly elongated, slender, and covered with a series of hard plates parallel to each other; the snout prolonged, with the mouth at the extremity; no ventral fins; and the males of many species have a pouch for the reception of the eggs, in which the young are hatched. Several species are known, five to ten inches long, all inhabiting tropical and temperate seas.

Fig. 231.

Pipe-Fish, *S. Peckianus*, Storer.

The Genus *Hippocampus* — Sea-Horses — has the body short, compressed, covered with angular and spinous plates, neither ventral nor caudal fin, a prehensile tail, and the head and neck have some resemblance to those of a horse. Several species are known, from three to six inches long, all marine. *H. hudsonius*, Dekay, of the Atlantic coast of the United States, is about five inches long.

Fig. 232.

Sea-Horse, *H. hudsonius*, Dekay.

SUB-SECTION VI.

THE ORDER OF PLECTOGNATHES.

THIS Order comprises fishes whose chief characteristic is that the maxillary bone is permanently attached to the intermaxillary, which alone constitutes the jaw.

GYMNODONTIDÆ. — This Family embraces those whose jaws are furnished with a bony substance resembling enamel, and divided internally into laminæ. These laminæ are really true teeth united, which succeed each other as fast as they are destroyed by trituration. About sixty species are known, all marine.

The Genus *Diodon* has all the teeth united into one in each jaw, and the surface of the body covered with spines.

The Genus *Tetrodon* has each jaw divided in the middle by a suture, thus giving the appearance of two teeth in each jaw; and the body is wholly or partly covered with spines. The members of this genus, and the Diodons, possess the faculty of inflating themselves like a balloon, by swallowing air. The Common Swell-Fish, or Puffer, *T. Turgidus*, Mitch., of the Atlantic, is from six to twelve inches long.

Fig. 233.

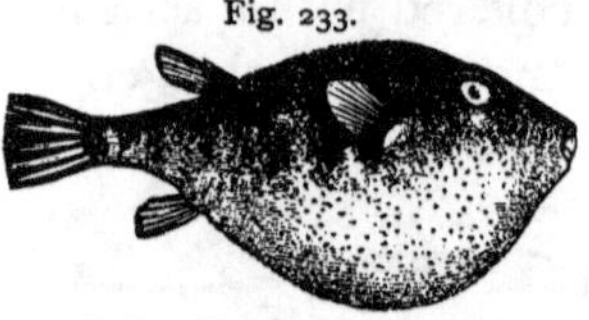

Puffer, *T. turgidus*, Mitch.

The Genus *Orthagoriscus* has the body short and compressed. The Short Sun-Fish, *O. mola*, Schr., of the Atlantic, attains the length of four feet, and the weight of five hundred pounds or more.

Fig. 234.

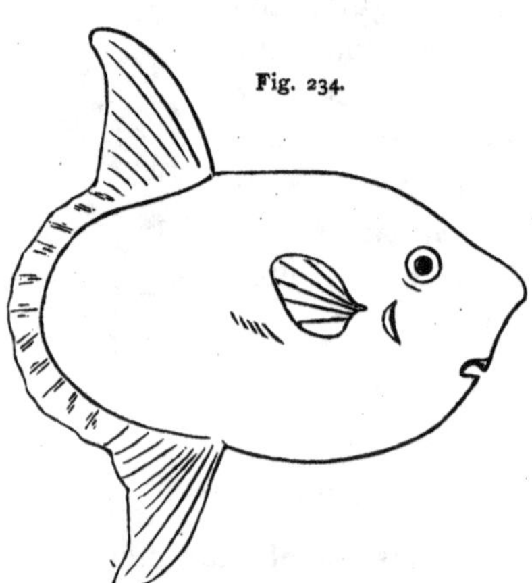

Sun-Fish, *O. mola*, Schr.

BALISTIDÆ, OR FILE-FISH FAMILY.— This Family embraces fishes with a conical or pyramidal snout, compressed body, eight teeth in a single row in each jaw, the skin scaly or granulated, two dorsals, the first composed of one or more spines, and the ventrals indistinct or wanting. They abound in the warm regions, and their colors are brilliant. Several species, from three to nine inches long, are found on the Atlantic coast of the United States.

OSTRACIONIDÆ, OR TRUNK-FISH FAMILY. — This Family comprises fishes which have the head and body covered with regular bony plates, soldered in such a manner as to form an inflexible shield, so that the mouth, tail, and fins are the only movable parts. Thirty or more species are known, inhabiting tropical and temperate seas. The Genus *Lactophrys* contains two or more species, from three and a half to fourteen inches long, which are found on the Atlantic coast of the United States.

Fig. 235.

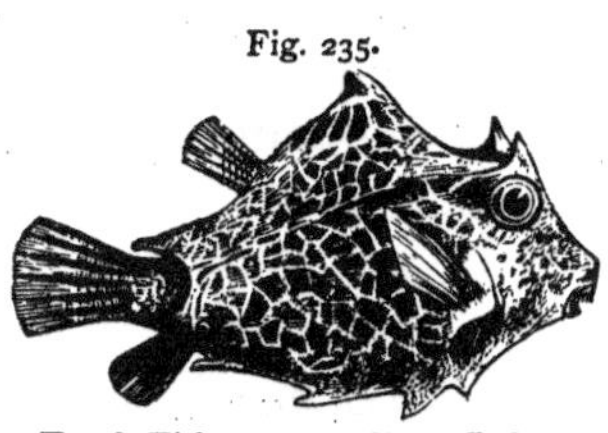

Trunk-Fish, *L. camelinus*, Dekay.

SUB-SECTION VII.

THE ORDER OF STURIONES, OR STURGEONS.

THIS Order embraces cartilaginous fishes with free gills, and one large external opening on each side, with a strong operculum.

STURIONIDÆ, OR STURGEON FAMILY.—This Family contains those which have large bony plates arranged in longitudinal rows, the mouth under the snout, without teeth, and very protractile, and the lobes of the tail unequal. The Genus *Acipenser* contains the Sturgeons proper. They inhabit lakes and the sea, and ascend the rivers of many countries. The Sharp-nosed Sturgeon, *A. oxyrhynchus*, Mitch., of the Atlantic coast of North America, is from four to eight feet long. The Lake Sturgeon, *A. rubicundus*, LeS., of the Great Lakes, is about four feet long, and of a ruddy hue. Several other species belong to North America. The Great Sturgeon, *A. huso*,

Fig. 236.

Sturgeon, *A. oxyrhynchus*, Mitch.

Linn., of Europe, attains a weight of twelve or fifteen hundred pounds. The isinglass of commerce is prepared from the swimming bladder of the Sturgeon.

SUB-SECTION VIII.

THE ORDER OF PLAGIOSTOMI, OR SELACHIANS.

THIS Order and the next comprise cartilaginous fishes called Fixed-Gilled Chondropterygians. Instead of having the gills free on the external edge, and opening at their intervals into a common chamber, as in all the preceding fishes, these have them adhering by the external edge in such a manner that the water escapes through as many holes in the skin as there are intervals between the gills; or else the holes terminate in a common duct, through which the water passes out.

SQUALIDÆ, OR SHARK FAMILY.—This Family embraces the Sharks, several species of which are viviparous, and others produce eggs invested with a yellowish transparent horny substance, the angles of which are prolonged into horny tubes.

The Genus *Alopias* contains the Long-tailed or Thresh-

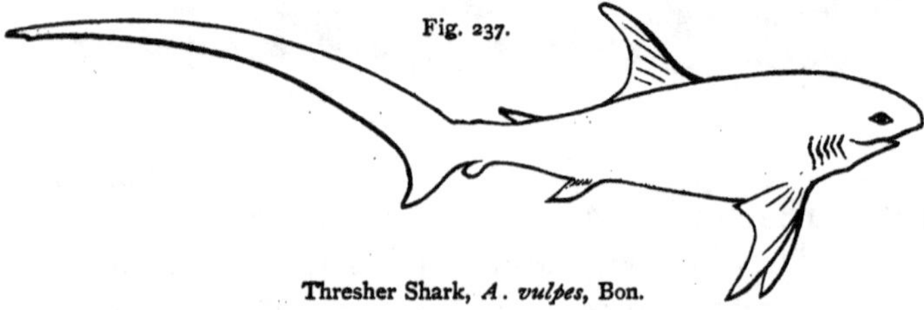

Fig. 237.

Thresher Shark, *A. vulpes*, Bon.

er Shark, *A. vulpes*, Bon., twelve to fifteen feet long, with the upper lobe of the tail about the length of the body. It inhabits the Atlantic.

The Genus *Lamna* has the snout pyramidal, with the nostrils under the base. The Mackerel Shark, or Green-

Fig. 238.

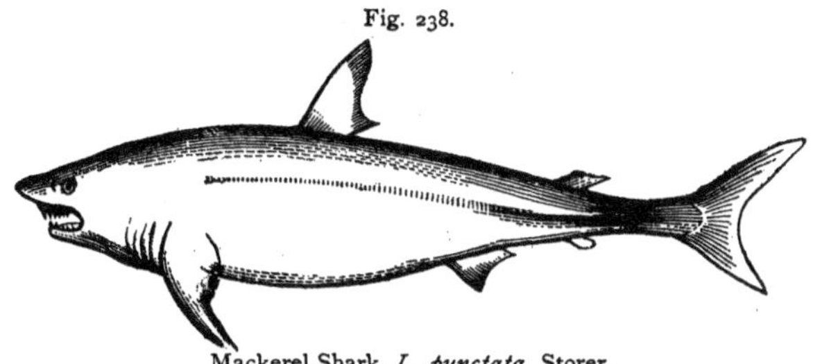

Mackerel Shark, *L. punctata*, Storer.

backed Shark, *L. punctata*, Storer, is from four to eight feet long, tail with a keel on its side, and the lobes not greatly unequal.

Fig 239.

Head of Mackerel Shark.

The Genus *Mustelus* has the teeth blunt, forming a closely compacted pavement in each jaw. The Dog Shark, *M. canis*, Dekay, is from two to four feet long.

The Genus *Selachus* has small, smooth teeth, the branchial apertures all before the pectorals, long, and nearly surrounding the neck. The Basking Shark, *S. maximus*, Yarrell, exceeds thirty feet in length. Although so large, it lacks the ferocity of other species.

The Genus *Somniosus* contains the Sleeper, or Nurse Shark, *S. brevipinna*, LeS., of the eastern coast of North America. It is so sluggish, that it often allows itself to be captured on a cod line.

The Genus *Acanthias* has a sharp, stout spine in front of each of the two dorsals. The Dog-Fish, *A. americanus*, Storer, is from one to three feet long. It is caught in great numbers for the sake of its oil.

Fig. 240.

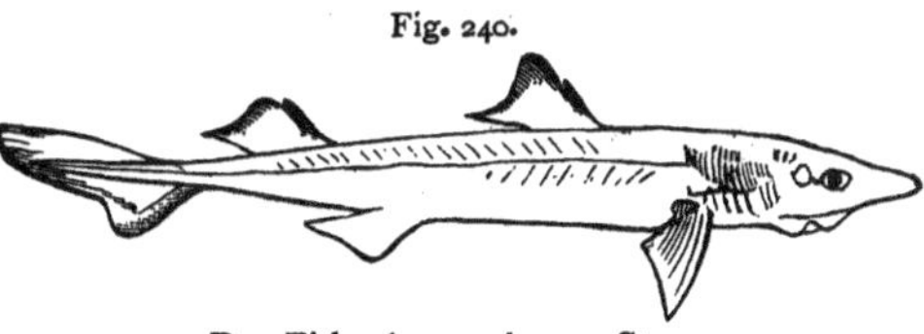

Dog-Fish, *A. americanus*, Storer.

The Genus *Zygæna* has the head flattened horizontally, with the sides much extended laterally. The Hammer-

Fig. 241. Fig. 242.

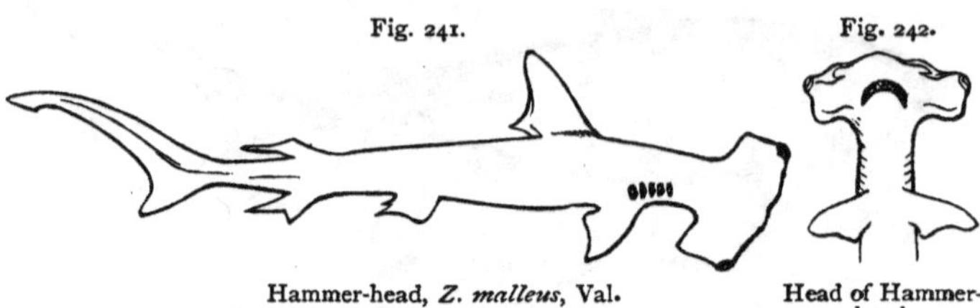

Hammer-head, *Z. malleus*, Val. Head of Hammer-head.

head Shark, *Z. malleus*, Val., attains the length of twelve feet, and is bold and ferocious.

The Genus *Pristis* has a very long, depressed snout, armed on each side with pointed spines, planted like teeth.

Fig. 243.

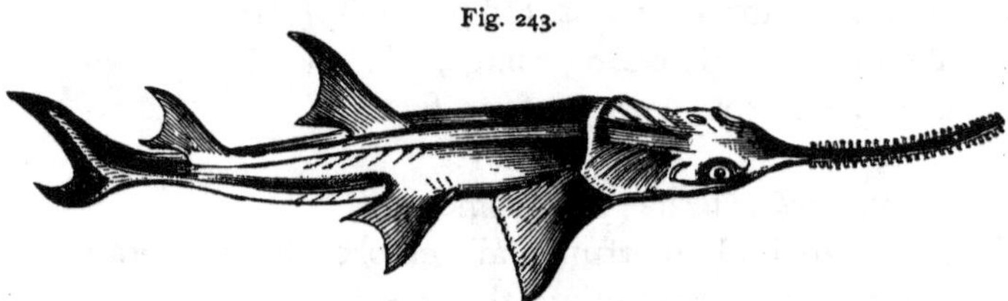

Saw-Fish, *P. antiquorum*, Lath.

The Common Saw-Fish, *P. antiquorum*, Lath., attains the length of fifteen feet.

Fig. 244.

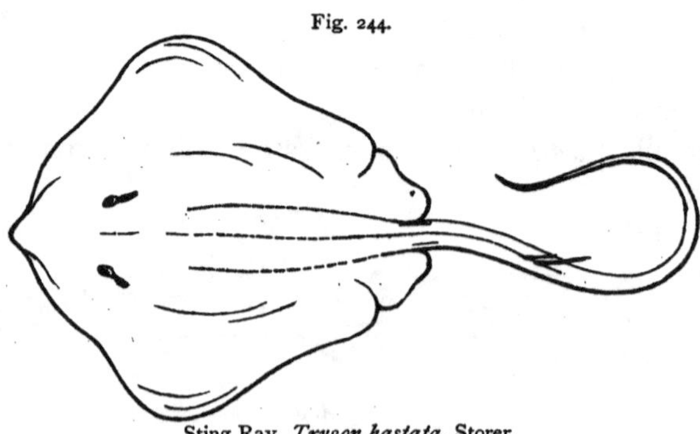

Sting Ray, *Trygon hastata*, Storer.

RAIIDÆ, RAY OR SKATE FAMILY. — This Family comprises fishes with the body very much flattened, the mouth, nostrils, and branchial openings below, and the dorsals, when present, upon the tail. The eggs are brown, coriaceous, and rectangular, with the angles extended into points. Members of this family are found in all seas, and more than a hundred species are known, from two to six feet or more in length.

The Genus *Torpedo* has the space between the pectorals, head, and the branchiæ filled on each side with a singular apparatus formed of little membranous tubes placed close together and subdivided by horizontal partitions into small cells filled with mucus, and traversed by nerves proceeding from the eighth pair. In this apparatus resides the electric or galvanic power which has made the Torpedo so celebrated. Violent shocks are received by coming in contact with it when alive.

Fig. 245.

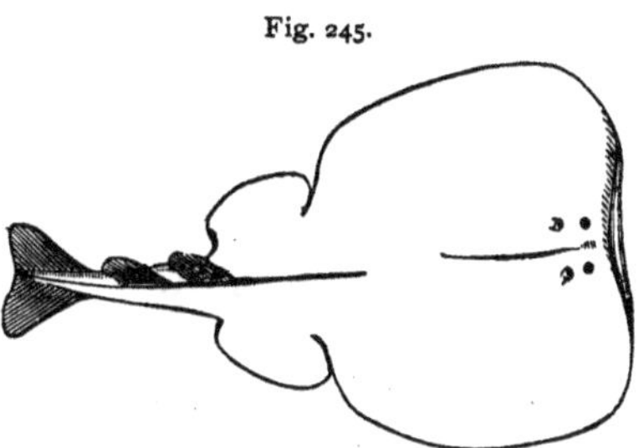

Torpedo, *T. occidentalis*, Storer.

The Genus *Cephaloptera* contains the Vampire of the Ocean, *C. vampirus*, Mitch., which attains the width of sixteen or eighteen feet, and ten feet or more in length, and weighs several tons. Dekay states that this monster of the deep has been known to seize the cable of a small vessel at anchor, and draw it several miles with great velocity!

SUB-SECTION IX.

THE ORDER OF CYCLOSTOMES, OR SUCKERS.

THIS Order comprises chondropterygians which, as regards the skeleton, are the most imperfect of all vertebrates, their vertebræ being simply cartilaginous rings scarcely differing from one another. But one of the most characteristic features of these animals is the tongue, which moves forwards and backwards like a piston, enabling them to produce a vacuum, and thus fix themselves to solid bodies as well as to fishes.

PETROMYZONIDÆ, OR LAMPREY FAMILY. — This Family comprises the Lampreys and their allies. The Genus *Petromyzon* has the maxillary ring armed with strong teeth. The American Sea Lamprey, *P. americanus*, LeS., is from two to three feet long, ending behind in a sharp tip. It ascends rivers, and piles up heaps of stones, among which it lays its eggs. The Bluish Sea Lamprey, *P. nigricans*, LeS., is five to seven inches long, and is found attached to cod, haddock, and other fishes.

Fig. 246.

Lamprey, *P. americanus*, LeS.

The Genus *Myxine* contains low forms, which Linnæus classed with worms. The Hag, *M. limosa*, Girard, is from six to eight inches long. It is common in the waters about Grand Menan.

Fig. 247.

Hag, *M. limosa*, Girard.

The Genus *Branchiostoma*, *Amphioxus* of authors, contains animal forms which are considered the lowest of all vertebrates. Several species are known, one of which is found on the Atlantic coast of the Southern States.

Fig. 248.

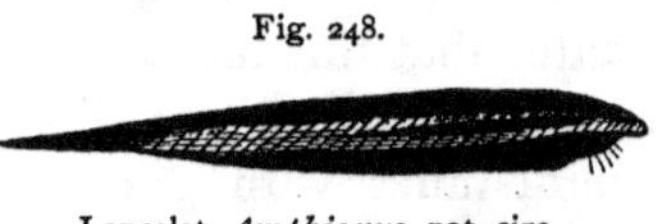

Lancelot, *Amphioxus*, nat. size.

Fossil fishes are found in all the systems of rocks from the Silurian upward, and their history, as written by Cuvier, Agassiz, and others, is full of instruction and interest.

CHAPTER III.

THE BRANCH OF ARTICULATA, OR ARTICULATES.

THE Branch of Articulata comprises all animals which possess bilateral symmetry, and which are divided transversely into rings or joints more or less movable upon each other, and whose hard parts are external. The alimentary canal lies in the centre of the body, and above it the dorsal vessel or heart. The nervous system consists of a sort of brain, which lies above the œsophagus, from which two threads, passing around the œsophagus, extend beneath the alimentary canal, along the floor of the general cavity of the body, and connect at certain distances small nervous centres or ganglia, whence arise the nerves of the body and limbs. Each of these nervous centres seems to fulfil the functions of a brain to the surrounding parts, and preserves their sensibility for a greater or less length of time after the animal has been divided. The number of these nervous centres generally corresponds to the number of the segments of the body. Articulates are divided into three classes, — Insects, Crustaceans, and Worms.

SECTION I.

THE CLASS OF INSECTS.

THE Class of Insects comprises articulates whose respiratory apparatus consists of air-holes, called stigmata, placed along the sides of the body, and connected with a system of air-tubes, called tracheæ, which branch throughout the interior of the body and carry air into every part. It includes three orders, — Insects proper, Spiders, and Myriapods.

Insects proper have the body divided into three plainly marked regions, — the head, chest or thorax, and hind body or abdomen. The head is furnished with antennæ, mouth, and eyes; to the thorax are appended the legs and wings, when these exist; and the abdomen contains the principal organs of digestion and other viscera, and to it belong the piercer and sting with which many insects are provided. The antennæ serve the purpose of feelers, and are also probably connected with the sense of hearing. The mouth-parts are modified in some groups for chewing purposes, in others as sucking organs. The eyes, though apparently only two in number, are really compound, each consisting of many single eyes closely united, and incapable of being moved in their sockets. Many winged insects have one, two, or three eyelets on the crown of the head. The legs are six in number, and are attached to the under side of the thorax, one pair to each of the three rings. The leg consists of the hip-joint, by which the leg is fastened to the body, the thigh, the shank, and the foot, — the latter consisting generally of five pieces, called tarsi, connected end to end, and armed at the extremity with one or two claws. The wings are two or four, or wanting. The piercer — more properly ovipositor — mentioned above, is in some cases a flexible or jointed tube, capable of being thrust out of the end of the body, and is used for conducting eggs into holes where they are to be deposited; in other cases it consists of a scabbard containing a central borer or saws, which are used in making holes in which eggs are to be placed. The sting — a modified ovipositor — consists of a sheath covering a sharp instrument for inflicting wounds, and connecting with it inside of the body is a sac of poison. Insects have three nervous centres, the largest in the head, a smaller in the thorax, and the smallest in the hind body. The breathing-holes, or stigmata, are generally nine in number on each side of the body. The heart consists of a long tube, lying just under the covering of the back, having small holes on each side for the admission of the blood, which is yellow or colorless, and which is prevented from escaping again by means of valves within. The heart is divided into several chambers by transverse partitions, in each of which there is a valve, which allows the blood to flow from the hinder part forwards, but not backwards. The blood does not circulate in arteries and veins, as in the higher animals, but is driven from the forward part of the heart into the head, whence it returns to the body and is mixed with the nutritive fluids that filter through the walls of the viscera, and, thus mingled and aerated by contact with the air-tubes, penetrates among the flesh and other internal parts, and nourishes the body and sustains life.

Insects are produced from eggs, and are never spontaneously generated. A very few insects do not lay their eggs, but retain them in the body till they are hatched, and thus such insects are ovoviviparous. Others always lay their eggs where the young, as soon as hatched, will find a plentiful supply of food. Most insects, in passing from the egg to the adult state, undergo great changes of form and habits. These changes are called transformations or metamorphoses, and are so great, that the same insect, at several different ages, may be mistaken for as many different animals. There are three more or less distinctly marked states in the life of an Insect, — the *larva*, the *pupa or chrysalis*, and the *imago or perfect state.* In the larva period, which is the one of

Fig. 249.

Larva.

Fig. 250.

Pupa or Chrysalis of Fig. 249.

Fig. 251.

Imago of Figs. 249, 250.

Fig. 252.

Larva.

Fig. 253.

Pupa of Fig. 252.

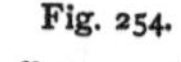

Fig. 254.

Imago of Figs. 252, 253.

infancy, and much the longest, Insects are always wingless, pass most of their time in eating, and grow rapidly, and the body, with some exceptions, is elongated, worm-like, and consists of fourteen segments, one of these being the head. For example, all caterpillars are butterflies and moths in the larva state, or state of infancy. When the larva has attained its full growth as a *larva*, it retires to some concealed spot, and, in many cases, spins a silken covering, called cocoon, casts its skin, and presents itself as a much shortened, oblong, oval, or conical body, and apparently lifeless ; in this form it is called a *pupa* or *chrysalis*, and the period during which the insect remains in this state is called the pupa or chrysalis period. At the end of this stage, which varies greatly in duration, the insect again sheds its skin, and comes forth fully grown, and, with few exceptions, provided with wings, and in this state is called a perfect insect, or *imago*. Thus, after insects enter upon the adult state, they no longer increase in size, but, having provided for a continuation of their kind, soon die. All insects which pass through the changes pointed out above are said to undergo a *complete transformation*. But there are some which, although differing greatly in the young from the adult state, do not pass through these changes ; but whose larvæ pass by more or less insensible gradations to the pupa state, and from the latter to perfect insects, all the while remaining active. All such are said to undergo only a *partial transformation*. For example, the grasshopper is hatched from the egg as a wingless insect. As it grows it casts its skin from time to time, becomes proportionally longer, and in due time wings begin to appear on the top of the back. It continues to eat voraciously, grows rapidly, hops without aid of wings,

Fig. 255.

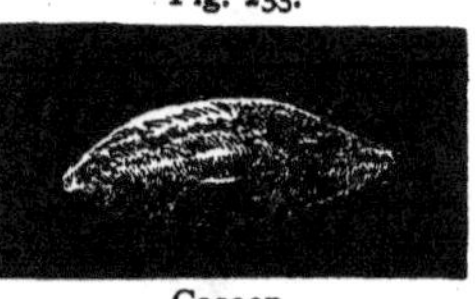

Cocoon.

repeatedly casts off its skin, and appears after each such process with longer wings and limbs more completely developed, until at length it ceases to grow, and, shedding its skin for the last time, it comes forth a fully grown insect, an adult grasshopper. The larvæ and pupæ of those insects which undergo only a partial transformation have six legs, the same number as adult insects. Of the larvæ that undergo a complete transformation, some have no legs, as maggots; others have six, a pair to each of the first three segments; others still, as caterpillars, have six true legs attached to the first three segments, and, besides these, several fleshy legs, sometimes numbering ten or more, placed beneath the abdominal segment, and known as prop-legs. The two sexes of insects differ in size, the female being larger than the male, and in many cases the former appears to have one ring less than the latter, since the terminal ring is obsolescent, or forms a small portion of the ovipositor.

Insects proper are divided into seven sub-orders, and, following the arrangement of A. S. Packard, Jr., in "Synthetic Types of Insects," these stand as follows:—

Hymenoptera, or Membranous-winged Insects, as Bees, Wasps, Ichneumons, Saw-Flies, Ants, and their allies;

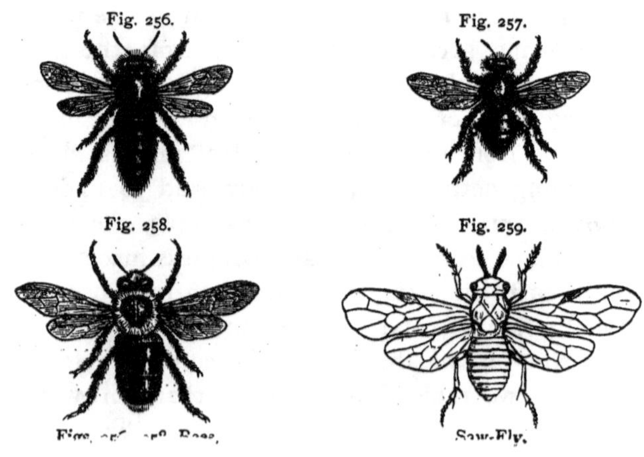

Fig. 256. Fig. 257. Fig. 258. Fig. 259.

Saw-Fly.

Lepidoptera, or Scaly-winged Insects, as Butterflies and Moths;

Fig. 260.

Butterfly.

Diptera, or Two-winged Insects, as Flies, Mosquitoes, and their allies;

Fig. 261. Fig. 262. Fig. 263.

Flies.

Coleoptera, or Sheath-winged Insects, as Beetles in their various forms;

Fig. 264. Fig. 265.

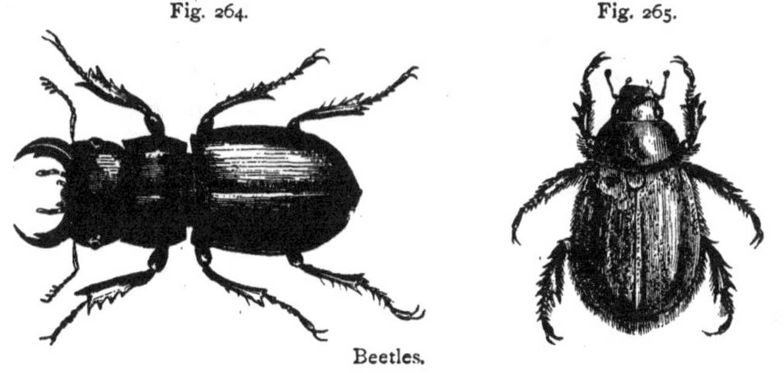

Beetles.

Hemiptera, or Bugs, Cicadas or Harvest-Flies, and the like;

Fig. 266. Fig. 267.

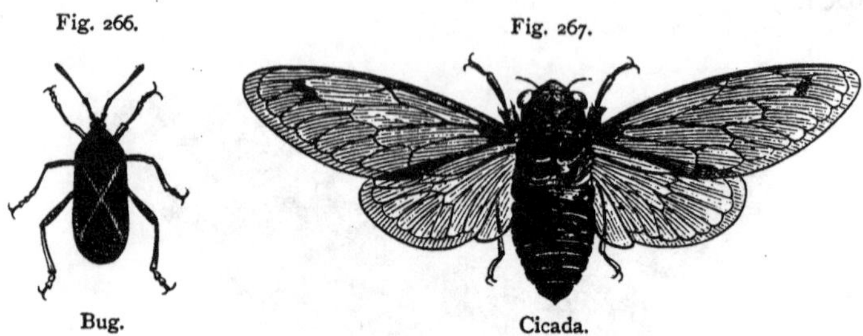

Bug. Cicada.

Orthoptera, or Straight-winged Insects, as Grasshoppers, Katydids, Cockroaches, Crickets, and their allies; and

Fig. 268.

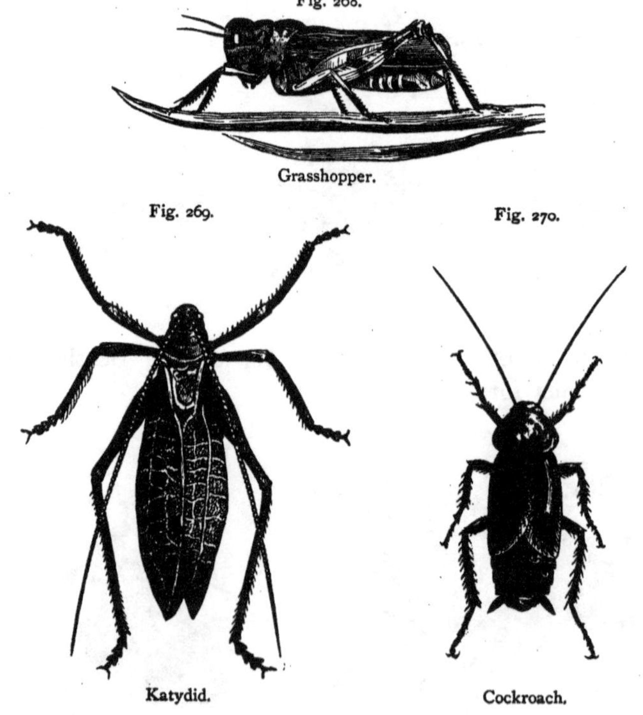

Grasshopper.

Fig. 269. Katydid.

Fig. 270. Cockroach.

Neuroptera, or Nerve-winged Insects, as Dragon-Flies and their allies.

Fig. 271.

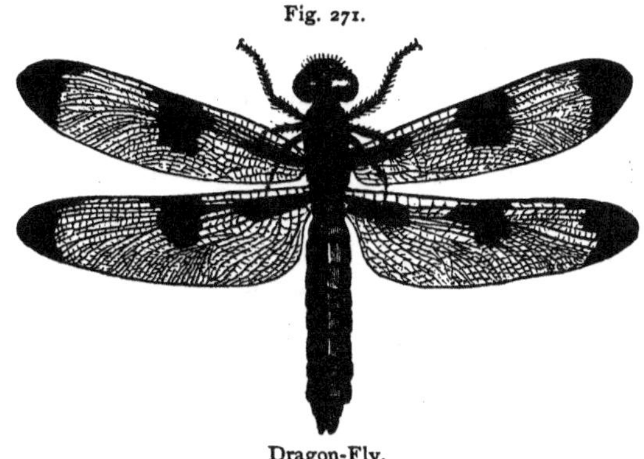

Dragon-Fly.

SUB-SECTION I.*

THE SUB-ORDER OF HYMENOPTERA, OR MEMBRANOUS-WINGED INSECTS.

THE Sub-Order of Hymenoptera comprises insects with four membranous and more or less transparent wings, the hind pair the smaller, and all traversed by a few irregularly branching veins. These insects have four jaws, the upper pair horny and fitted for biting or cutting, and the lower pair longer, softer, and with the lower lip adapted for collecting honey. The males have no weapons except their jaws; but the females are provided with either a piercer or a sting in the hind extremity of the body. They all undergo a complete transformation in coming to maturity; but the Piercers differ from the Stingers in their early stages. The young of the latter

* Contrary to the general plan of this book, a Sub-Section is devoted to each Sub-Order of Insects proper. This is done mainly for the sake of convenience, and to give greater prominence to these groups, which, until recently, have been treated as Orders.

are soft, maggot-like, and destitute of legs. Some of the piercers have this form, but others more nearly resemble grubs and caterpillars. Nearly all of the larvæ spin silken cocoons, in which they undergo their transformations. The Hymenoptera are all diurnal, fly swiftly, and in the number and variety of their instincts they surpass all other insects. The number of species is very great. None are aquatic. The families are arranged in this work according to the classification proposed by A. S. Packard, Jr.* It is the reverse, with several alterations, of the classification by Latreille.

APIARIÆ, *Latr.*, OR BEE FAMILY. — This Family comprises hymenoptera which have the body densely hirsute, the mouth-parts lengthened and partially united to form a sort of proboscis that can be folded up under the head, and the first joint of the two hind legs often very large, flattened, and fitted for collecting and carrying the pollen of flowers. With some exception they are eminently social in their habits, and the species often consist of *males or drones*, *females or queens*, and *imperfect females*

Fig. 272. Queen. Fig. 273. Drone. Fig. 274. Worker.
Hive-Bee, *A. mellifica*, Linn.

or workers. The last are smaller than the others, and are often improperly called *neuters*.

The Genus *Apis* contains the Hive-Bees. The Common Hive-Bee, *A. mellifica*, Linn., is known by every one. In a single community of this bee there are at least

* "How to Observe and Collect Insects." — Maine Scientific Survey Report for 1862.

two thousand males, fifty thousand workers, and only one queen. Of all insects no other has so excited the interest and admiration of mankind in every age. A volume might well be devoted to its intensely interesting and fascinating history, as traced by Réaumur, Huber, and others. The Hive-Bee is indigenous to the Eastern hemisphere.*

The Genus *Bombus* embraces the Humble-Bees, of which there are many species, over forty belonging to North America, and ten of these to New England, and which are at once known by their large and very hirsute bodies. They build nests in the ground or under loose stones, and their cells are large, oval, and partially separate. There are generally one hundred of these bees in a community, sometimes four hundred. A single female that has survived the winter, founds a colony in the spring. About the middle of May the workers begin to hatch. Late in the summer there is a brood of males and females.

The Genus *Xylocopha* contains the Carpenter-Bees, which are of large size, and which form a tube or burrow a foot or more in length, curved, and open at each end, in a wooden post or stump, and deposit therein their eggs, arranging them in successive layers in masses of pollen.

The Genus *Megachile* comprises the Leaf-Cutters, which cut circular pieces from leaves, and with these make a honey-tight cell, which they build in holes excavated in trees or decayed wood, or in the earth.

The Genus *Osmia* includes the Mason-Bees, which are bluish or green, and have a circular, much incurved abdomen. They make their nests with sand in crevices. *Andrena* resembles the Hive-Bee, but is smaller, and its members burrow in the ground. *Cœleoxys* has the abdomen triangular. Its species lay their eggs in the nests of other bees. *Nomada* is not hirsute, and its slender

* When not otherwise stated, the Insects described in this book belong to the United States.

form and gay colors remind us of the wasps. It enters the nests of Andrena, and feeds upon its food.

VESPARIÆ, *Latr.*, OR WASP FAMILY.—This Family comprises hymenoptera which fold their wings longitudinally.

The Genus *Vespa* contains those that live in colonies, composed of males, females, and workers. They construct complex nests, either under ground or attached to bushes, trees, fences, or buildings. These nests consist of several tiers of hexagonal cells, with their mouths downward, supported by pedicels, and all surrounded by a paper-like substance gnawed from wood or the bark of trees, and reduced to a paste by the action of the jaws. The cells in a single nest sometimes number sixteen thousand. Unlike the Hive-Bee, the communities of Wasps are dissolved annually on the approach of winter. The males die and the females disperse, seeking a sheltered winter retreat. Each female that survives the winter lays the foundations of a new colony in the spring, building a small nest in which she lays her eggs, from which hatch a community composed entirely of workers. These assist the parent; and at length, in autumn, three generations have been produced, the last composed of males and females, the community has become very large, and the few-celled nest has grown to one containing thousands of cells. The Hornet, *V. crabro*, Linn., was introduced into this country from Europe.

The Genus *Polistes* contains wasps which build an open nest of comparatively few cells, arranged in one tier, and attached by a short pedicel. The Genera *Odynerus* and *Eumenes* comprise the Solitary Wasps, which build nests of sand, and store them with other insects.

Fig. 275.

Wasp, *P. pallipes*, Lapel.

Crabronidæ, *Latr.*, or Wood-Wasp Family. — This Family comprises hymenoptera which have the head cuboidal, the thorax spherical, somewhat flattened, and a flattened abdomen, rarely pedicelled. They are often found resting on leaves in the sunshine.

The Genus *Crabro* — Wood-Wasps proper — bores into posts and stumps, where it makes its nest, and stores it with insects.

The Genus *Philanthus* burrows in sandy places, and stores its nests with hive-bees, a single individual of which, after being stung, is deposited with an egg.

Bembecidæ, *Latr.*, or Bembex Family. — This Family embraces hymenoptera which have the head large, body flattened, and the labrum large, long, and triangular. The Genus *Bembex* burrows in the sand, and stores its nest with diptera for the future larvæ.

Sphegidæ, *Latr.*, or Mud-Wasp Family. — This Family comprises hymenoptera which have long antennæ, pedicelled abdomen, and long hind legs. They are large, black and red, brown and red, or wholly blue or black. They are very active, and their sting powerful.

Scolietæ, *Latr.*, or Scolia Family. — This Family contains hymenoptera which have the body long, rather narrow, and hirsute, the abdomen sessile, with two prominent terminal spines in the males, and with short, spiny, fossorial legs. They are black, with bright yellow spots along the sides of the abdomen.

Formicariæ, *Latr.*, or Ant Family. — This Family comprises hymenoptera which have the head triangular, long geniculate antennæ, the pedicel which connects the abdomen with the thorax in the form of a knot or scale, and the legs slender. They live in communities, which are often large, and each species consists of males, females, and workers, the two former furnished with long and deciduous or loosely attached wings, and the last destitute of wings. The workers have the care of the nest or habitation, and of rearing the young. Some kinds of ants make their nest in the ground, others raise large ant-hills, and others live in stumps and trunks of trees. The workers go abroad in search of food, appear to communicate with each other, and to assist each other in their labors. They feed the larvæ, take them into the sunshine in fine weather, and back again on the approach of bad weather, or at night, and watch over them with great fidelity and earnestness. Most ant-hill communities are composed of individuals of the same species; but in some cases the workers procure auxiliaries by visiting the ant-hills of other species, and forcibly taking the larvæ and pupæ and bringing them back to the domicil of the invaders, where they are tended and reared by other workers of their own species which have either undergone metamorphosis there, or which have been stolen from their original home. The male ants have the body small, and the antennæ and legs long and slender; the females are much larger, with antennæ and legs shorter and thicker.

CHRYSIDIDÆ, *Latr.*, OR CHRYSIS FAMILY.—This Family comprises hymenoptera which are oblong and compact, the abdomen sessile, and with only three to five rings visible; the remaining ones being drawn within, forming a long and large ovipositor, which can be thrust out like the joints of a telescope. The abdomen beneath is concave, and can be flexed upon the breast so as to make the insect appear globular. The richness of the colors of these insects vies with that of the humming-birds, and they are often called Golden Wasps. They are constantly in motion, flying or running about on walls, fences, and sand-banks, in the hot sunshine; and their antennæ are in constant vibration. They lay their eggs in the nests of other hymenoptera, that their young may feed upon the larvæ of those upon whom they have intruded.

PROCTOTRUPIDÆ, *Latr.*—This Family comprises a vast number of minute hymenoptera with rather long, slender bodies, wings without nervures, and covered with minute hairs, and the antennæ often haired on the joints. Scarcely any of these insects exceed a quarter of an inch, and most are very minute, hardly distinguishable by the untrained eye. They are generally black, varied with brown. They prey upon other small insects by ovipositing in their eggs or in their larvæ.

The Genus *Platygaster* contains a species which lays its egg in those of the canker-worm moth. It is only one twenty-fifth of an inch long.

CHALCIDIDÆ, *Spinola*, OR CHALCIS FAMILY.—This Family also comprises a great number of parasitic hymenoptera of small size, but of brilliant colors. Their antennæ are geniculate, and wings often deficient in nervures. Some species prey upon plant-lice, others lay their eggs in the nests of bees and wasps, and others consume the larvæ of the Hessian Fly, and of those insects which produce galls. Some species are parasites on other parasites; as, for instance, Aphidius of the next family, which is parasite on the plant-lice, has itself parasites which are members of this family.

ICHNEUMONIDÆ, *Latr.*, OR ICHNEUMON FAMILY.—This Family, the most extensive in numbers of the hymenoptera, comprises insects with the body long and narrow, the antennæ long, ovipositor generally long, and protected by two threads or sheath-pieces of the same length, and the anterior pair of wings always exhibiting perfect cells upon their disk. The color is generally black, varied with red, yellow, or white. They attack almost all other in-

Fig. 276.

Ichneumon, *I. suturalis*, Say.

sects, depositing their eggs in the eggs, larvæ, or pupæ, upon which the young feed when hatched. Those genera which have the ovipositor short deposit their eggs in exposed larvæ, while those which are provided with longer ones penetrate into holes and under bark; and for this end some species have the ovipositors two or three times the length of the body. Ichneumons prey especially upon the Lepidoptera, and it is thus that those destructive insects are held somewhat in check. When the eggs are laid upon the surface of the larvæ, the parasites, as soon as hatched, eat their way into their victim. When deposited inside, the young ichneumons feed on the tissues of the body, gradually consuming its life, till the parasite goes into the pupa state, and the insect dies. There may be only one ichneumon thus feeding within, or many of them, which at length fill the inside of the body with little cocoons placed vertically next to one another. Some ichneumons do not destroy their victim in the larva state, but allow it to become a pupa in the body of which they undergo their transformations, and come forth perfect insects. Most of the insects of this family spin a silken cocoon; and these cocoons are sometimes found in a mass together enveloped in a general covering of glossy silk.

Fig. 277.

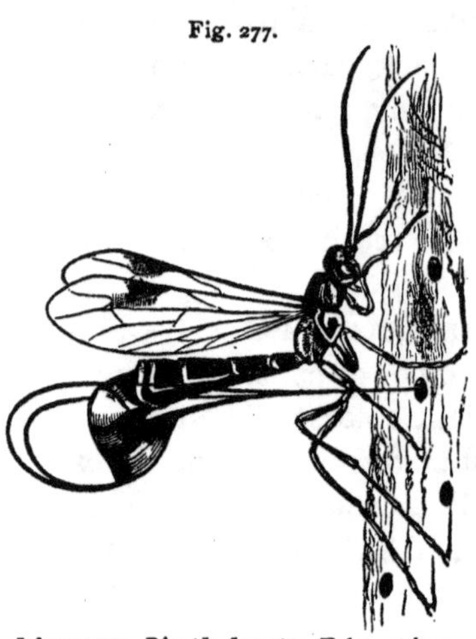

Ichneumon, *Pimpla lunator*, Fabr., ovipositing in holes bored by Tremex.

EVANIALES, *Latr.* — This Family contains parasitical insects which have the abdomen attached by a peduncle, nearer to the back than usual; the ovipositor straight and often exserted; hind legs longest; antennæ filiform or setaceous, and thirteen or fourteen jointed.

CYNIPSERA, *Latr.*, OR GALL-FLY FAMILY.—This Family comprises small hymenoptera which have the head short and broad, thorax thick and oval, abdomen much compressed and attached to the thorax by a very short peduncle, and the wings few veined. The females have a long, slender ovipositor, with which they insert their eggs into leaves and other parts of plants. These punctures cause excrescences called galls, the form and solidity of which vary according to the nature of the plant or parts of the plant that receive the wounds, and according to the species of gall-fly that make them. The eggs introduced into the punctures increase in size, and at length hatch, and the larvæ feed upon the vegetable matter in which they find themselves imbedded. With some exceptions, they undergo their transformations within the galls, and, gnawing through the shell, fly away. Some species gnaw through at the end of their larval life, and enter the ground to go into the pupa state. There are members of this family which produce no galls themselves, but are parasitic in galls produced by others; and they are called Guest Gall-Flies.

The Genus *Cynips* comprises those species which attack Oaks, and *Rhodites* those which are confined to the Rose. The nut-galls of commerce, used in making ink, in coloring, and in medicine, are caused by the punctures of gall-flies on a species of oak growing in Western Asia. The Rose-bush Gall-Fly, *R. dichlocerus*, Harr., about one eighth of an inch long, is bred in great numbers in the woody galls or long excrescences of the stems of rose-bushes.

Fig. 278.

Rose-bush Gall-Fly.

UROCERATA, *Latr.*, OR BORING SAW-FLY FAMILY. — This Family comprises comparatively rare hymenoptera which are among the largest of the sub-order. They have the body elongated, nearly cylindrical, and the blunt abdomen ending in a horny point. From beneath the abdomen projects a long, saw-like, and powerful borer, with which they bore holes into trees, in which to lay their eggs. The larvæ are borers in the trunks of trees.

The Genus *Tremex* contains *T. columba* of authors, which is an inch or more in length besides the borer, and that an inch long, projecting beyond the body. The head and thorax are rust-colored and black; the abdomen black, with seven ochre-yellow transverse bands. The Genus *Urocerus* comprises those which especially infest the pines.

Fig. 279.

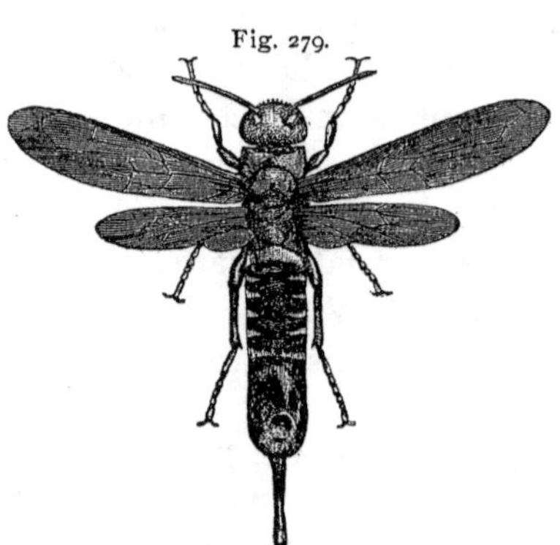

Saw-Borer Fly, *T. columba* of authors.

TENTHREDINETÆ, *Latr.*, OR SAW-FLY FAMILY. — This Family comprises hymenoptera in which the females have an ovipositor consisting of double saws lodged under the body, and covered by two pieces which serve as a sheath. They are sluggish in their habits, and fly only in the warmest days. Their larvæ have from eighteen to twenty-two legs, and are found in communities on the leaves of birch and alder, holding on by their true legs, while the rest of the body is curved curiously upwards; or appearing like slugs on the leaves of the pear and other fruit-trees, and those of the rose; while others feed upon the stems of plants, or roll up a leaf, or construct cases of particles of leaves to hide in.

The Genus *Selandria* embraces the Rose Saw-Fly and its allies. The Rose Saw-Fly, *S. rosæ*, Harr., is over three twentieths of an inch long, the color is deep shining black, the wings smoky, with a brown spot near the middle of the first pair. This species comes out of the ground from the last of May to the middle of June. The females fly little, remaining most of the day on the rose leaves, and, when touched, draw up their legs and fall. The female thrusts her saws obliquely into the skin of the leaf, and deposits in each incision a single egg, which hatches in about ten days. The larvæ are soft, head yellowish, with a black dot on each side of its body, green above and yellowish below. These gelatinous larvæ eat the upper surface of the leaf in large patches, leaving the veins and skin beneath untouched. They cast their skin several times, and after the last moulting are opaque yellowish. At length they go into the ground an inch or more, make an oval cell, where they go through their transformation, and come out of the ground early in August and lay eggs for a second brood, which in turn perform their work of destruction, go into the ground in autumn, and appear in the perfect form the following spring and summer.

The Vine Saw-Fly, *S. vitis*, Harr., is about a quarter of an inch long, jet black, except the upper side of the thorax, which is red, and the fore legs and under side of the others, which are light. This fly rises from the ground in spring, and lays its eggs on the lower side of the terminal leaves of the vine. In July the larvæ of various sizes are seen in swarms beneath the leaves, feeding side by side. They begin at the edges, eat the whole leaf, then take another, and thus proceed down the branch. When full grown they are about five eighths of an inch long, head and tip of the tail black, body light green above, with two transverse rows of minute black dots across each ring; under surface yellowish. After the last moulting they become yellow, burrow in the ground, form small oval cells lined with silk, and in about a fortnight come forth in the perfect form and lay their eggs for a second brood, which are not transformed to flies until the next spring.

The Genus *Lophyrus* contains the Fir-tree Saw-Fly, *L. abietis*, Harr., which feeds upon the fir-tree. The male is about one fourth of an inch, and the female about three tenths of an inch long.

Fig. 280.

Saw-Fly (enlarged), *L. abietis*, Harr.

The Genus *Cimbex* contains large Saw-Flies. The Elm Saw-Fly, *C. ulmi*, Peck, is about three fourths of an inch long, head and thorax shining black, abdomen steel-blue, with three or four oval yellowish spots on each side; legs blue black, feet yellow. The male has the body longer, narrower, and no spots on the sides, and appears so different from the female, that it has been described as distinct. These flies appear in the latter part of May and June, and the female lays her eggs on the elm, upon the leaves of which the larvæ feed. These come to their growth in August, and then are an inch and a half to two inches long, thick, and covered by a firm skin, with numerous transverse wrinkles; color pale greenish-yellow, with a black stripe of two black lines from the head to the tail, and black spiracles. When at rest, they lie upon the side, and look somewhat like a snail-shell. It makes an oblong cocoon, very closely woven, and tough, about an inch long, in which it remains unchanged, under leaves and rubbish, till the next spring.

SUB-SECTION II.

THE SUB-ORDER OF LEPIDOPTERA, OR BUTTERFLIES AND MOTHS.

THE Sub-Order of Lepidoptera comprises insects which have four wings covered on both sides with scales that are removed by the slightest touch; a tongue consisting of two tubular threads placed side by side and forming a channel by their junction, and thus adapted for suction, and, when not in use, rolled up like a watch-spring beneath the head, and more or less concealed on each side by a little palpus or feeler; six legs, the first pair being very short; and feet which are five-jointed and terminated by a pair of claws. They undergo a complete transformation. In the larva state they are called caterpillars, and have from ten to sixteen legs. The first three pairs of legs are covered with a shelly skin, are jointed and tapering, and the extremity is armed with a little claw. The remaining ones — prop-legs — are thick, fleshy, without joints, elastic, and at the extremity are surrounded by a ring of minute hooks. The jaws are placed at the sides of the mouth, move laterally, and are quite strong. The middle of the lower lip contains a conical tube, from which the larva spins a silken thread. The material from which the silk is made is contained in two long, slender bags in the interior of the body, and ending in the spinning-tube just mentioned. This material is a viscid fluid, which, as it flows from the tube and comes to the air, hardens. Some caterpillars make but little silk; others produce it in great abundance. Most of them feed upon the leaves of plants. Some eat buds, blossoms, seeds, and roots; others eat the solid wood; others devour fabrics and furs; and others leather, meat, lard, and wax. Some species are gregarious; others are solitary in their habits. In coming to their growth as larvæ, caterpillars usually change their skins four times. At length they cease eating, and most of them spin a silky covering for themselves, called a cocoon; some suspend themselves in various ways without making a cocoon; and others enter the ground. In all these situations the larva soon bursts open the skin upon the back, sloughs it, and thus passes into the pupa or chrysalis state, when, at first sight, it appears destitute of head and locomotive appendages; but closer examination reveals traces of head, tongue, antennæ, wings, and legs. In due time the pupa-skin is rent on the back, and the winged insect emerges, soft and weak, and the wings shrivelled; but soon the superfluous moisture of the body evaporates, the wings expand to their full dimensions, the limbs acquire firmness, and the perfect insect goes forth to feed upon water and the sweet fluids of flowers, which constitute its only food. Butterflies are at once distinguished from all the rest of the sub-order by their knobbed antennæ, erect wings when at rest, and diurnal habits. Their caterpillars do not spin cocoons, but the pupæ are bare. Moths have the antennæ variously formed, and are mainly nocturnal in their habits. The caterpillars of many of them spin cocoons.

The following families are arranged according to Latreille, with some modifications.

PAPILIONIDÆ, *Latr.*, OR PAPILIO FAMILY. — This Family embraces lepidoptera which are the largest of our butterflies, and which generally have their hind wings extended into a tail-like appendage.

The Genus *Papilio* contains the Papilios proper. The Asterias Butterfly, *P. asterias*, Drury, expands three and a half to four inches, and is black, with a double row of yellow dots upon the back, a broad band of yellow spots across the wings, a row of yellow spots near the hind margin, seven blue spots on the hind wings between the yellow band and outer row of yellow spots, and an eye-like spot of orange, with a black centre, near their hind angle. The female differs in having only few yellow spots

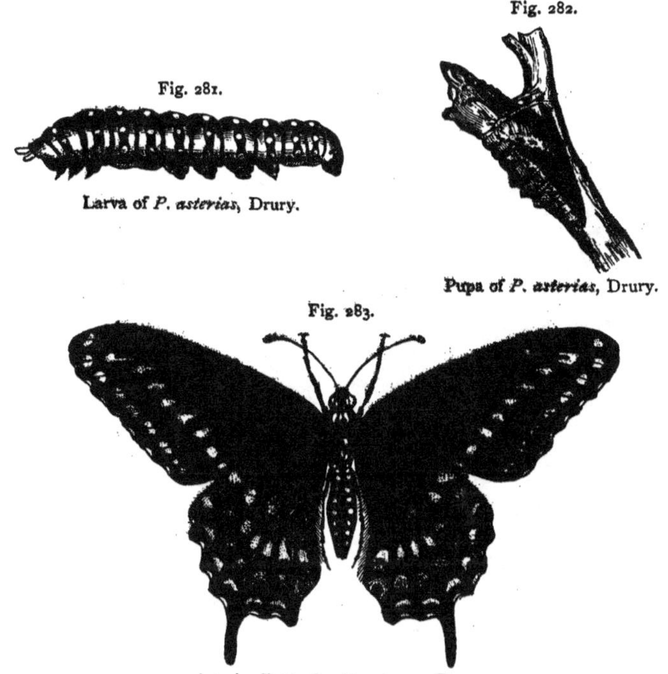

Fig. 281. Larva of *P. asterias*, Drury.

Fig. 282. Pupa of *P. asterias*, Drury.

Fig. 283. Asterias Butterfly, *P. asterias*, Drury.

on the upper surface. The caterpillar feeds upon parsley, parsnips, carrots, and celery, and, when touched, protrudes from a slit in the first segment a pair of soft, orange-colored, V-shaped organs, which diffuse a disagreeable odor.

The Turnus Butterfly, *P. turnus*, Linn., expands from four and a half to five inches, and is of a beautiful yellow color, marked with black, and the hind wings have an orange-red spot near their hind angle. The larva feeds upon the leaves of the apple and wild-cherry trees, folding them so as to make a sort of case for itself. It is about two inches long, green above, with rows of blue dots, a yellow eye-spot with a black centre on each side of the third segment, and a yellow and black band across the fourth segment. It becomes a chrysalis early in August, and comes out a butterfly the next summer.

Fig. 284.

Turnus Butterfly, *P. turnus*, Linn.

PIERIDÆ, *Boisd.*, WHITE, OR SULPHUR BUTTERFLY FAMILY. — This Family comprises butterflies which have the hind wings rounded, and also forming a gutter for the reception of the abdomen. Their prevailing colors are white, orange, and sulphur.

The Genus *Pieris* contains the White Butterfly, *P. oleracea,* Boisd., seen in May and June, and again in July and August. It expands about two inches.

The Genus *Colias* contains the Common Yellow Butterfly, *C. philodice,* Godart, seen in great numbers in fields and by road-sides. The caterpillars, found upon clover and allied plants, are green, slightly downy, and attain an inch and a half in length, then suspend themselves by a loop from the stem of a plant, and form a straw-colored chrysalis.

NYMPHALIDÆ, *Boisd.*, OR NYMPHALIS FAMILY. — This Family comprises butterflies which are remarkable for their beautiful colors and splendid ornamentation. The fore legs are rudimental.

The Genus *Limenitis* has the knob of the antennæ long, straight, and slender, the edges of the wings, especially the hind ones, scalloped, but not tailed. The Misippus Butterfly, *L. misippus*, Harr., expands from three

Fig. 285.

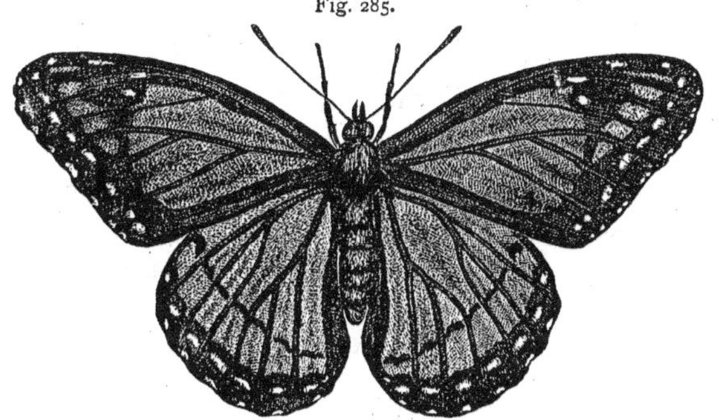

Butterfly, *L. misippus*, Godart.

to three and a half inches, and is tawny yellow above, paler below, the wings veined with black and surrounded by a broad black border, spotted with white, and there is a triangular patch spotted with white near the tips of the fore wings, and a curved black band on the hind ones. The caterpillar lives on the willow and poplar. The butterfly appears in June and September.

The Arthemis Butterfly, *L. arthemis*, Boisd. & LeC., expands about three inches, and is at once distinguished by the broad white, curved band, beginning just beyond the middle of the front edge of the fore wings, and crossing both wings. The male has a row of orange-colored spots on the hind wings, next to the border.

The Genus *Danais* has the knob of the antennæ long and curved. The Archippus Butterfly, *D. erippus*, Doubl., expands from three and three fourths to four and a half inches; the wings are tawny orange above, nankin-yellow beneath, veins black, and have a black border, spotted with white. The males have an elevated black spot near the middle of the hind wings. It flies in the latter part of summer. The caterpillar lives upon the silk-weed.*

The Genus *Argynnis* has the wings without indentations, and the hind ones are generally ornamented beneath with silvery or pearly spots. The Idalia Butterfly, *A. idalia*, Godt., expands about three and a half inches, and the fore wings are deep tawny-orange, spotted with black, and with a broad black hind border, around which, in the female, there is a row of white spots; hind wings bluish-black above, with two rows of spots behind, both of which are of cream color in the female; but in the males the spots of the outer row are deep tawny-orange. The fringes of all the wings are spotted with white; all have a row of pearly-white crescents beneath, and the under surface of the hind wings have each seventeen more white spots. It flies in July and August.

The Aphrodite Butterfly, *A. aphrodite*, Fabr., expands two and three fourths to three and a half inches, the

* When ready to go into the chrysalis state, the caterpillar of this species, and others of its family, spins a tuft of silk on the under surface of a piece of timber, board, branch, or leaf, and, fixing the hooks of its hind feet in the threads, suspends itself, and curves the forward part of the body upward. In a short time the skin is rent on the back, and the chrysalis appears through the fissure. The chrysalis now performs a feat which seems incredible until it is witnessed; for it has to release itself entirely from the caterpillar-skin, which has been worked back to its hind extremity, and to fasten itself to the silken tuft by its hooks at the tail. By means of two of the movable rings near the middle of the body, the chrysalis seizes a portion of the empty caterpillar-skin, and, thus supporting itself, withdraws wholly

wings tawny in the male, ochre-yellow in the female, both with a black line near the hind margins, within which is a row of black crescents, and within the latter a row of round black spots, the rest of the surface having irregular black spots. Under the tips of the fore wings are seven or eight silvery spots, and on the under surface of the hind wings are more than twenty large silvery-white spots. It flies in July and August.

The Bellona Butterfly, *A. bellona*, Godt., expands an inch and three quarters, and the wings are tawny, with two rows of black spots around the hind margins, and at a distance from these is a row of round black spots; basal half of the wing with blackish blotches.

The Myrina Butterfly, *A. myrina*, Godt., expands an inch and three quarters to nearly two inches, the wings tawny, bordered above with black, with a row of black crescents adjoining the border, and a row of round black spots at a distance from it; basal half of the wings with irregular black spots. It flies in May and June, and again in August and September.

The Genus *Melitæa* resembles Argynnis, but has the under surface of the wings marked with various colors, not silvery spots, and the caterpillars are covered with blunt tubercles, beset with very short, stiff bristles.

The Phaeton Butterfly, *M. phaeton*, Boisd. & LeC., expands two inches or more, and the wings are black, with

from the skin. The skin is now suspended to the tuft, and the chrysalis to the skin; but the chrysalis is some distance from the tuft of silk, to which it must climb in order to cling to it by the hooks of its tail. To accomplish this, it extends the rings of the body as much as possible, then, bending together two of them above those by which it is clinging to the dry skin, it catches hold of the skin higher up, at the same time letting go below; and by repeating this process with different rings in succession, at length it reaches the tuft of silk, and fixes its hooks therein. Both the chrysalis and the dry caterpillar-skin are now suspended from the tuft; but the former by its motion soon dislodges the latter, and is thus left suspended alone. The whole of this operation is performed in a very few minutes.

a row of orange crescents around the hind margin, within which are rows of cream-colored spots, and there are two orange-red spots on the fore wings. It flies in June.

The Tharos Butterfly, *M. tharos*, Boisd. & LeC., expands about an inch and a half, and the wings are tawny orange above, with a broad black hind border, bearing a row of tawny crescents, and before these a row of round black spots; the basal half of the wings is marked with black, running together like net-work; and on the fore wings is a large black spot. It flies most of the summer.

The Genus *Pyrameis, Cynthia* as restricted by Harris, has the wings more or less scalloped, but not indented or tailed. The larvæ are solitary.

The Thistle Butterfly, *P. cardui*, Doubl., expands two and a half inches or more, and the wings are tawny above, with a rosy tinge, and spotted with black and white; hind wings marbled beneath, and with five eye-like spots near the hind margin, and a triangular white spot in the middle. The caterpillars are found on thistles. It flies from May to the end of summer.

Hunter's Butterfly, *P. Huntera*, Doubl., expands about two and a half inches, and the wings are tawny above, variegated and spotted with black and white; hind wings marbled beneath, and with two large eye-like spots near the hind margin. It flies in the latter part of summer and in early autumn.

The Atlanta Butterfly, *P. atlanta*, Hübn., expands two and a quarter to three inches, and the wings are black above, with an orange-red band across the middle of the forward ones, and white spots near their tips; hind wings with a marginal red band bearing a row of black dots, the two nearest the hind angle with a blue centre. Harris says it was probably introduced from Europe. In the caterpillar state it feeds upon the nettle.

The Genus *Junonia* contains the Lavinia Butterfly,

J. cœnia, Hübn., which expands from two to two and a half inches; the wings are dark brown above, each with a large and a small eye-like spot on both sides; the fore wings have two orange-red spots near the middle of the front margin, and a whitish band enclosing the eye-like spots, and the hind wings a reddish band near their hind margin. It is common in the Southern States all summer; rarer northward.

The Genus *Vanessa* has the wings tailed or jagged on their hind edges. The caterpillars are armed with branching spines, and live in company.

The Antiopa Butterfly, *V. antiopa*, Ochs., expands about three and a half inches, and the wings are purplish brown above, with a broad buff-yellow margin, near the inner edge of which is a row of blue spots. This butterfly first appears in midsummer, and a second brood appears in autumn, and some of the latter may be found either flying or in sheltered places throughout the winter. The caterpillars are spiny, black, minutely dotted with white, with a row of eight dark brick-red spots on the back.

The White J-Butterfly, *V. j-album*, Boisd. & LeC., expands two and a half to three inches, and has on the under side of the hind wings a j-shaped, silvery white mark.

Milbert's Butterfly, *V. Milberti*, Godart, expands over two inches, and is black above, with a broad orange-red band near the hind margin of all the wings, behind which, on the hind wings, there is a row of blue crescents, and on the fore wings a white spot near the tips, and two orange-red spots near the middle of the front margin.

The Genus *Grapta* has the wings more incised than Vanessa. The Semicolon Butterfly, *G. interrogationis*, Doubl., expands two and a half inches or more, and is distinguished by the pale golden semicolon on the middle of the under surface of the hind wings.

The Comma Butterfly, *G. comma*, Doubl., expands over two inches, and is distinguished by the silvery-white comma on the under surface of the hind wings.

The Progne Butterfly, *G. c-argenteum*, Kirby, expands about two inches, and has an angular silvery mark on the middle of the under surface of the hind wings.

SATYRIDÆ, *Boisd.*, OR SATYRUS FAMILY.—This Family comprises butterflies which have the wings broad and more or less rounded. The larvæ are pale green.

The Genus *Satyrus*, *Hipparchia*, Fabr.,—Hipparchians,—has the wings of a most delicate brown color, with eye-like spots near the outer margins. The species expand two or three inches.

Fig. 286.

Mt. Butterfly, *C. semidea*, Ed.

The Genus *Chionobas* contains species which are restricted to Arctic and alpine regions.

The Mountain Butterfly, *C. semidea*, Edw., is found as yet only on Mount Washington, in New Hampshire.

LYCÆNIDÆ, *Leach*, OR AZURE-BUTTERFLY FAMILY.—This Family embraces very small and very beautiful butterflies, which, in the caterpillar state, much resemble wood-lice, and whose legs are so short that they seem to glide over surfaces, and whose chrysalids are short, thick, with the under side flat and the upper very convex.

The Genus *Chrysophanus* contains the Copper Butterflies. The American Copper Butterfly, *C. americana*, D'Urban, expands over one inch; the fore wings are coppery-red above, with about eight square black spots, and the hind margins bordered with dusky brown; hind wings with a few small black spots on the middle, and a coppery-red band on the hind margin. It flies all summer. The caterpillar is green, and lives upon sorrel.

The Genus *Lycæna* contains the Azure Butterflies, small and delicate species, which expand about an inch, and which are generally of some shade of blue or brown above, and grayish dotted with black below.

The Genus *Thecla* has generally two thread-like tails on each hind wing. In some cases the hind wings are merely notched. The species expand over an inch.

HESPERIDÆ, *Latr.*, OR SKIPPER FAMILY. — This Family comprises butterflies which have the body short and thick, head large, eyes prominent, antennæ short, with the knob curved like a hook or bent to one side, legs six, and the four hindmost shanks armed with two pairs of spurs. Skippers fly with a jerking motion, and hence their name. They are generally of a rich brown, marked with yellow spots. The species are quite numerous, and expand from one inch and a half to two inches and a half.

Fig. 287.

Skipper, *Hesperia*.

SPHINGIDÆ, *Latr.*, OR HAWK-MOTH FAMILY. — This Family comprises very large lepidoptera which have the antennæ thickened in the middle and more or less tapering at each end, and generally hooked at the tip. The wings are narrow in proportion to their length, and fastened together by a bunch of stiff hairs situated on the shoulder of each hind wing, which connect with a hook on the under side of each fore wing; and when at rest, the wings are more or less inclined like a roof. They fly with great power and rapidity, and, with some exceptions, frequent flowers in the morning and evening twilight, and are easily mistaken for humming-birds. The caterpillars are remarkable for their great size, general appearance, and curious attitudes. Supporting themselves on their hind legs, they elevate the forward part of the body, and remain for hours in this sphinx-like position, as in Fig. 288. They have sixteen legs, in pairs beneath the first, second,

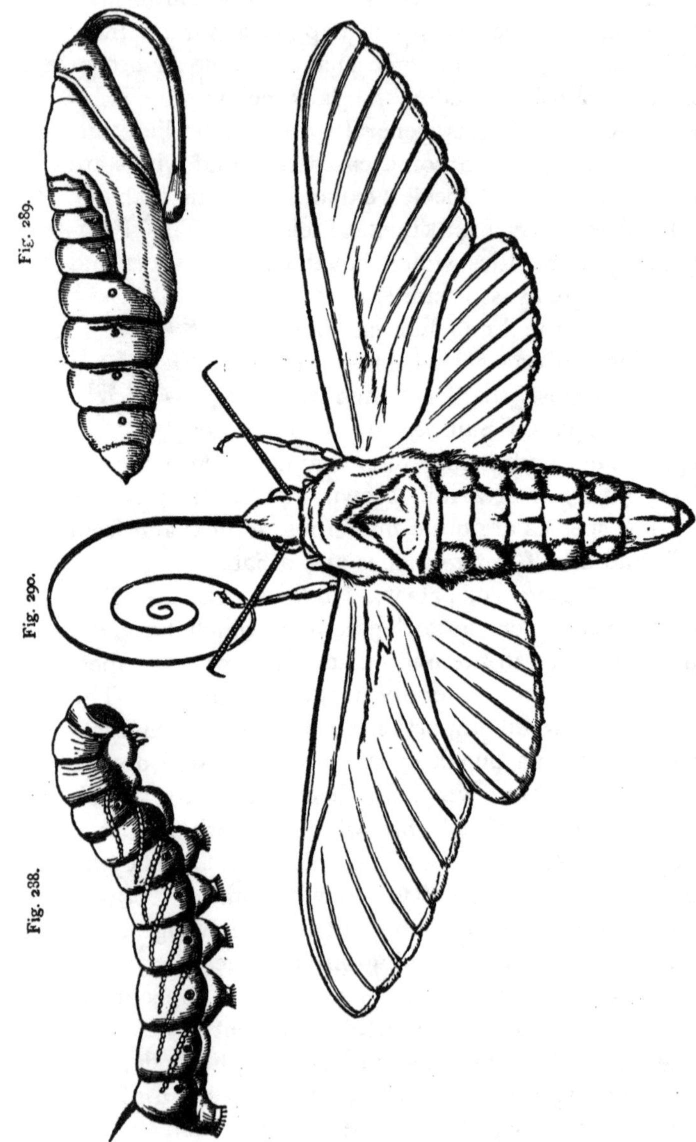

Five-spotted Sphinx, *S. quinquemaculatus* of authors. Fig. 288, Larva; Fig. 289, Pupa; Fig. 290, Imago.

third, sixth, seventh, eighth, ninth, and last segments, and, with some exceptions, they have either a horn or tubercle on the top of the last segment.

The Genus *Sphinx*, as formerly limited, contains the Sphinges proper, which by modern writers are referred to several different genera.

The Five-spotted Sphinx, *S. quinquemaculatus* of authors, expands about five inches, and is gray, variegated with blackish, and on each side of the body there are five round orange-colored spots encircled by black. Its tongue is five or six inches long when fully unrolled, but when not in use is coiled up nearly out of sight. The larva is green, feeds upon the potato-vine, and attains the length of three inches. At the close of summer it enters the ground, becomes a bright brown chrysalis, with a long, slender tongue-case bent over from the head to the breast, like a pitcher-handle. The Carolina Sphinx, *S. carolina*, Linn., closely resembles the preceding one.

The Genus *Ceratomia* has four short, fleshy horns on the thorax of the larva. The Four-horned Sphinx, *C. quadricornis*, Harr., expands about five inches, and the wings are light brown, variegated with dark brown and white; hind body marked with five longitudinal brown lines. The caterpillar feeds upon the elm, attains the length of three inches and a half, has four horns on the thorax, and a stiff spine at the hind extremity.

The Genus *Philampelus* embraces sphinges whose larvæ feed upon the grape-vine. The Satellitia Hawk-Moth, *P. satellitia*, Linn., expands four or five inches, and is light olive, variegated with patches of darker.

The Achemon, *P. achemon*, Drury, expands three or four inches, and is reddish ash, with two triangular patches of deep brown on the thorax, two square ones on each fore wing, and the hind wings pink, with a broad ashy border.

The Genus *Ellema* is smaller than the preceding. *E. Harrisii*, Clemens, flies the last of June.

The Genus *Smerinthus* has the wings scalloped or notched, and the tongue very short. The caterpillars are rough, with a stout thorn on the tail.

The Blind Smerinthus, *S. excæcata* of authors, expands about two and a half inches, is fawn-colored, clouded with brown, except the hind wings, which are rose in the middle, and have a black eye-spot with a blue centre.

Fig. 291.

Sesia, *S. thysbe*, Fabr.

The Genus *Sesia* contains the Clear-winged Sphinges, which are distinguished by their transparent wings and broad tails. They hover over flowers like humming-birds in the daytime, in July and August.

ÆGERIDÆ, *Steph.*—This Family comprises moths which have the body large, wings mainly transparent, and, in a state of repose, about half erect. The larvæ bore the stems of trees and shrubs.

The Genus *Trochilium* is hornet-like in appearance, with the antennæ thickened nearly to the end, which is curved but not hooked. The Peach-tree Borer, *T. exitiosa*, Say, expands about an inch and a half, the fore wings are blue and opaque, the hind ones transparent, and the middle of the abdomen has a broad orange-colored belt. The male has all the wings transparent, bordered and veined with steel-blue, which is the general color of the body in both sexes. The female deposits her eggs in summer on the

Fig. 292.

Peach-tree Borer, *T. exitiosa*, Say.

trunk of the tree, near the roots; when hatched, the borers penetrate and devour the inner bark and sap-wood. When about a year old they make their cocoon, become chrysalids, and from June to October come forth in the perfect state. It also attacks the cherry.

The Currant-bush Ægerian, *T. tipuliforme*, Harr., expands about three fourths of an inch, wings transparent, veined, fringed with black, and with a copper-tinged band across the tips of the first pair; body blue. The eggs are laid singly near the buds of the currant-bush, and, when hatched, the caterpillar penetrates to the pith, which it devours, forming a burrow several inches long, causing the bushes to sicken, and frequently to break off. The moth comes forth in early summer.

The Genus *Ægeria* contains the Squash-vine Ægeria, *Æ. curcurbitæ*, Harr., which expands an inch to an inch and a half, the body orange color, spotted with black, and hind legs fringed with long orange and black hairs; only the hind wings transparent. In the caterpillar state it devours the squash and other vines in the month of August, perforating the stem near the ground.

ZYGÆNIDÆ, *Latr.*—This Family comprises lepidoptera which have the body slender, wings rather narrow and covered with powdery scales. They are called Glaucopidians, from the glaucous appearance of some of them.

The Genus *Eudryas* comprises the Beautiful Wood-Nymphs. The Wood-Nymph, *E. grata*, Fabr., expands one inch and a half to one inch and three fourths, the fore wings pure white, with a broad stripe along the front edge for more than half its length, and a broad band around the outer hind margin, of a deep purple brown, the band edged on the inside with olive green, and marked towards the edge with a wavy white line; under side of the fore wings yellow, with a round and kidney-shaped black spot. The hind wings are yellow, with a broad purplish brown hind border above, on which there is a wavy white line; below they have a central black dot. The caterpillar, which infests the grape-vine, attains one inch and a quarter in length, is blue, transversely banded with deep orange, the bands dotted with black; the top of the eleventh ring is humped.

The Genus *Harrisina* contains *H. americana*, which expands about an inch, and is blue black, with a saffron-colored collar, and a notched tuft at the hind extremity. The caterpillar feeds upon the vine and woodbine.

BOMBYCIDÆ, *Latr.*, OR SILK-WORM FAMILY. — This Family comprises mainly thick-bodied moths which have the head small and sunken, antennæ generally feathered or pectinated, mouth-parts short, thorax woolly, and the fore legs very hairy. The caterpillars have sixteen legs, and, with few exceptions, spin cocoons. Some genera are small; others are the largest of all the Lepidoptera. The members of this family supply the world with silk.

The Genus *Lithosia* has the body slender, and the very narrow fore wings lie flat upon the top of the back.

The Striped Lithosian, *L. miniata*, Kirby, expands over an inch, and is deep scarlet, with three broad lead-colored stripes on the fore wings; hind pair with a broad lead-colored border behind; abdomen with broad scarlet stripe below.

The Genus *Deïopeia* contains the Beautiful Deïopeia, *D. bella*, Drury, which expands about an inch and three fourths, its fore wings deep yellow, crossed by about six white bands, each bearing a row of black dots; hind wings scarlet, with an irregular black border.

Fig. 293.

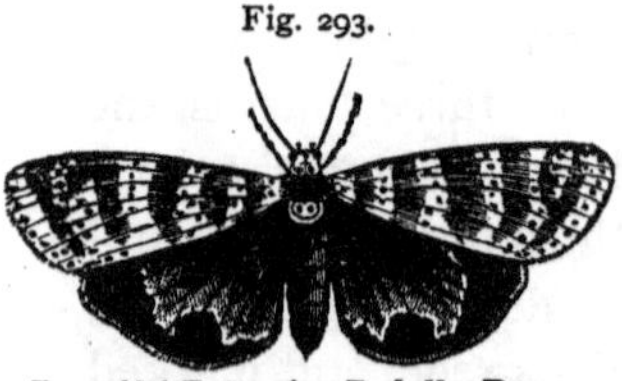

Beautiful Deïopeia, *D. bella*, Drury.

The Genus *Callimorpha* contains the Soldier-Moth *C.*

militaris, Harr., which expands about two inches, and the fore wings are white bordered with brown, and with an oblique band of the same color from the inner margin to the tip; hind wings white, without spots.

The Genus *Crocota* contains pale red species. It is like a geometrid moth in form, and may easily be mistaken for one.

The Genus *Arctia* has the body thick, and the larvæ have whorls of long hairs and are called Woolly Bears.

The Virgin Tiger-Moth, *A. virgo*, Sm. Abb., expands two and a half inches, and the fore wings are flesh-red fading to reddish buff, and marked with stripes and spots of black; hind wings vermilion blotched with black.

The Harnessed Moth, *A. phalerata*, Harr., expands from one inch and a half to one inch and three fourths, and the color is pale buff, the hind wings next to the body and the sides of the body reddish, and the fore wings with two longitudinal black stripes and four triangular black spots.

The Isabella Tiger-Moth, *A. isabella*, Hübner, expands two inches or more, and the color is tawny, with a few black spots on the wings, and a row of black dots on each side and above the body. The caterpillar is very thickly clothed with short, stiff, even hairs, which are black on the first four and last two segments, and tan-red on the intermediate ones. If taken up it immediately rolls itself into a ball. It remains torpid through the winter, makes an oval blackish cocoon in the spring, and comes out a moth in summer.

The Genus *Spilosoma* has the color white, gray, or yellow, with black dots or stripes.

The Virginia Ermine-Moth, or White Miller, *S. virginica*, Fabr., expands from an inch and a half to two inches, color white, with a black point on the fore wings and two black dots on the hind ones.

The Salt-Marsh Moth, *S. acræa*, Drury, expands about two inches, and the fore wings are generally white, hind wings and abdomen yellow, the wings with black spots, and the abdomen with a row of black spots above, two

Fig. 294.

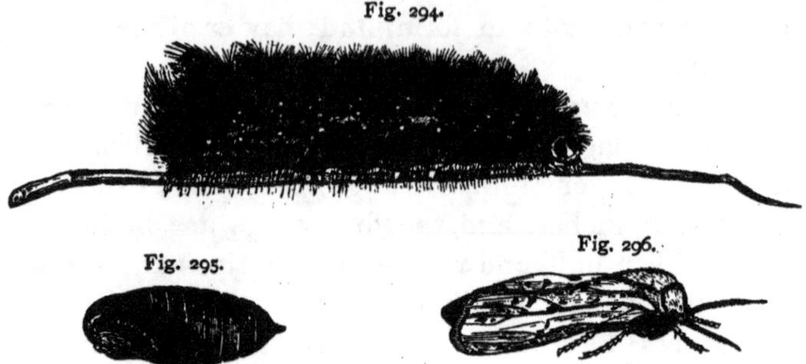

Fig. 295. Fig. 296.

Salt-Marsh Moth, Larva, Pupa, and Imago, *S. acræa*, Drury.

rows on the sides, and one row below. The female expands about two inches and a half, and differs from the male in having the hind wings white, or in having all the wings ashy gray, with the usual black spots. The caterpillars, produced from eggs laid on the grass of salt-marshes, attain the length of an inch and three quarters, and are clothed with brown hairs in spreading tufts.

The Weaver, *S. textor*, Harr., expands an inch and a quarter, and is white, without spots. The larvæ weave large webs over the branches of fruit and other trees in the fall, and devour the upper skin and pulpy part of the leaves.

The Genus *Halesidota*, or *Lophocampa*, Harr., comprises moths whose larvæ are short, thick, and have a crest of tufts along the back. It contains the Hickory Tussock-Moth, *H. caryæ*, Harr., which expands about two inches and a quarter; and the Checkered Tussock-Moth, *H. tessellaris*, Sm. Abb., which expands about two inches.

The Genus *Orgyia* embraces Tussock-Moths which in

the caterpillar state have long pencils of hair projecting before and behind the body.

The White-marked Orgyia, or Tussock-Moth, *O. leucostigma*, Smith, expands one inch and three eighths, and the wings are ashen gray crossed by wavy darker bands on the forward pair, on which there is also a black spot near the tip, and a minute white crescent near the outer hind angle; females lighter, and apparently wingless.

The Genus *Notodonta* contains moths which in the larva state are singularly humped, and have the last pair of prop-legs prolonged in many cases, and, when at rest, elevated over the back.

The Unicorn Moth, *N. unicornis*, Sm. Abb., expands an inch and a quarter to an inch and a half, and the fore wings are light brown, with patches of greenish white, and with wavy dark brown lines, two of which enclose a whitish space near the shoulders; hind wings of the male dingy white, with a dusky spot near the inner hind angle. The caterpillar, found in August and September on plum and apple trees, attains an inch in length, and is distinguished by the horn arising from the top of the fourth ring.

The Genus *Lagoa* has the body very stout and woolly, and the wings short, broad, and wrinkled transversely.

The Common Lagoa, *L. crispata*, Packard, lives in the raspberry, in the larva state.

The Genus *Limacodes* has the wings rather broad and deflexed, hind tibiæ with four spurs, and abdomen slightly tufted. The larvæ are slug-like, and the under side of the body is smeared with a sticky fluid, which leaves a slimy track wherever they go. Their cocoons are round and parchment-like, and fastened to the twigs of plants.

The Genus *Bombyx* contains the celebrated Silk-worm Moth, *B. mori*, Linn. The caterpillar, known as the Silkworm, feeds upon the leaves of the mulberry. It spins a cocoon containing about one thousand feet of silk. It is a native of China.

The Genus *Attacus*, as limited by Linnæus, is the prominent group of the family. It contains exceedingly large and magnificent species, with large, eyed wings, and antennæ broadly feathered on both sides in both sexes. The larvæ bear tubercles tipped with bristles. Recent writers refer the species to several genera.

A. atlas, of China, expands eight inches.

The Cecropia Moth, *A. cecropia*, Linn., expands from five inches and three fourths to six inches and a half, and the wings are grizzly dusky brown, with clay-colored hind margins, and near the middle of each wing there is an opaque kidney-shaped dull red spot, having a white centre and a narrow black edging, and beyond the spot a wavy dull red band, internally bordered with white; the fore wings are dull red next to the shoulders, with a curved white band, and near their tips is an eye-like black spot within a bluish-white crescent. This magnificent moth appears in June. The caterpillar attains the length of three inches or more, and is of a light green color, and has red and yellow warts armed with short bristles. The cocoon is three inches long, and fastened to the side of a twig. The outer coat is wrinkled, and resembles strong brown paper, and inside of this is loose yellowish-brown strong silk, surrounding an inner oval cocoon, which is composed of the same kind of silk very closely woven. The moth comes out of the small end, and the threads of silk so converge afterwards as almost to close the opening. It is the largest moth in North America. This species and the next are now referred to *Samia*.

Fig. 297.

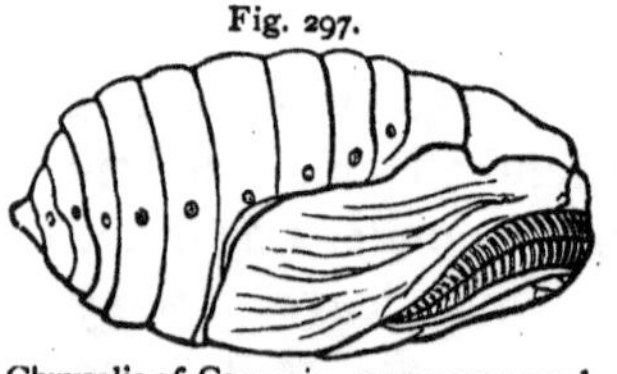

Chrysalis of Cecropia, cocoon removed.

The Promethea Moth, *A. promethea*, Drury, expands from three inches and three fourths to four inches and a quarter, and the male is deep smoky brown above, the female light reddish-brown; in both the wings are crossed by a wavy whitish line near the middle, and have a clay-colored border, which is marked by a wavy reddish line; and near the tips of the fore wings there is an eye-spot within a bluish-white crescent. This moth appears in July. The caterpillar feeds upon the sassafras-tree. Before making its cocoon, it fastens to the twig with silken threads the leaf that is to cover its cocoon, so that it shall not fall in autumn, and then spins its cocoon on the leaf, bending over the edges of the latter to cover it.

The Luna Moth, *A. luna*, Fabr., expands from four inches and three quarters to five inches and a half, and the wings are of a delicate light green, the hinder angle of the posterior ones prolonged into a tail of an inch and a half or more in length, and each wing with an eye-like spot, which is transparent in the centre and encircled by rings of white, red, yellow, and black. The caterpillar lives on the walnut and hickory, attains its growth, two or three inches in length, in July and August, is pale bluish-green, with a yellow stripe on each side of the body, and the back is crossed between the rings by the same color. On each ring are about six pearly-colored warts tinged with purple, and at the extremity of the body are three brown spots edged above with yellow. It draws together two or three leaves, and spins its cocoon inside of them; the cocoons fall with the leaves in autumn, and the next June the beautiful Luna appears. This species is now referred to *Tropæa.*

The Polyphemus Moth, *A. polyphemus*, Fabr., expands from five and a quarter to six inches; wings ochre-yellow, and on each there is a transparent eye-spot, divided by a slender line, and encircled by yellow and black rings. Before, and adjoining the eye-spots of the hind wings, there is a large blue spot shading into black. It appears in June. It is now referred to *Telea.*

The Genus *Saturnia* — Saturnians — has the antennæ widely feathered only in the males, and the larva has small warts crowned with long prickles or branching spines, and these prickles sting severely.

The Io, *S. Io*, Sm. Abb., expands from two inches and three fourths to three inches and a half, and the male is deep yellow, the fore wings marked and spotted with purplish-red, and the hind wings bordered next the body and banded near the hind margin with the same color, and within this band there is a curved black line; middle of the wing with a large, round, blue spot with a broad black border and a central white mark; fore wings of the female purplish-brown mingled with gray, with a brown spot surrounded by an irregular gray line. It is now referred to *Hyperchiria.*

The Proserpina, *S. maia*, Drury, expands from two inches and a half to three inches, and the wings are thin, black, and both pairs crossed by a broad yellow-white band, near the middle of which, on each wing, there is a kidney-shaped spot of black with a whitish crescent.

The Genus *Ceratocampa* embraces moths which in the caterpillar state are armed with horns, and which eat the leaves of forest-trees, and enter the ground to undergo their transformations. The species are of gigantic size.

The Regal Walnut-Moth, *C. regalis*, Fabr., expands from five to six inches, the fore wings olive-colored, with several yellow spots, and veined with broad red lines; hind wings orange red, with two large irregular yellow patches before, and a row of wedge-shaped, olive-colored spots between the veins behind. The larva feeds upon the walnut-tree, and attains four or five inches in length. It is now referred to *Citheronia.* It flies in July and August.

The Genus *Dryocampa* has the antennæ deeply pectinated to much beyond half the length, and thence minutely serrated to the tips in the male, but simple in the female. The larvæ enter the ground to go through their transformations.

The Imperial moth, *D. imperialis*, Harr., *Eacles imperialis*, Hübner, expands from four and a half to five inches, and the wings are yellow, thickly sprinkled with purplish-brown dots; a large patch at the base, a spot near the middle, and a band towards the hind margin of each wing, light purplish-brown. The males have additional purple, covering much of the outer hind margin of the fore wings. It flies in June and July.

The Senatorial Dryocampa, *D. senatoria*, Fabr., expands from an inch and three quarters to two inches and a half, and is ochre-yellow, the wings tinged with purplish-red, and crossed by a narrow purplish-brown band. The fore wings are dotted with blackish, and have a small round white spot near the middle.

The Genus *Hepialus* has the antennæ very short, slender, almost threadlike.

The Silver-spotted Hepialus, *H. argenteo-maculatus*, Harr., expands two inches and three quarters, and the fore wings have a triangular spot and dot of silvery white near the base.

The Genus *Cossus* has the wings long, thickly veined, and the antennæ with a double row of short teeth along the under side.

The Locust-tree Carpenter-Moth, *C. robiniæ*, Peck, expands about three inches, the color gray, the fore wings thickly covered with dusky netted lines and irregular spots, and the hind wings more uniformly dusky. The male is much smaller and darker, and has a large ochre-yellow spot near the hind margin of the hind wings. The caterpillar bores the locust-trees and the red oak.

The Genus *Psyche* comprises moths which in the larva state live in cases open at both ends. They are called Sack-bearers. *Perophera* is a closely allied genus, and contains Melsheimer's Sack-bearer, *P. Melsheimerii*, Harr., a very interesting species. Its case, as seen by Harris, consists of two oblong oval pieces of leaf fastened together by their edges.

The Genus *Clisiocampa* has a stout woolly body, short, stoutly pectinated antennæ, and short, broad wings.

The American Tent-Caterpillar Moth, *C. americana*, Harr., expands from an inch and a quarter to an inch and a half, and is reddish-brown, the fore wings crossed by two oblique, dingy-white lines. The caterpillars of this species abound in neglected orchards and nurseries, and upon wild cherry-trees, and are familiar to every one. The eggs from which they hatch are situated in a cluster nearly surrounding the small branches towards their extremity, and are covered with a sort of water-proof varnish. They hatch at the time of the unfolding of the leaves of the cherry and apple tree. The little caterpillars soon form a small tent between the forks of the branches, a little below the position of the eggs. Here they remain when not engaged in eating. In crawling from one twig to another, they spin a fine silken thread, which serves to conduct them back to their tent. As they grow larger, they enlarge the tent, surrounding it from time to time with new layers. They feed at stated times, and at once return to their tents when they have finished eating. They always rest in their webs at noon. In stormy weather they do not come out at all. At maturity, which occurs from the first to the middle of June, they begin to leave the trees, separate, wander about for a time, and at

Fig. 298.

Tent-Caterpillar Moth, *C. americana*, Harr.

length, in some sheltered place, spin their cocoons. These are regular, oval, and loosely woven, and the meshes are filled with a thin paste, which on drying becomes a yellow powder. After remaining in the chrysalis state from fourteen to seventeen days, they come forth in the winged form.

Fig. 299.

Cocoon, *C. americana*, Harr.

NOCTUÆLITÆ, *Latr.*, OR OWLET-MOTH FAMILY. — This Family comprises thick-bodied, swift-flying moths which have the antennæ long and tapering, rarely pectinated, the thorax thick and often crested, and each of the fore wings marked behind the middle of the front edge with two spots, one round and small, the other larger and kidney-shaped. They are exceedingly alike in general appearance, and are mostly some shade of gray or brown. A few fly by day, but most of them only at night. They are greatly attracted by light, and thus enter houses in great numbers on summer evenings. The larvæ taper towards each end, and make thin, earthen cocoons.

The Genus *Leucania* contains the Army-worm Moth and its allies. The species are yellowish-white, and the larvæ naked.

The Genus *Agrotis* — Dart-Moths — has the antennæ in the males generally pectinated on the under side, wings nearly horizontal when closed, and the fore legs often spiny. The larvæ are well known as Cut-Worms, which do great damage in the fields and gardens, by cutting off the leaves of plants, or by cutting down the tender plants close to the ground. They are very destructive in early and middle summer, after which they go into the chrysalis state in the ground, and later come forth as moths.

The Genus *Mamestra* has the wings rather broad, thorax slightly crested, and a W- or M-shaped character near the outer hind margin of the fore wings; larvæ brilliant.

The Genus *Gortyna* contains the Spindle-worm Moth and its allies, which in the larva state live in the roots and stems of plants.

The Spindle-worm Moth, or Corn Gortyna, *G. zeæ*, Harr., expands about an inch and a half, and the fore wings are rust-red mottled with gray, with a tawny patch near the tip, and a few black dots on the veins; hind wings yellowish-gray, with two indistinct dusky bands and a dusky spot. The caterpillar, known as the Spindle-Worm, attains one inch in length, is smooth, naked, yellowish, the head and top of the first and last rings black, and a darker row of shining black dots across each of the others. It bores into the stems of Indian corn.

PHALÆNIDÆ, *Latr.*, OR GEOMETRID FAMILY. — This Family comprises moths whose larvæ seem to measure the surfaces over which they pass. The necessity of this sort of movement results from the fact that they have only ten legs; six true ones under the fore part of the body, and four prop-legs at the hind extremity; three intermediate pairs of prop-legs being wanting. Some, however, have twelve or fourteen legs, but in such cases the additional prop-legs are too short to assist much in creeping, so that these also creep like those above described. Geometrids live upon trees, and in most cases undergo their transformation upon or in the ground, which they reach by letting themselves down by a silken thread, which they spin from their mouth while descending. They are generally smooth, and when at rest, many of them stand on the two hind pairs of legs, with the body extended, and thus may be easily mistaken for a little twig. Often, when disturbed, they let themselves down, and, when no danger is apprehended, return to the tree again by the same thread by which they descended. In the perfect state these insects are mainly slender-bodied

Fig. 300.

Geometer, or Span-worm.

moths, with tapering antennæ and large and delicate wings. The females in some cases are wingless.

The Genus *Geometra* contains the Chain-dotted Geometer, *G. catenaria*, Harr., represented in the larva state by Fig. 300. Its cocoon is made with meshes, through which the insect may be seen. It is now *Bupalus*.

The Genus *Anisopteryx* has the wings in the two sexes very unequal, or the females are wingless.

The Canker-worm Moth, *A. vernata*, Peck, expands about an inch and a quarter, and the wings are large, thin, and silky. The females are wingless. The larvæ, called Canker-worms, make their appearance upon the trees about the time the leaves of the apple-tree begin to start from the bud. They hatch from clusters of eggs which have been placed upon the branches at various times in and since the preceding autumn. They immediately commence their depredations. The leaves are found at first to be pierced with small holes, but as the caterpillars grow they enlarge these holes, and at length little more is left than the midrib and veins. The Canker-worms vary greatly in color at different ages, and individuals of the same age differ in this respect. When not eating, they are stretched out at full length beneath the leaves. When about four weeks old they reach their full size, and are then about an inch long. They now quit eating, descend to the ground, and, entering to the depth of two to six inches, each makes a little cavity by repeated turnings, and soon passes into the chrysalis state. They remain in this state till after the first frosts of autumn, when they begin to come forth in the moth state, and continue to do so, whenever the weather is mild enough, throughout the remainder of the autumn and the winter. They rise in the greatest numbers, however, in the spring. They come out of the ground mainly in the night, and often make their appearance in immense numbers. The females crawl up the nearest trees, where they are joined by the males, and soon begin to lay their eggs, which they place in rows, forming separate clusters of sixty to a hundred or more, each cluster being the product of a single female. Canker-worms are among the most destructive of all the insects, and it is not till they have nearly ruined the foliage of a tree that we are fully aware of their presence.

The Genus *Hibernia* contains the Lime-tree Winter-Moth, *H. tiliaria*, Harr., which expands an inch and three fourths, and the fore wings are rusty buff, with fine brownish dots and two transverse wavy brown lines; hind wings paler, with a brownish dot in the middle.

PYRALIDÆ, OR DELTA-MOTH FAMILY. — This Family comprises those called Delta-Moths because of their triangular form when the wings are closed.

The Genus *Hypena* contains the Hop-vine Moths, which have their feelers long, wide, and held close together and projecting like a snout, and the antennæ bristle-like. The caterpillars are false loopers, bending up the back a little when they creep. They are about eight tenths of an inch long when fully grown, and green. When disturbed they bend their bodies with a jerk, first on one side, then on the other, each time throwing themselves to a considerable distance. They eat large holes in the hop-leaves.

The Genus *Aglossa* contains the Grease-Moth, *A. pinguinalis*, Harr., which has the wings narrow, glossy, smoky gray, and crossed by wavy lighter-colored bands. The larva lives in fatty substances.

The Genus *Pyralis* contains the Meal-Moth, *P. farinalis*, Harr., which expands about one inch, the fore wings light brown, crossed by two curved white lines, and there is a dark chocolate-brown spot on the base and tip. The caterpillar is found in old flour-barrels.

The Genus *Galleria* contains the Bee-Moth, *G. cereana*,

Fabr., which expands from one inch to one inch and a half, and the fore wings slope steeply and turn up at the end; the male is grayish, the fore wings scalloped on the hind margin, glossed and streaked with purplish brown on the outer edge, and marked with a few dark spots near the inner margin; hind wings light yellowish gray with light fringes. The female has the fore wings longer in proportion, less deeply scalloped, and the color darker. By day bee-moths remain quiet about the bee-houses, but at night they hover around the hives, into which they enter and lay their eggs; or, not succeeding in this, they deposit their eggs upon the outside. There are two broods in a year. The larvæ feed upon wax; they enter the hive as soon as they are hatched, and work their way in all directions through the waxen cells, and thus destroy them. During the day they remain concealed in silken tubes, which they begin to make for themselves as soon as hatched. They enlarge these tubes as they increase in size, and cover them with a coating of wax as a defence against the stings of the bees, and thus they are able to go on with their work of destruction with impunity.

The Genus *Crambus* has the feelers long, wings oblong, white and buff-yellow, sometimes ornamented with golden spots. The species fly in the grass in great numbers, and rest on the stems with the head downwards.

TORTRICIDÆ, *Leach*, OR LEAF-ROLLING FAMILY. — This Family comprises moths which in the larva state roll the edges of leaves of plants into cylindrical rolls open at each end. The moths are mostly small, few expanding more than one inch, the antennæ naked, fore wings generally ornamented with spots and bands, hind wings without ornaments, and the inner edges folded against the side of the body.

Fig. 301.

Leaf-roller, *Tortrix*.

The Genus *Penthina* contains the Apple-worm Moth,

P. pomonella, Harr., which expands three fourths of an inch, and the fore wings are crossed by gray and brown lines, and near the hind angle is a large dark spot with bright copper-colored edges; hind wings and abdomen light yellowish-brown with a satin lustre. The larva is the well-known apple-worm, which every child has seen. During the latter part of June and in July the moths fly about apple-trees every evening and lay their eggs in the cavity at the blossom end of the little apples. The eggs hatch in a few days, and the little caterpillars immediately burrow into the fruit. There is generally but one to each apple, and so small at first, that it is detected only by the reddish powder it throws out in eating its way through the calyx. In the course of about three weeks it reaches its full size, and meanwhile has burrowed to the core, and through the apple in various directions. It bores a hole through the side of the apple, out of which it thrusts its chips, and through which the insect ultimately escapes. Soon after the half-grown apples fall, the caterpillars leave them, crawl into crevices, and each spins a delicate white cocoon, in which, in most cases, it remains till the next summer, when it comes forth in the perfect form.

TINEIDÆ, *Leach*, OR TINEA FAMILY. — This Family comprises moths which in the larva state gnaw winding paths in the substances upon which they subsist. They devour some of the fragments, and fasten together others with silken threads, thus making a covering for their tender bodies. Some thus make cylindrical burrows; others make cases, which they bear about with them. They are the smallest of the Lepidoptera, and are generally very beautiful.

The Genus *Anacampsis* contains the Angoumois Grain-Moth, *A. cerealella*, Olivier, which expands half an inch, and is pale cinnamon-brown above, with a satin lustre;

hind wings lead-color; the antennæ are thread-like, and consist of numerous beaded joints, and two tapering feelers are turned over the head. It lays from sixty to ninety eggs in clusters of about twenty on a single kernel of grain. In four to six days these eggs produce little worm-like caterpillars not thicker than a hair. Each burrows in a single kernel, and devours the mealy substance, and the work of destruction goes on so unseen, that it is only detected by the softness of the grain or the loss of its weight. When fully grown the Angoumois caterpillar is not more than one fifth of an inch long, white, with brownish head, six small jointed legs, and ten extremely small prop-legs. It goes into the chrysalis state within the kernel. Before this it has gnawed a hole nearly through by which to escape when it has finished its transformations. The insects of the summer brood come to the full larva growth in about three weeks, remain for a time in the chrysalis state, and in autumn they appear in the winged form, and may be found in the evening in great numbers laying their eggs upon stored grain. The moth-worms of the second brood remain in the grain through the winter, change into winged moths in the summer, and lay their eggs on the ears of the growing grain.

The Genus *Tinea* — Tineans — contains a large number of very small moths, found in houses, stores, granaries, and mills, and which in the larva state are very destructive, devouring almost all kinds of substances. The winged moths enter through the cracks into closets, drawers, chests, or get under the edges of carpets, or into the folds of curtains and garments, and deposit their eggs, which hatch into caterpillars in about fifteen days, and immediately begin to gnaw the substances within reach, and cover themselves with the frag-

Fig. 302.

Tinean.

ments, shaping them into hollow rolls and lining them with silk. They generally live in these rolls through the summer, enlarging them as they grow, and carrying on their work of destruction, but in the autumn become torpid, change to chrysalids in the spring, and in twenty days come forth winged moths, which in turn lay their eggs for a new brood.

PTEROPHORII, *Latr.*, OR FEATHER-WINGED MOTH FAMILY. — This Family comprises moths which have the wings divided lengthwise into narrow-fringed branches, resembling feathers, and the body and legs very long and slender.

SUB-SECTION III.

THE SUB-ORDER OF DIPTERA, OR TWO-WINGED INSECTS.

THE Sub-Order of Diptera comprises insects which have only two wings and two knobbed threads, called balancers, in the place of the hind wings, and a mouth formed for sucking or lapping. The sucker or proboscis is composed of from two to six bristles, in some cases as sharp as needles, and either enclosed in the upper groove of a proboscis-like sheath, terminated by two lips, or covered by one or two laminæ, which constitute a sheath for it. They undergo a complete transformation in coming to maturity; the larvæ are without feet, and are called maggots, and have their breathing openings generally at the hind extremity. The pupæ or chrysalids are in most cases enclosed in the dried skin of the larvæ, though sometimes naked. Diptera are exceedingly numerous both as regards species and the swarms of individuals of each species. They are mainly small, and many are very minute. In the classification we follow Latreille, with modifications by Westwood, Loew, and others, aided by Loew's Monograph of North American Diptera.

Culicidæ, *Latr.*, or Gnat Family. — This Family comprises diptera which have the body and legs much elongated and very delicate, antennæ densely pilose, and which have a long piercing sucker consisting of five bristles. Such are the Gnats and Mosquitoes which abound in all parts of the world, and which are represented by a large number of species. The female lays her eggs on the surface of the water, and the larvæ may be seen in all stagnant pools throughout the warm and hot months, for there are several generations in a season. The larvæ rest with the head downwards, and the hind extremity, which contains the respiratory openings, is at the surface of the water. They are very lively, and swim or wriggle with great agility, dive from time to time, but soon come again to the surface in order to breathe. At length they cast their skins and enter the pupa state, in which they still live at the surface of the water, and move by means of the tail; but they now take a different position, and breathe through two tubes placed on the thorax. After being in the pupa state a few days, the skin splits on the back between the breathing-tubes, the winged insect appears, and, after resting awhile on the empty skin as it floats upon the surface of the water, spreads its wings, and, humming its peculiar note, flies away in search of a victim whom it may pierce for blood, which is its choicest food. These insects discharge a poisonous fluid into the wounds which they inflict, and this is the cause of the irritation and swelling which results from their attacks.

Tipulariæ, *Latr.*, or Crane-Fly Family. — This Family embraces diptera which have the body slender, legs very long, and long antennæ and palpi. The Genus *Cecidomyia* has the antennæ verticillate.

The Hessian Fly, *C. destructor*, Say, is one tenth of an inch long, and expands about one fourth of an inch; head, antennæ, and thorax black; wings blackish, except the tawny base, and fringed with short hairs. The hind body is tawny, with black on each ring, and clothed with fine grayish hairs; legs brownish and feet black. Two broods of this fly appear in a year, — one in spring and one in autumn. The females lay their eggs on the young blades of wheat, both in spring and fall. The eggs are about one fiftieth of an inch in length, four thousandths of an inch in diameter, cylindrical, translucent, and pale red, and hatch in about four days, producing pale red maggots. The larvæ immediately crawl down the leaf, and work their way between the latter and the main stalk, till they come to a joint. Here they rest a little below the surface of the ground, with the head downwards, and do not move from the place till they have undergone their transformations. They injure the plant, not by eating it or boring into it, but by sucking its sap; and when several are fixed upon one stem, the plant becomes exhausted, and withers. The larvæ come to their full size in five or six weeks, and are then three twentieths of an inch long, and covered with a hardening brown or chestnut-colored skin, and the insect is then said to be in the flax-seed state, so called from its resemblance to a flax-seed. In April and May they complete their transformations, and come forth in the winged state, and soon begin to lay their eggs upon the spring wheat, and upon that sown the autumn before. The maggots hatched from these eggs pass down the stem as before stated, take the flax-seed form in June or July, and in autumn most of them are transformed into winged insects; but others remain through the winter, and are transformed in the spring, as observed above. These

Fig. 303.

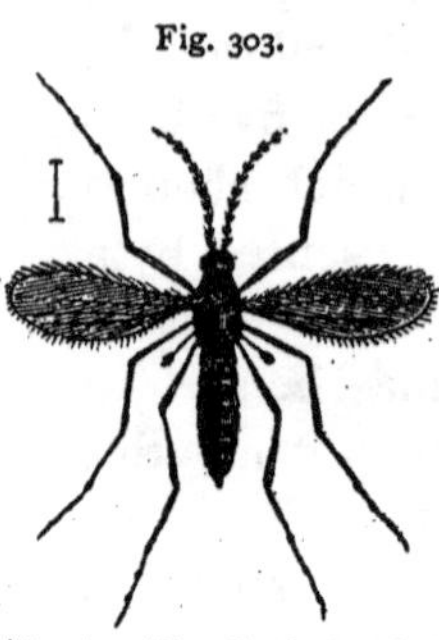

Hessian Fly, *C. destructor*, Say.

flies sometimes move in immense swarms in search of fields of their favorite grain where they may lay their eggs. The Hessian fly received its popular name from the belief that it was brought to this country in straw by the Hessian troops under the command of Sir William Howe.

The American Wheat-Fly, *C. tritici*, Kirby, is one tenth of an inch or less in length, orange-colored, the legs long, slender, and pale yellow, wings transparent and iridescent, eyes black and prominent, and antennæ long and blackish, those of the male being twice as long as the body, and consisting of twenty-four joints, and those of the female about as long as the body, and consisting of twelve joints. The wheat insects, in their perfect form, appear some time between the first of June and the last of August, according to locality or state of the season. They often move in immense swarms, taking wing in the morning and evening twilight, and in cloudy weather, at which times they lay their eggs in the opening flowers of the grain of barley, rye, and oats, as well as wheat. The eggs hatch in about eight days, producing little yellow larvæ or maggots, which are found within the chaffy scales of the grain. The eggs are laid, and consequently hatched, at different times, so that all do not come to maturity together; but they appear to come to their growth in about fourteen days. These insects prey upon wheat in blossom and in the milk, and do not touch the kernel after it has become hard. They prevent the kernel from filling at all, which happens when they attack the grain soon after the blossoming; or they cause the kernels to wither, which happens when they attack it somewhat later. At length they cease eating, become sluggish and torpid, and soon after moult their skins, after which they become active again, and in a few days descend to the ground of their own accord, or are shaken down by the wind. Here they burrow, remain through the winter in the larva form, pass into the pupa state in early summer, and in a few days afterwards come forth in the winged state.

The Genus *Chione* contains the Snow-Gnats, which are very small, wingless, and look like spiders, and are found in great numbers in the last of winter upon the snow.

The Genus *Simulium* comprises the Black Flies and their allies. The larvæ are aquatic.

TABANIDÆ, *Leach*, OR HORSE-FLY FAMILY. — This Family comprises large diptera which in the female have a proboscis enclosing six sharp lancets, and in the male, four; the eyes are very large and cover nearly the whole head, thorax oblong, and abdomen triangular. They are among the largest of the Diptera, and are notorious for their attacks upon horses and cattle, piercing them and sucking their blood, and causing them great pain. The larvæ live in the ground.

Fig. 304.

Horse-Fly, *T. lineola*, Fabr.

The Genus *Tabanus* contains the Black Horse-Fly, *T. atratus*, Fabr., which is seven eighths of an inch long, and expands nearly two inches; the Orange-belted Horse-Fly, *T. cinctus*, Fabr.; and the Lined Horse-Fly, *T. lineola*, Fabr., at once distinguished by the whitish dorsal line.

ASILICI, *Latr.*, OR ASILUS FAMILY. — This Family comprises diptera which are of large size, with the body long, slender, and clothed with stiff bristles. They are rapacious, seizing and bearing away other insects. The larvæ live in the roots of plants.

Fig. 305.

Asilus, *A. æstuans*, Linn.

The Genus *Asilus* contains the principal species.

A. sericeus, Say, is about an inch long, brownish yellow,

wings smoky brown, with brownish-yellow veins. The larvæ feed upon the roots of the rhubarb.

BOMBYLIARII, *Latr.*, OR BEE-FLY FAMILY.—This Family comprises diptera which have a long, slender proboscis, and the body covered with hairs. They fly with great swiftness, and are found in sunny paths in woods in spring.

Fig. 306.

Bee-Fly, *Bombylius æqualis*, Fabr.

SYRPHIDÆ, *Leach*, OR SYRPHIAN FAMILY. — This Family comprises diptera which bear a great resemblance to the Hymenoptera, many of them looking so much like bees and wasps as to be easily mistaken for them by the casual observer. The head is hemispherical, antennæ three-jointed, the third joint largest, eyes large, and body rather flattened. They are ornamented with yellow bands and spots. They fly with amazing rapidity, and many delight to hover immovably over certain spots. In the larva state they feed upon plant-lice.

DOLICHOPIDÆ, *Latr.* — This Family embraces small, brilliant, metallic-colored flies, with the abdomen compressed and incurved at the tip, legs long, slender, and armed with bristles. They are found solitary in damp situations, or in numbers flying and running upon pools and streams in spring. *Dolichopus* is generally green.

ŒSTRIDÆ, *Latr.*, OR BOT-FLY FAMILY. — This Family comprises diptera which have their antennæ very short and inserted in two little holes upon the forehead, head large, eyes small, with a large space between them, wings large, covering the balancers, and the hind body of the females with a conical tube bent under the body, and with which they deposit their eggs while flying. The larvæ inhabit various parts of the body of herbivorous animals. They are thick, fleshy, without feet, tapering towards the head, which in most cases is armed with two hooks, and the segments of the body are also armed with hooks or prickles. More than twenty species are known, and several are found in this country.

The Genus *Gasterophilus* comprises three species, which infest the horse. The Large Bot-Fly, *G. equi*, Linn., lays her eggs upon the fore legs of the horse; the Red-tailed Bot-Fly, *G. hæmorrhoidalis*, Linn., lays her eggs upon the lips; and the Brown Farrier Bot-Fly, *G. veterinus*, Green, under the throat. By biting the parts where the eggs are laid, the horse gets the larvæ into his mouth, swallows them, and, clinging to the walls of the stomach, they remain there till fully grown.

Fig. 307.

Bot-Fly, *G. equi*, Linn.

The Genus *Œstrus* contains the Ox Bot-Fly, *O. bovis*, Fabr., which deposits her eggs in the skin of the backs of cattle, and the larvæ live in large open sores.

The Genus *Cephalemyia* contains the Sheep Bot-Fly, *C. ovis*, Linn., which lays its eggs in the nostrils of sheep, and the larvæ crawl into the cavities in the bones of the forehead, and in many cases produce death.

Muscidæ, *Latr.*, or Fly Family. — This Family comprises diptera which have short antennæ that end with an oval joint and a lateral bristle, a short, soft proboscis ending with large fleshy lips, enclosing a sucker composed of only two bristles, and capable of being entirely retracted into the oral cavity. The larvæ are fleshy, whitish maggots, and never cast their skins; but when they pass to the pupa state, they shorten, become oblong-oval, dry, hard, and brown on the outside. This family includes about one third of all the Diptera, and its members are known under the names of House-Flies, Blow-Flies, Flesh-Flies, Flower- and Fruit-Flies, Cheese-Flies, &c. Meigen has described about seventeen hundred European species.

The Genus *Sarcophaga* contains the Flesh-Flies, which are viviparous, and which deposit their larvæ upon animal matter and other substances in a state of decomposition. Reaumur found twenty thousand larvæ in a single flesh-fly.

The Genus *Musca* contains the common House-Fly, *M. domestica*, Linn., the larvæ of which are found in manure; the Blue-bottle Fly, *M. Cæsar*, Linn.; and the Meat-Fly, *M. vomitoria*, Linn., which deposits its eggs, known as *fly-blows*, upon meat and other dead animal matter.

The Genus *Stomoxys* comprises flies which frequent our apartments, and pierce our flesh to obtain blood. *Anthomyia* contains species whose larvæ attack the radish and other roots. *Ortalis* and its allies produce galls in plants, or lay eggs in raspberries and other fruit. *Tephrites* causes the swellings in the stems of asters. *Oscinus* lays its eggs in the flowers of grain, and the young consume the grain. *Scatophaga* contains the Dung-Flies, which are almost always seen on manure. The males are bright ochre-yellow.

HIPPOBOSCIDÆ, *Leach*, OR SPIDER-FLY FAMILY.—This Family comprises small flat-bodied flies, which infest quadrupeds and birds. They are not produced from eggs in the usual way, but are brought forth in the pupa state. The pupæ are soft and white at first, but soon become hard and brown.

The Genus *Hippobosca* nestles in the hair of the horse; *Ornithomyia* lives in the plumage of birds; *Nycteribia* contains the Bat-ticks, and *Mellophaga* the Sheep-ticks.

PULICIDÆ, *Steph.*, OR FLEA FAMILY. — This Family embraces the fleas, which are wingless flies, with hard, compressed bodies, a sucker-like arrangement of the mouth-parts, and large hind legs, formed for leaping. Different species inhabit different animals.

SUB-SECTION IV.

THE SUB-ORDER OF COLEOPTERA, OR BEETLES.

THE Sub-Order of Coleoptera comprises insects whose anterior or upper wings are represented by a pair of horny cases, called elytra, meeting in a straight line upon the top of the back, and often having a small triangular or semicircular piece, called the scutellum, wedged between their bases. The posterior or under wings are thin, membranous, and, when at rest, longitudinally and transversely folded. Beetles are provided with two pairs of lateral-moving jaws, and in the larva state are grubs, and undergo complete transformation in coming to maturity. There are probably a hundred thousand species in all. In the classification of the Coleoptera, we follow LeConte.

CICINDELIDÆ, *Leach*, OR TIGER-BEETLE FAMILY.—This Family comprises brilliant-colored beetles with large head, globose eyes, long antennæ, and very long and dentated mandibles. They prefer warm and sandy places, run with swiftness, take wing on the slightest alarm, but soon alight again. They are carnivorous, and very voracious, devouring other insects in great numbers. The larvæ are soft, white, and provided with powerful jaws, and, like the adults, carnivorous. They dig vertical holes in the ground, in which they remain, the head just closing the entrance; and, when some insect passes near enough, they seize the victim and drag it into their retreat.

Fig. 308.

Fig. 309.

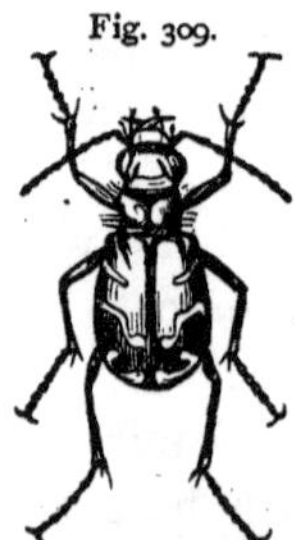

Tiger-Beetles.

Fig. 310.

Tiger-Beetle, Larva.

The Genus *Cicindela* contains the principal species. The common Tiger-Beetle, *C. vulgaris*, Say, is well represented by Fig. 309, and the Hairy-necked Tiger-Beetle, *C. hirticollis*, Say, by Fig. 308.

CARABIDÆ, *Leach*, OR PREDACEOUS GROUND-BEETLE FAMILY. — This Family comprises a large number of predaceous beetles which have the upper jaws very powerful and hooked, body oblong and firm. Some are without wings, having only elytra. They prey upon the larvæ of other beetles, upon herbivorous beetles, and other insects. The Genus *Calosoma* contains large and splendid species, which prey upon canker-worms. The Caterpillar-Hunter, *C. scrutator*, Fabr., is green. The Glowing Calosoma, *C. calidum*, Fabr., is black, with six rows of sunken brilliant red metallic spots.

Fig. 311.

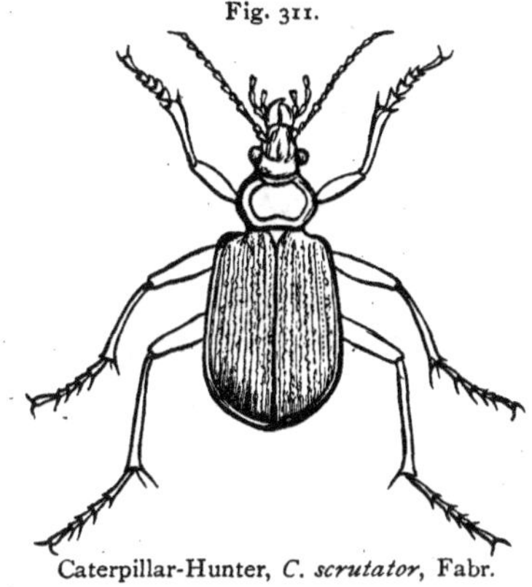

Caterpillar-Hunter, *C. scrutator*, Fabr.

DYTICIDÆ, *McLeay*. — This Family embraces aquatic beetles of an oval or rounded form, with the posterior legs longest, and strongly fringed to aid in swimming. They are excessively voracious both in the adult and larval state, devouring not only other insects, but also young fishes.

Fig. 312.

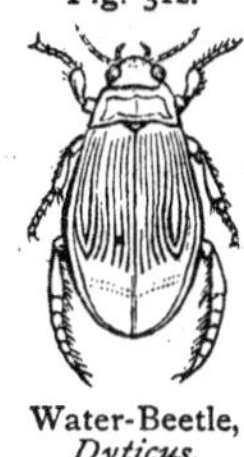

Water-Beetle, *Dyticus*.

GYRINIDÆ, *Latr.*, OR WHIRLIGIG-BEETLE FAMILY. — This Family comprises aquatic beetles which have an oval, generally glossy and brilliant body, of small or of moderate size, and which are found on the surface of still waters, where they appear like brilliant spots gliding in all sorts of curves and gyrations. They swim by means of the four hinder legs.

HYDROPHILIDÆ, *Leach*. — This Family comprises aquatic beetles which are ovate, hemispherical, and with the thorax broader than long, the edges and the tibiæ slightly spined, but terminated by strong spurs, and the tarsi are commonly ciliated so as to aid in swimming. They are less agile in swimming than the Dyticidæ, and move their posterior legs alternately. They stay in the water by day, but take wing at night.

SILPHIDÆ, *Leach*, OR CARRION-BEETLE FAMILY.—This Family embraces beetles which have the body broad and depressed, thorax shield-like, mandibles strong and exserted, and the antennæ thickened towards the tips, and made up of several pieces. These beetles live together in great numbers in the bodies of decaying animals, and thus greatly assist in removing such noxious substances. Some species have the habit of burying all the small dead animals which they find, and they find out with astonishing quickness where such animals are; for although a carrion beetle may not be found in a given locality, let but a dead frog, mouse, or bird be thrown upon the ground, and in a few hours these beetles will be seen about it, and soon they begin to dig beneath it, and will continue their digging till they have sunk it into the ground out of sight. The female then lays her eggs in it, so that, when the young hatch, they find themselves in the midst of suitable food.

Fig. 313.

Carrion-Beetle, *Silpha*.

STAPHYLINIDÆ, *Leach*, OR ROVE-BEETLE FAMILY.—This Family comprises beetles which are long, narrow, and depressed, with the head large and flattened, stout mandibles, short antennæ, thorax as wide as the abdomen, and the abdomen much longer than the elytra. When touched, or when they run, they elevate the abdomen and flex it in every direction. They run swiftly, take wing quickly, are very voracious, and revel in all kinds of decayed vegetable and animal substances. The

Fig. 314.

Rove-Beetle, *Staphylinus*.

larvæ closely resemble the perfect insects both in appearance and habits.

HISTERIDÆ, *Leach*, OR MIMIC-BEETLE FAMILY. — This Family comprises small beetles of a square or oblong-quadrate form, very hard, and with a highly polished surface. The antennæ are short, elbowed and terminated by a large, solid club, and the elytra short and truncate. These beetles scarcely exceed one third of an inch in length, the color is generally black and shining, and they have the habit of counterfeiting death when disturbed.

DERMESTIDÆ, *Leach*, OR SKIN-BEETLE FAMILY. — This Family embraces beetles which have an ovoid or oblong thick body, covered with scales or hairs, the head short and deeply inserted into the cavity of the thorax, and five-jointed tarsi. In the larva state they attack the skins and bodies of dried animals of all kinds, feathers, furs, dried meats, bacon, horns and hoofs, books, papers, and cork.

The Genus *Dermestes* contains the Bacon-Beetle, *D. lardarius*, Fabr., which is oblong-oval, black, the base of the elytra grayish-buff.

The Genus *Anthrenus* contains *A. musæorum*, Fabr., which is met with in its perfect state in flowers, and which feigns death the moment it is disturbed. The larvæ are very destructive to zoölogical collections, and are distinguished by their elongate-ovate thick form, strong horny jaws, small bundles of hairs arranged along the sides, and the larger tufts of hair upon the tail.

BYRRHIDÆ, *Leach*. — This Family comprises beetles which have the body short, oval, or rounded, very convex, and generally covered with a short pile.

LUCANIDÆ, *Latr.*, OR HORN-BUG FAMILY. — This Family comprises beetles which have the body hard, oblong, and rounded behind, head large and broad, thorax short and as wide as the abdomen, upper jaws large, and

Fig. 315.

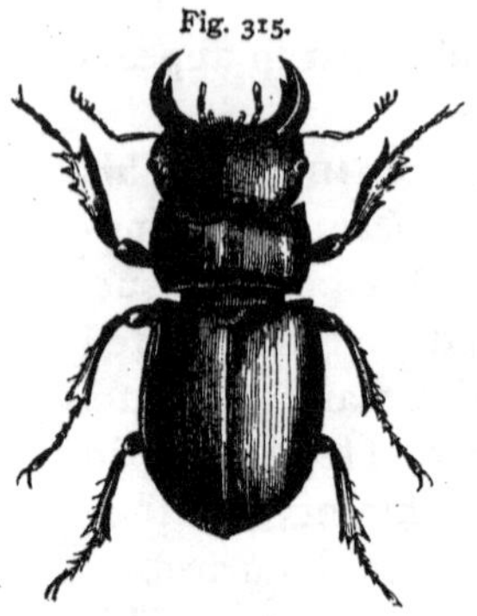

Horn-Bug, *Lucanus dama*, Fabr.

in some cases curved, in others branched, antennæ bent in the middle, and composed of ten joints, the last three or four of which are leaf-like, and project on the inside. They fly only at night, spending the day upon the trees, and feeding upon the leaves. The grubs of the larger kinds are six years in coming to their growth, living all the time in the trunks and roots of trees.

SCARABÆIDÆ, *Latr.*, OR SCARABÆIAN FAMILY. — This very extensive Family embraces beetles which, though differing in many respects, agree in having the antennæ ending in a knob composed of three or more leaf-like pieces; a projecting plate or clypeus, which extends forwards over the face like a visor; a short, broad, thick, and convex form; legs fitted for digging, being toothed on their outer edges; and feet five-jointed. Some live mainly upon or beneath the surface of the earth, and are hence called Ground-Beetles; others in the winged state are found upon trees, the leaves of which they devour, and are called Tree-Beetles; and others, which in the perfect state feed upon the juices of flowers, are called Flower-Beetles. This group has been much divided, thus appearing in some works as many distinct families.

Fig. 316.

Scarabæian, *Phanæus*.

The Genus *Copris* contains Dung-Beetles, that enclose their eggs in pellets of manure, which they roll along with their hind feet, and at length bury them.

The Genus *Phanæus* comprises very brilliantly colored species, the males of which have horn-like prominences on the head and thorax.

The Genus *Geotrupes* embraces large green or purplish species, known as Earth-borers. *Melolontha* and its allies include the Leaf-eaters, which have powerful and horny jaws, fitted for cutting and grinding leaves.

The Genus *Macrodactylus* contains the Rose-chafers.

The Common Rose-chafer, *M. subspinosa*, Fabr., is seven twentieths of an inch in length, and is covered with very short and close ashen down; legs pale red.

The Genus *Lachnosterna* contains the May-Beetles.

The May-Beetle, *L. quercina*, Knoch, is nine tenths of an inch long, chestnut-brown, smooth, but finely punctured, and each wing-case has three slightly elevated longitudinal lines; breast clothed with yellowish down. The grub is white, with a brownish head, attains almost the size of our little finger, and feeds upon grass roots.

The Genus *Pelidnota* contains the Spotted Pelidnota, *P. punctata*, Fabr., which is about an inch long, oval, reddish-yellow, with three black spots on each elytron, and the thorax with a black dot on each side.

The Genus *Cotalpa* contains the splendid Goldsmith-Beetle, *C. lanigera*, Linn., which is nearly an inch long, lemon-color above, and glittering like burnished gold on the head and thorax; under side copper-colored, and covered with whitish wool. It appears in spring and early summer. It flies in the morning and evening twilight, feeds upon tender leaves, and clings to the under side of the leaves during the day.

Fig. 317.

Goldsmith Beetle, *C. lanigera*, Linn.

The Genus *Dynastes* contains the Hercules Beetles of South America, which are five inches long.

The Genus *Cetonia* — Flower-Beetles — has the form oblong oval, lower jaws soft on the inside, and often provided with a flat brush of hairs, upper jaws without a grinding-plate, and the antennæ ten-jointed. Beetles of this genus feed upon the juices of flowers, and pollen.

The Genus *Osmoderma* embraces the Scented Beetles.

The Rough Osmoderma, *O. scabra*, Beauv., is about one inch long, broad, oval, and flattened, purplish black, with a copper lustre, the head hollowed on top, and the wing-cases thickly, deeply, and irregularly punctured.

BUPRESTIDÆ, *Leach*, OR BUPRESTIAN FAMILY. — This Family comprises elongated, flattened, very solid beetles, with the head sunk into the thorax to the eyes, the antennæ eleven-jointed, serrate, and legs short. Their colors are brilliant, often metallic. The Buprestians are diurnal, found on trunks of trees, and when disturbed fold their legs and feign death. The larvæ are wood-borers, are several years in coming to their full growth, and their transformations take place within the trees. Species differing from one another bore the peach, plum, oak, pine, and hickory.

Fig. 318.

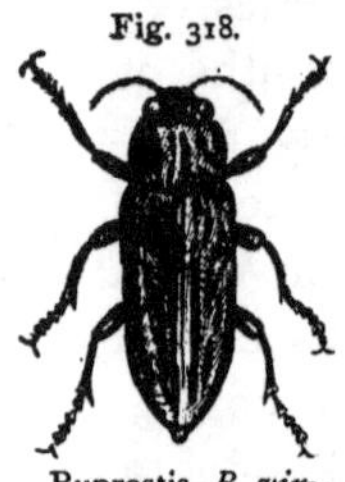

Buprestis, *B. virginica*, Drury.

ELATERIDÆ, *Leach*, OR SPRING-BEETLE FAMILY. — This Family comprises beetles which have a hard body, usually tapering behind, the head sunk to the eyes in the thorax, and the latter as broad at the base as the wing-covers, rounded before, and the hind angles short and prominent. They are at once distinguished by their power of throwing themselves upward with a jerk after they have been placed upon the back. They perform this feat by means of a spine, the point of which fits into a cavity behind it, situated on the under side of the breast, between the bases of the first pair of

Fig. 319.

Spring-Beetle, *E. oculatus*, Linn.

legs. When once upon the back, their legs are so short that they cannot turn themselves like other insects; but, folding their legs close to the body, and bending back the head and thorax, and thus unsheathing the spine, they straighten the body suddenly, the point of the spine strikes forcibly upon the edge of the sheath, and, acting like a spring, throws the insect upwards. In the adult state they feed mainly upon flowers. The larvæ devour wood and roots, and are called wire-worms. *Elater* is the principal genus.

LAMPYRIDÆ, *Leach*, OR FIRE-FLY FAMILY. — This Family comprises beetles which resemble the Elaters, but are shorter, broader, and softer. In some species the females are wingless, and in others furnished with only short elytra. Some are carnivorous, preying in the larva state upon snails. Many of the species are phosphorescent, the luminous matter occupying the under side of a few of the terminal rings; and it appears that the insect can vary the intensity of the light.

MALACHIDÆ, *Redt.* — This Family contains small beetles which have the body furnished with extensible vesicles.

CLERIDÆ, *Kirby.* — This Family comprises beetles which are long, often cylindrical, with the thorax narrower than the elytra, and head prominent. They are fast runners, handsomely variegated in their colors, and feed upon juices of flowers. The larvæ are carnivorous.

The Genus *Clerus* has the elytra of a bright red color, ornamented with purple. The larvæ are destructive to bees and wasps, in the nests of which the female deposits her eggs.

LYMEXILLIDÆ, *Leach.* — This small Family embraces beetles which have the body long, nearly cylindrical, softer than the Elaters, head broad before, narrowed behind, and not sunk into the thorax. The grubs penetrate timber. One species has been very destructive to ship-timber in the North of Europe.

PTINIDÆ, *Leach.*—This Family comprises small dull-colored beetles, obtuse at each end, with the head small and immersed in the thorax to the eyes, and the antennæ long, and constantly in motion when the insect is walking. When disturbed they feign death, withdrawing their head and antennæ, and contracting their legs, and will suffer themselves to be pulled to pieces rather than show signs of life.

The Genus *Anobius* contains the Death-Watches, which make a ticking noise, that is regarded by the ignorant with great superstition. This ticking is produced during the pairing season by striking their jaws upon the object upon which they are standing, and is a signal which is replied to by the mate.

TENEBRIONIDÆ, *Latr.*, OR MEAL-WORM FAMILY.—This large Family comprises beetles which have the body oblong or ovate, depressed, or slightly elevated, antennæ clavate, feet short, and the colors black or brown.

The Genus *Tenebrio* contains the species whose larva is the well-known meal-worm.

MORDELLIDÆ, *Leach.*—This Family comprises small, wedge-shaped, glistening, pubescent, black beetles, which are found upon flowers, and which, when disturbed, leap like fleas. The larvæ live in the pith of plants.

MELOIDÆ, *Gyllenh.*, OR BLISTERING-BEETLE FAMILY.—This Family comprises beetles which are mainly soft, and celebrated for secreting cantharidine, a blistering property, which has caused them to be extensively used in pharmacy. They are also remarkable in the successive forms of the larvæ, in the first of which they are very small active parasites, infesting bees. The adults are called Cantharides. They have the head broad, antennæ long, wing-covers soft and more or less bent downward, and they feign death when alarmed. The Genus *Cantharis*, or *Lytta*, contains the principal species.

The Striped Cantharis, *C. vittata*, Harr., is about half an inch long, and light yellowish-red above, with two black spots on the head, and two black stripes on the thorax and on each wing-cover; under parts black, covered with a grayish down. The Margined Cantharis, *C. marginata*, Olivier, is over half an inch long, wing-covers black, with a narrow gray margin. The Ash-colored Cantharis, *L. Fabricii*, LeC., and the Black Cantharis, *L. atrata*, Fabr., each about half an inch long, are also common species. The Spanish Fly, *C. vesicatorius* of authors, inhabits the South of Europe, and is golden green.

The Genus *Meloe* contains the Narrow-necked Oil-Beetle, *M. angusticollis*, Say, an inch long, color Prussian-blue.

STYLOPIDÆ, *Kirby*. — This Family comprises small beetles which at first sight seem to bear no resemblance to the other coleoptera, and which were formerly regarded as a distinct order, named Strepsiptera. They are less than a quarter of an inch long, the elytra pad-like, but the hind wings are greatly developed. They are parasite in various aculeate hymenoptera.

CURCULIONIDÆ, *Latr.*, CURCULIO OR WEEVIL FAMILY. —This Family embraces hard-shelled beetles which have the fore part of the head prolonged into a broad muzzle or a longer and slender snout, at the extremity of which is the mouth, armed with small horny jaws. They are exceedingly numerous in genera and species, and in many cases very minute. They are timid, and quickly feign death when disturbed. The larvæ are white, thick grubs.

Fig. 320.

Pea-Weevil, *B. pisi*, Linn.

The Genus *Bruchus* contains the Pea-Weevil, *B. pisi*, Linn., which lays its eggs on the pea when in flower, and the larva enters the pea through the green pod, and remains there till the following spring, when it emerges as an imago. Harris says that the Baltimore Oriole splits open the green pods for the sake of

Fig. 321.

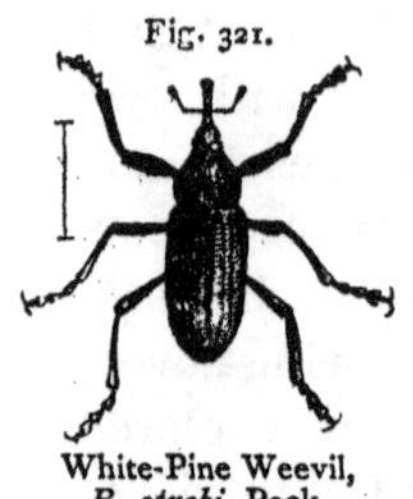

White-Pine Weevil, *R. strobi*, Peck.

obtaining the grubs contained in the peas.

The Genus *Rhynchœnus* has the snout long and slender, and the antennæ near the middle of it.

The White-Pine Weevil, *R. strobi*, Peck, bores the pine when in the larval state.

The Long-snouted Nut-Weevil, *R. nasicus*, Say, attacks nuts while it is in the grub state.

Fig. 322.

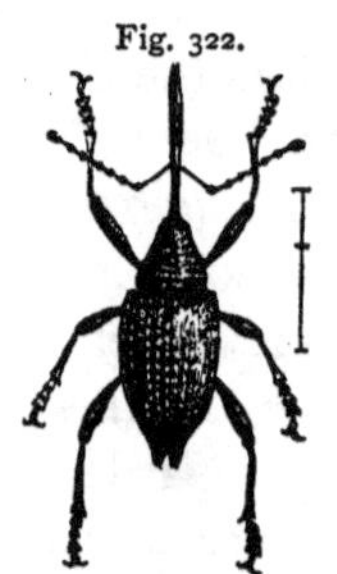

Long-snouted Nut-Weevil, *R. nasicus*, Say.

The Genus *Curculio* contains the Plum-Weevil, *C. nenuphar*, Herbst, about one fifth of an inch long exclusive of the snout, rough, dark-colored, and, when shaken from the tree, appearing like a dried bud. This beetle makes a crescent-shaped incision with its snout in the surface of the plum, and then lays an egg in the wound; and it goes from one plum to another till the stock of eggs is exhausted, so that where the beetles are plentiful not a plum escapes stinging. From the egg there hatches a whitish grub, destitute of feet, and with a rounded light brown head, which immediately burrows into the fruit, and ultimately penetrates to the stone.

Fig. 323.

Plum-Weevil, *C. nenuphar*, Herbst.

The Genus *Calandra* comprises beetles which attack stored wheat, corn, and rice. They lay their eggs upon the grain, and the larvæ, as soon as hatched, burrow into the kernel, destroying everything except the hull, which is left entire, so that the injury done is not perceived till the lightness of the kernel reveals it.

The Grain Weevil, *C. granarius*, Linn., of Europe, is about one eighth of an inch long, and pitchy-red. A single pair may have six thousand descendants in a single year.

Fig. 324.

Rice Weevil, *C. oryzæ*, Linn.

The Rice Weevil, *C. oryzæ*, Linn, is about one tenth of an inch long, with two red spots on each wing-cover. It not only attacks rice, but wheat and Indian corn. In the Southern States it is called Black Weevil.

CERAMBYCIDÆ, *Leach*, OR CAPRICORN-BEETLE FAMILY. — This Family comprises beetles which have the antennæ very long, tapering, and generally curved like the horns of a goat. When caught, they make a squeaking noise by rubbing the joints of the thorax and abdomen together. In the larva state they are the most destructive of all wood-eating insects, and are known as *borers*. They are long, whitish, and fleshy, and provided with short, powerful jaws, by which they bore a cylindrical passage through the hardest wood. Some species always keep one end of their burrows open, out of which they cast their chips; others, as fast as they proceed, fill their passages behind them with their cuttings, which are the well-known powder-post. They remain in the larva state from one to three or more years, then go into the pupa state at the extremity of their burrows, and at length appear as beetles. They are popularly known as Long-horns.

Fig. 325.

Prionus, *P. laticollis*, Drury.

The Genus *Prionus* has the antennæ composed of flattened joints, which project on the inside like the teeth of a saw.

The Broad-necked Prionus, *P. laticollis*, Drury, is an inch and three quarters long, black, and the larva lives in the trunk of the poplar.

The Genus *Stenocorus* has the wing-covers narrow, and notched or armed with two little thorns at the tip, and the antennæ very long.

The Oak-Pruner, *S. villosus*, Fabr., is about half an inch long, slender, dull brown sprinkled with gray spots, scutel yellowish-white, the third and fourth joints of the antennæ tipped with a small spine. It lays its eggs in July, placing each one in the joint of a leaf-stalk, near the extremity of a branch. As soon as the larva is hatched, it penetrates to the pith, and then moves towards the body of the tree, devouring the pith as it goes. At the close of the summer, it has moved several inches; and now, having arrived at its full growth, it cuts out all the wood at the lower extremity of its burrow, leaving only the bark to sustain the branch; then, retiring a little, it stops up the downward end of its burrow, and awaits the fall of the branch, which takes place during the first strong wind. Branches an inch in diameter and several feet in length are thus cut off. The larva goes into the pupa state in the spring, and comes out a beetle in June.

The Genus *Clytus* contains capricorn-beetles which are beautifully colored with black and yellow. They are seen in great numbers upon flowers and upon the trunks of the locust-trees in the early autumn.

The Painted Clytus, *C. flexuosus*, Fabr., is about three quarters of an inch long. The Beautiful Clytus, *C. speciosus*, Say, is an inch long, and lays its eggs on the trunk of the maple in July and August. It is the largest known Clytus.

Fig. 326.

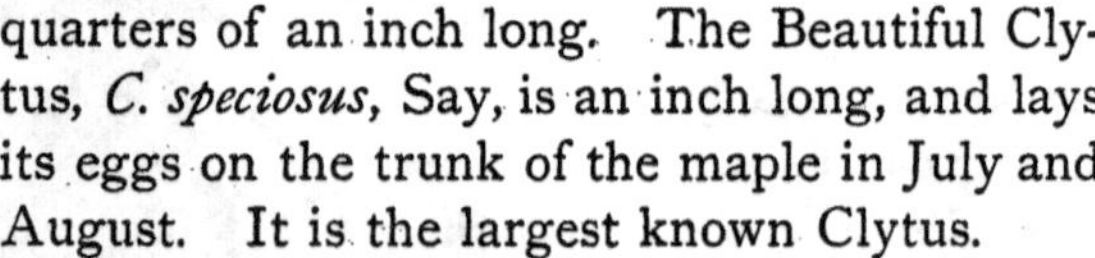

Painted Clytus, *C. flexuosus*, Fabr.

The Genus *Lamia* contains the Tickler, *L. titillator*, Fabr., which is about an inch and a quarter in length, brownish, mottled with spots of gray, and the antennæ of the male are about twice as long as the body. It gets its name from its habit of gently touching with the tips of its long antennæ the surface over which it walks.

The Genus *Saperda* contains the Apple-tree Borer, *S. bivittata*, Say, which in the beetle state is about three fourths of an inch long, brown above, and marked with two white longitudinal stripes. It lays its eggs on the bark, near the roots of the tree. The larvæ are whitish, nearly cylindrical, and tapering from the first ring to the hind extremity. The head is small, horny, and of a brown color. Their jaws are strong, and with them they cut a cylindrical passage through the bark, pushing their cuttings backward, and in the course of the two or three years in which they remain in the larva state, they penetrate eight or ten inches upward into the trunk of the tree, and end their burrow just under the bark. Here they go through their transformation, which is completed in June, and the beetle gnaws through the bark, which is all that covers it.

Fig. 327.

Apple-tree Borer, larva, *S. bivittata*, Say.

Fig. 328.

Apple-tree Borer, *S. bivittata*, Say.

The Genus *Leptura* contains capricorn beetles which have the body, in most cases, narrowed behind, and the antennæ implanted in the middle of the forehead.

The Genus *Desmocerus* contains the Cloaked Lepturian, *D. palliatus*, Harr., which is of a Prussian-blue color, with one half of the fore part of the wing-covers orange yellow. The larvæ live in the lower parts of elder-stems.

CHRYSOMELIDÆ, *Leach*, OR CHRYSOMELA FAMILY. — This Family embraces beetles which have a hemispherical or ovate form, small and sunken head, and antennæ inserted wide apart. They are blue, green, and golden.

The Genus *Lema* contains the Three-lined Leaf-Beetle, *L. trilineata*, Olivier, which is one fourth of an inch long, rusty buff, with two black dots on the thorax, and three black stripes on the wing-covers. Beetles of this species appear in June on the leaves of the potato, gnawing large and irregular holes through them. Soon they begin to lay their oblong-oval and golden-yellow eggs, gluing them to the leaves in groups of six or eight. Their filthy grubs are hatched in about a fortnight, and after feeding upon the potato leaves for about fifteen days, enter the ground, pass into the pupa state, remain two weeks, and then come forth perfect beetles, which lay eggs for a new brood.

Fig. 329.

Cucumber Beetle.

The Genus *Galeruca* contains the Squash or Cucumber Beetle, *G. vittata*, Fabr., less than one fifth of an inch long, light yellow, with a black head and three black longitudinal stripes; black below.

The Genus *Chrysomela* is very gayly colored.

Fig. 330.

Ladder Beetle, *C. scalaris*, LeC.

The Ladder Chrysomela, *C. scalaris*, LeC., is about three tenths of an inch long, the head, thorax, and under part of the body dark green; wing-covers silvery-white, with green spots on the sides, and a broad green jagged stripe along their inner edges; antennæ and legs red.

The Blue-winged Chrysomela, *C. cœruleipennis*, Say, is about one seventh of an inch long.

The Gilded Eumolpus, *E. auratus*, Fabr., is golden green, and about three eighths of an inch long.

The Genus *Cassida* contains the Tortoise Beetles, which are broad, oval, or rounded, nearly flat, and with the thorax projecting over the head.

The Genus *Hispa* contains little Leaf-Beetles, which are rough above, and whose larvæ burrow under the skin of the leaves of plants, especially of the apple and its allies.

The Genus *Haltica* embraces the Flea-Beetles, little shiny black leaping species, which injure leaves.

Coccinellidæ, *Latr.*, or Lady-bug Family. — This Family comprises hemispherical beetles, which are black, red, or yellow, with round or lunate spots. Both in the perfect and larva state they devour plant-lice. When about to become chrysalids, they attach themselves to a leaf.

Fig. 331.

Lady-bug, *C. novemnotata*, Harr.

SUB-SECTION V.

THE SUB-ORDER OF HEMIPTERA, OR BUGS, CICADAS, ETC.

THIS Sub-Order embraces insects which have the mouth-parts in the form of a slender horny beak, consisting of a horny sheath, containing three stiff and intensely sharp bristles. When not in use, this beak is bent under the body, and lies upon the breast.

The Hemiptera comprise two great groups, — the true Hemiptera, which are designated as Hemiptera heteroptera, and include all insects properly called Bugs; and Hemiptera homoptera, which contain all insects properly called Harvest-flies, Plant-lice, and Bark-lice.

The Hemiptera heteroptera, or Bugs, have the wing-covers thick in their basal portion, thin towards their tips, and lying flat on the top of the back, and the thin portions crossing each other.

The Hemiptera homoptera have the wing-covers of uniform thickness throughout, and not lying flat upon the back, nor crossing each other at their extremities, and both the wings and wing-covers are more or less sloping at the sides of the body. The families of the two groups are arranged according to Westwood, with the addition of Thripsidæ and Pediculi to the sub-order.

CICADARIÆ, *Latr.*, CICADA, OR HARVEST-FLY FAMILY. — This Family comprises homopterous insects which have very broad heads, large and convex eyes on each side, and three eyelets on the crown. Their wing-covers and wings are both transparent, and distinctly veined. The males are provided with an apparatus by which they are enabled to produce an exceedingly loud and shrill buzzing sound, which in some species may be heard at the distance of a mile; and the females are furnished with a unique kind of piercer for perforating the limbs of trees, in which they lay their eggs. The organs which enable the male to produce its music consist of a pair of kettle-drums, one in each side of the abdomen, formed of convex pieces of parchment finely plaited, and played by means of muscular fibres fastened to the inside. By the rapid contraction and relaxation of these fibres, the drum-heads are alternately tightened and loosened, and thus the sounds are produced; while other cavities in the body, separated by thin, transparent, and brilliant membranes, assist greatly in increasing the intensity of the sounds. The piercer of the female consists of three pieces, two outer ones grooved on the inside and toothed on the outside like a saw, and a central one, a spear-pointed borer, which plays between the other two. The Greeks were charmed with the singing of the Cicadæ, and often kept them in cages, that they might enjoy their music. They ate both the pupæ and the perfect insect.

The Genus *Cicada* contains all the species, of which there is a score or more in this country.

The Seventeen-year Cicada, *C. septendecim*, Linn., is

Fig. 332.

Seventeen-year Cicada, *C. septendecim*, Linn.

about an inch long, black, the wing-covers and wings transparent, with the forward edges and larger veins and eyes orange-red, and near the tips of the wing-covers there is a dusky zigzag line. This species is popularly known as the Seventeen-year Locust, a name which should be at once abandoned, as the present species does not even belong to the same sub-order as the *Locust*. It is believed that this insect appears in the same locality only at intervals of seventeen years, and hence its specific name. It makes its appearance in the early part of summer. Sometimes the cicadas of this species come in such immense swarms as to bend, and even break, the limbs of the forest upon which they alight, and the woods are filled from morning till night with the noise of their rattling drums. After pairing, the females proceed to lay their eggs, which they accomplish as follows: they select small branches, which they clasp with their legs, and then repeatedly thrust their piercer obliquely into the bark and wood in the direction of the fibres, and at the same time, by means of their lateral saws, they detach little splinters of the wood at one end, which serve as a fibrous cover to the perforations. By repeated thrusts they form a longitudinal fissure capable of holding from ten to twenty eggs, which are conveyed into the nest by means of the grooved side-pieces of the piercer, and deposited in pairs, but separated from each other by woody fibre, and placed so that one end points upward. When one fissure is filled, the insect makes another close by on the same limb; and when one limb is sufficiently stocked with eggs, she takes another, and thus continues till her store of eggs, consisting of four or five hundred, is laid. Soon afterwards she dies. The eggs are one twelfth of an inch long, pearly white, and hatch in fourteen, forty-two, or fifty-two days, authorities differing in regard to the number. Soon after they are hatched the young fall to the ground, where they immediately bury themselves, burrowing by means of their broad, strong fore-feet, which are perfectly adapted for digging. They follow the roots of plants, upon whose juices they feed. They continue this sort of life till their time of transformation approaches, when they gradually ascend towards the surface, making circuitous, cylindrical passages, about five eighths of an inch in diameter. When the time comes, they leave the ground in the night, crawl up the trunks of trees, and, after resting awhile, prepare to shed their skins, which have now become dry. After some effort, they open a longitudinal fissure in the skin of the back, and through this opening the full-grown and perfect cicada comes forth, leaving its empty pupa-skin still attached to the tree.

The Dog-day Harvest-Fly, *C. canicularis*, Harr., is over an inch long, the body black above, ornamented with olive-green, and the under side covered more or less with

Fig. 333.

Dog-day Harvest-Fly, *C. canicularis*, Harr.

a white substance resembling flour. It makes its appearance with the beginning of dog-days, and its singing may be heard among the trees through the middle of the day. The pupæ of this species, as well as of the preceding, as they come out of the ground and crawl up the trees, look like large beetles.

FULGORIDÆ, *Leach*, OR LANTERN-FLY FAMILY. — This Family comprises insects which are distinguished by a curious prolongation of the forehead, which in some cases equals all the rest of the body, and is the part asserted by writers to emit a strong light by night. They belong mainly to tropical and sub-tropical regions. They produce a waxy secretion, and the Chinese collect this for the manufacture of the white wax so highly esteemed in the East Indies.

CERCOPIDÆ, *Leach*, OR TREE-HOPPER FAMILY. — This extensive Family comprises hemiptera homoptera of small size, well fitted for leaping, and which are found amongst plants and on trees, upon the sap of which they subsist, imbibing such quantities, in many cases, that it oozes out of their bodies continually in the form of little bubbles, and covering the insect entirely in a mass of frothy matter or foam. Many of them are remarkable for their singular and even grotesque shapes. They are known as Tree-hoppers and Frog-hoppers. Figs. 334 and 335 represent one species, the first being an enlarged profile view, the other of the natural size.

Fig. 334. Fig. 335.

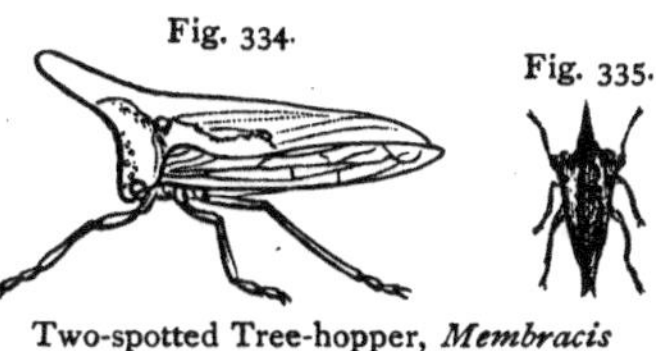

Two-spotted Tree-hopper, *Membracis bimaculata*, Fabr.

APHIDÆ, *Leach*, OR APHIS FAMILY. — This Family comprises hemiptera homoptera which have the body short, and furnished at the hind extremity with two little tubes or pores, from which exude minute drops of a very sweet fluid. Their upper wings are much longer than the body, about twice as large as the lower ones, nearly triangular, and, when at rest, almost vertical. Aphides, or Plant-lice, inhabit all kinds of plants, the leaves and softer portions being often completely covered with them. The young are hatched in the spring, and soon come to maturity, and, what is remarkable, the whole brood consists of wingless females; and what is still more remarkable, these females bring forth living young, each female producing fifteen or twenty in a day. These young are also wingless females, and at maturity bring forth living young, which are also all wingless fe-

Fig. 336.

Aphis, *A. mali*, Harr.

males, and in their turn bring forth living young; and in this way brood after brood is produced, even to the fourteenth generation, in a single season, and this without the appearance of a single male. But the last brood in autumn contains both males and females, which at length have wings, pair, stock the plants with eggs, and then perish. Réaumur has proved that a single aphis in five generations may become the progenitor of about six thousand millions of descendants. Wherever plant-lice abound, ants collect to feed upon the honey-like fluid produced by them; and the most friendly relations exist between these two kinds of insects. The ants even caress the plant-lice with their antennæ, apparently soliciting them to give out the sweet fluid, and the plant-lice yield to these solicitations; and a single aphis has been known to give in succession a drop to each of a number of ants waiting to receive it! In return, the ants take the kindest care of the plant-lice, warding off or removing anything that might be injurious to them. Plant-lice are kept in check by the beetles called Lady-bugs, already described.

The Genus *Eriosoma* contains Downy Plant-lice, or those which have a sort of woolly or cottony covering.

COCCIDÆ, *Fallen*, OR BARK-LICE FAMILY.—This Family comprises small insects, which, in the form of oval, rounded, or other shaped scales or shields, cover the bark of the stems and branches, and, in some cases, the leaves and roots of plants. The males alone are winged, and pass through the usual changes, while the females increase in size, always keeping the scale-like form. The Genus *Coccus* is the principal one.

C. ilicis, Linn., lives on a low shrub of the Levant, and is the insect which supplied the famous dye κόκκος of the Greeks, Coccus of the Romans, Kermes of the Arabs, Cocchi of the Italians, and Alkermes of the Persians. The Scarlet Grain, of Poland, *C. polonicus*, Linn., was also employed as a dye. The Cochineal, *C. cacti*, Linn., of Mexico, lives upon the cactus, and is collected in such quantities, that 800,000 pounds of this insect have been shipped to Europe in a single year.

C. lacca, Kerr, is the species which furnishes the Indian material called *lac*. The female attaches itself to the twigs of various trees, and in this state is called *stick-lac;* when separated and pounded, and the greater part of the coloring matter extracted, it is called *seed-lac;* when formed into cakes, *lump-lac;* and when strained and formed into thin leaves, *shell-lac*.

C. manniparus, Ehren., is found on Tamarix, a large tree growing on Mount Sinai, the shoots of which are covered with females, which puncture them, and thus cause them to discharge a gummy secretion, which quickly hardens and drops from the tree, and is collected by the natives, who regard it as the real manna.

NOTONECTIDÆ, *Latr.*, OR BOAT-FLY FAMILY. — This Family comprises hemiptera proper which are aquatic, being specially formed for swimming, the hind pair of legs being greatly elongated and strongly ciliated. They are remarkable for the habit of swimming on their backs. They prey upon other insects.

Fig. 337.

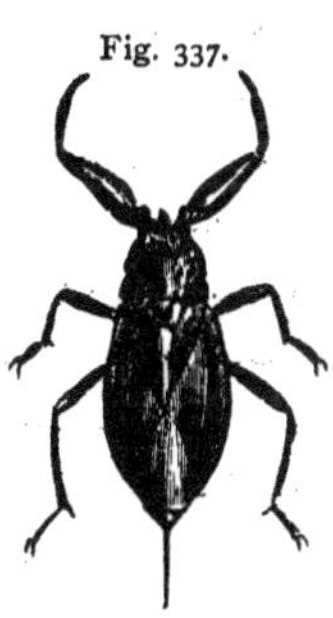

Scorpion-Bug, *Nepa apiculata*, Harr.

NEPIDÆ, *Leach*, OR SCORPION-BUG FAMILY. — This Family embraces aquatic bugs which have the body oval and much depressed. They are rapacious, and seize their prey with the fore legs, which flex upon themselves and act as piercers. They can sting severely.

HYDROMETRIDÆ, *Leach*, OR WATER-MEASURER FAMILY. — This Family embraces hemiptera proper which have the body long and narrow, and which differ in their habits from all the rest of the sub-order, being always found upon the surface of standing or running waters, upon which they move backwards or forwards with the greatest facility. The Genus *Gerris* is the most active. It is long and narrow, and wherry-shaped.

COREIDÆ, *Leach*, OR SQUASH-BUG FAMILY.—This Family comprises bugs which have the body oblong oval. The Genus *Coreus* contains the common Squash-Bug, *C. tristis*, DeGeer, which is six tenths of an inch long, rusty black above, dingy ochre-yellow beneath. It passes the winter in crevices and holes, in a torpid state; and when the vines of the squash put forth a few rough leaves, it collects beneath them, and soon begins to lay eggs, which it fastens in clusters to the under side.

Fig. 338.

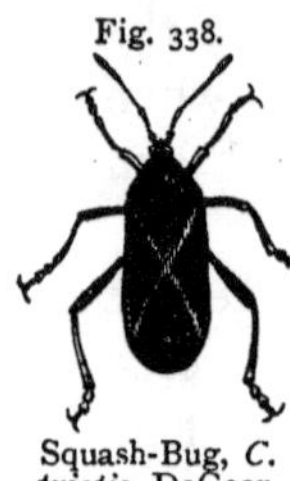

Squash-Bug, *C. tristis*, DeGeer.

The Genus *Lygæus* contains the Chinch-Bug, or White-winged Lygæus, *L. leucopterus*, Say, three twentieths of an inch long, the wing-covers white, and each with a central line and a marginal oval spot of black.

THRIPSIDÆ, *Westwood.*—This Family contains very minute insects with the body long and depressed, and eyes large. They are very agile, leaping when disturbed, and they attack leaves, melons, cucumbers, and beans, causing them to be covered with decayed patches.

CIMICIDÆ, *Westwood*, OR CIMEX FAMILY.—This Family comprises bugs which have the body very flat, and their antennæ terminating abruptly in the form of a seta. The Genus *Cimex* contains the Bed-Bug, *C. lectularius*, Linn., a representative of this family too well known to need description. It is found in all countries and is not confined to houses, though there unfortunately we oftenest find it. It is also found in the pine forests of both hemispheres; but from what country it originally came is unknown. Bed-Bugs flourish in a warm temperature, but are not killed by freezing, and they have been kept alive six years without food.

PEDICULIDÆ, *Leach*, OR LICE FAMILY. — This Family comprises degraded wingless hemiptera, which have the mouth-parts in the form of a fleshy retractile sucker.

The Genus *Pediculus* contains those which are parasitic on man, and some of the brutes. Different varieties are found on the different races of men. The Genus *Mallophaga* includes those that live on birds.

SUB-SECTION VI.

THE SUB-ORDER OF ORTHOPTERA, OR STRAIGHT-WINGED INSECTS.

THE Sub-Order of Orthoptera embraces insects whose wings lie straight along the top or sides of the back, the upper ones being somewhat thick and opaque, and sometimes slightly overlapping, and the under ones larger, thin, and folded in plaits like a fan. They do not undergo a complete transformation in coming to maturity, but the young are constantly active, feeding and growing, and differ from the adults only in size, and in having only the rudiments of wings; and in frequently changing their skins. At length, having shed their skins for the sixth and last time, they come forth perfect insects, without having passed through the inactive phase of the pupa state. The families are arranged according to Latreille.

FORFICULARIÆ, *Latr.*, OR EARWIG FAMILY. — This Family comprises orthopterous insects which have the body long, somewhat flattened, and armed at the hind extremity with a pair of slender sharp-pointed blades or nippers, which open and shut horizontally. They prefer cool and damp places, collect under stones and the bark of trees, creep into crevices, fly at night, devour fruits, and defend themselves with their pincers. It has been said that they crawl into the ear.

Fig. 339.

Earwig, *Forficula*.

BLATTARIÆ, *Latr.*, OR COCKROACH FAMILY. — This

Fig. 340.

Cockroach, *B. orientalis*, Linn.

Family contains orthopterous insects which have the body oval, flattened, the hind extremity of the abdomen furnished with conical articulated appendages, and the antennæ long and many-jointed. Cockroaches are nocturnal, and are found not only in forests, but some species infest kitchens, store-rooms, and closets, devouring all kinds of provisions, and even fabrics. The Genus *Blatta* contains the species, of which we have several that are indigenous, and one, *B. orientalis*, Linn., which originated in Asia.

PHASMIDA, *Leach*, OR WALKING-STICK FAMILY. — This

Fig. 341.

Walking-stick, *D. femorata*, Scudd. Reduced one half.

Family comprises orthoptera which are at once distinguished by their very close resemblance to vegetable structures. Some appear like dry twigs; others have wings which almost exactly resemble green or dry leaves. They are sluggish in their movements, and are found principally in warm regions, though several species belong to temperate climes. Three or more are found in North America. Some of the tropical species are very large, even a foot in length.

The Genus *Diaphomera* contains *D. femorata*, Scudd., four inches long, which is one of our most common species. It is the *Spectrum femoratum* of Say.

Mantidæ, *Latr.*, or Mantis Family.—This Family embraces orthoptera which are much elongated, and whose fore legs are formed for seizing and holding prey. They are found upon plants and trees, where they sit for hours with the front part of the thorax elevated, and the

Fig. 342.

American Mantis, *M. carolina* of authors.

fore legs held up together like a pair of arms, prepared to seize any insect which may come within reach. Some of the superstitious inhabitants of Eastern countries believe that the Mantis in this attitude is engaged in devotion.

The Genus *Mantis* contains our only species, Fig. 342.

Gryllides, *Latr.*, or Cricket Family.—This Family comprises orthoptera which have an oblong depressed body, long antennæ, long anal stylets, and the female often has an ovipositor nearly as long as her body. The male chirrups to attract his mate, the apparatus being a specialization of the membrane and nervures at the base of the wings, so that the rubbing of the wings upon one another produces sound.

Fig. 343.

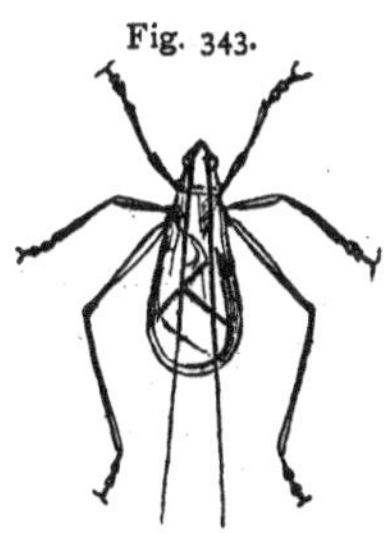

White Climbing-Cricket, *Œ. niveus*, Serville.

Fig. 344.

Mole Cricket, *G. borealis*, Burm.

The Genus *Gryllotalpa* contains the Mole Cricket, *G. borealis*, Burmeister, which is about one inch and a quarter long, and at once recognized by its stout fore legs, which are admirably adapted for digging. It burrows in the moist ground, raising ridges in its search for insects, on which it preys.

The Genus *Gryllus* contains the Field Crickets, which are dark colored or black; and *Œcanthus*, the Climbing Crickets, as in Fig 343, which are light colored.

LOCUSTARIÆ, *Latr.*, OR LOCUST FAMILY. — This Family embraces grasshopper-like orthoptera which have very long, slender antennæ, four-jointed tarsi, and the females have a long ovipositor. Many of them produce a stridulating sound by rubbing their wing-covers together.

The Genus *Ceuthophilus* contains those which are wingless, and live in concealment under stones, a dozen species of which are enumerated by Scudder.

The Genus *Cyrtophyllus* has the wing-covers much widened in the middle, and concave.

Fig. 345.

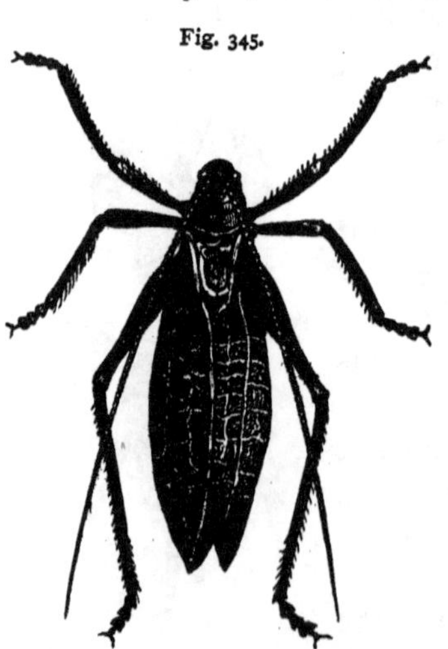

Katydid, *C. concavus*, Scudd.

The Katydid, *C. concavus*, Scudd., is one inch and a half long from the head to the end of the wing-covers, which enclose the body somewhat like the valves of a pod. This insect is silent during the day, hiding among the leaves; but at early twilight, in autumn, its notes come from the trees of the garden and groves, and continue till the dawn of day; and such a resemblance do they have to the words "Katy did," that this has become its name. The sounds are produced by a pair of taborets, one in the overlapping portion of each wing-cover, and formed by a thin transparent membrane, stretched in a strong frame. The friction of the frames of the taborets against each other, as the insect opens and shuts its wings, produces the sounds.

The Genus *Phylloptera* has the wing-covers shorter than the wings, and not concave.

The Oblong Leaf-winged Grasshopper, *P. oblongifolia*, Burm., is from an inch and three quarters to three inches long from the head to the tips of the wings. It is found in the perfect state in autumn upon trees, and when it flies makes a whizzing noise. The female lays her eggs on twigs of trees and shrubs.

The Genus *Phaneroptera* has the ovipositor curved sharply upward.

P. curvicauda, Serville, is about one inch and three quarters from the forehead to the tips of the wings, the wing-covers of uniform width, and shorter than the wings.

The Genus *Conocephalus* has the head ending in a conical projection.

The Sword-bearer, *C. ensiger*, Harr., is from an inch and three fourths to two inches long from the point of the head to the tips of the wing-covers, and is pale green. The piercer of the female is over an inch long.

The Genus *Xiphidium* is of small size, and the ovipositor is nearly straight.

The Slender Meadow-Grasshopper, *X. fasciatum*, Serville, is about eight tenths of an inch long from the head to the tips of the wing-covers.

The Genus *Orchelimum* has the ovipositor sabre-like in form.

ACRYDII, *Latr.*, OR MIGRATORY LOCUST FAMILY. — This Family contains orthopterous insects which have a large head, short and stout antennæ, very strong hind legs, three-jointed tarsi, and no projecting ovipositor. The genera and species are very numerous. In some cases they move in swarms so great as to darken the sky, and the places where they alight at once become destitute of a green leaf or a blade of grass. Many of them produce a stridulating noise by rubbing their thighs against their wing-covers.

The Genus *Chloëaltis* has the hind legs long and slender, and wings extremely short. The species are generally less than one inch long.

The Genus *Stenobothrus* contains the most common so-called Grasshoppers.

The Genus *Tragocephala* contains the Goat-headed Locusts, which have the antennæ shorter than the thorax, and slightly thickened towards the end; face oblique.

The Genus *Caloptenus* contains the Red-legged Locust and its immediate allies.

The Red-legged Locust, *C. femur-rubrum*, Burm., is about an inch long, grizzled with dingy olive and brown, and the hindmost shanks and feet blood-red, with black spines. The Yellow-striped Locust, *C. bivittatus*, Uhler, is about an inch long, dull green or orange, with a yellowish line on each side from the forehead to the tips of the wing-covers; the hindmost shanks and feet blood-red; the spines tipped with black.

The Genus *Acrydium* embraces the largest members of the family, including the celebrated Migratory Locusts of the East. Some tropical species are four inches long.

The Genus *Œdipoda* contains the most common large species of the United States.

The Carolina Locust, *Œ. carolina*, Burm., is about one inch and a half long, pale yellowish-brown, the under wings black with a broad yellow hind margin.

The Coral-winged Locust, *Œ. phœnicoptera*, Germ., is about one inch and a half long, and is light brown spotted with dark brown on the wing-covers, and the wings are coral-red with an external dusky border.

The Yellow-winged Locust, *Œ. sulphurea*, Burm., is about one inch long, and is dusky brown, the wings deep yellow next the body, dusky at the tip, the yellow portion bounded beyond the middle by a broad dusky brown band.

Fig. 346.

Clouded Locust, *Œ. nebulosa*, Erichs.

The Clouded Locust, *Œ. nebulosa*, Erichs., is about one inch long, dusky brown, with pale wing-covers clouded and spotted with brown, and the wings transparent.

The Genus *Tettix* — Grouse Locusts — has the thorax greatly prolonged over the entire abdomen, and the wing-covers exceedingly minute. The species are small, generally less than half an inch long, and extremely agile.

THYSANOURA, OR SPRING-TAIL FAMILY. — This Family comprises insects which are wingless, and which remind us of the Myriapods.

The Genus *Podura* has the body rather broad, hairy, abdomen with setæ converted into a forked tail bent beneath the body, and used to aid in leaping. They are found in gardens, hot-beds, and on the surface of quiet pools.

The Genus *Lepisma* is long, with silvery scales, and the abdomen has three long bristles. It is found among old books and woollens, and under bark.

SUB-SECTION VII.

THE SUB-ORDER OF NEUROPTERA, OR NET-WINGED INSECTS.

THE Sub-Order of Neuroptera embraces insects which have four membranous net-veined wings, the hinder ones largest, the mouth furnished with jaws, and the abdomen destitute of sting and piercer. In some the transformation is complete, in others only partial. The families are arranged according to Hagens's Synopsis.

TERMITIDÆ, *Latr.*, OR TERMITE FAMILY. — This Family comprises neuroptera which have the body depressed, wings when present longer than the body and laid horizontally on the back, head rounded, thorax nearly square or semicircular, abdomen with two small conical points at the extremity, and the legs short. Termites inhabit warm countries mainly, and are known by the name of White Ants. They live in communities, whose numbers are great. They are among the most destructive of all insects, particularly in the larva state, devouring all kinds of wooden furniture, boards, timber, and all the wood-work of houses, excavating galleries in all directions in these materials, leaving only a thin surface-crust or shell untouched, which on the slightest shock crumbles to pieces. A beautiful edifice in the Isle of France was thus entirely destroyed in a few months after its completion. Some species of this family raise their nests or domiciles above the surface of the ground, in the form of pyramids or turrets, sometimes surmounted with a solid roof, and are so high — ten or twelve feet sometimes — and numerous, that they resemble a little village. Some species make their nests in the form of a globular mass upon trees. Having become perfect insects, Termites leave their retreats and fly off at night in innumerable numbers.

PSOCIDÆ, *Latr.* — This Family includes minute neuroptera which resemble Aphides. They frequent the trunks of trees, old books and papers, and neglected collections of plants or insects.

The Genus *Atropos* contains *A. divinatorius* of authors, the little wingless louse-like insect always seen running over the leaves of dusty books and papers.

PERLARIÆ, *Latr.*, OR PERLA FAMILY. — This Family

comprises neuroptera which are oblong, depressed, with very long and many-jointed antennæ, and the abdomen furnished with two long articulated appendages. In the larva and pupa state they are found in streams under stones, and are active.

Fig. 347.

Stone-Fly, *Pt. regalis*, Newman. Reduced one half.

The Genus *Pteronarcys* retains its branchiæ of the larval state. *Perla* shows disparity between the sexes, the females being the smaller.

EPHEMERIDÆ, *Latr.*, OR MAY-FLY FAMILY. — This Family comprises very short-lived neuroptera which have the body long, slender, soft, the wings of very unequal size, antennæ minute, mouth-parts obsolete, and the abdomen with long articulated appendages. Though these insects live only for a few hours or a day in the perfect state, their existence in the larva and semi-pupa state extends through two or three years, and all this time they live in water. The larvæ have long antennæ, mandibles for chewing, ciliated filaments along the sides of the body for breathing, and three caudal appendages. When about to go through their final changes, the pupæ crawl to the surface, cast off the pupa-skin, and appear at first sight to be fully developed, with the wings expanded to the full size; this is the sub-imago state; they then fly with difficulty to the shore, affix themselves to plants and trees, and cast off a very delicate pellicle. After this the wings are brighter, and the tails greatly increase in length. May-

Fig. 348.

May-Fly, *Ephemera*.

Flies appear in such immense swarms in some parts of Europe, that the people collect their dead bodies into heaps, as dressing for the land. They are common in this country. One of our species is shown in Fig. 348.

ODONATA, *Fabr.*, OR DRAGON-FLY FAMILY. — This Family comprises neuroptera which are known as Devil's Darning-needles, and are distinguished by their long body, large, lustrous, gauze-like wings, large head, large lateral compound eyes, and three ocelli. They are among the most conspicuous of insects, and at once arrest the attention by their size, light and graceful figure, variegated colors, and the great velocity with which they speed their way over fields and meadows, or skim the surfaces of the pools or ponds in search of flies, mosquitoes, and

Fig. 349.

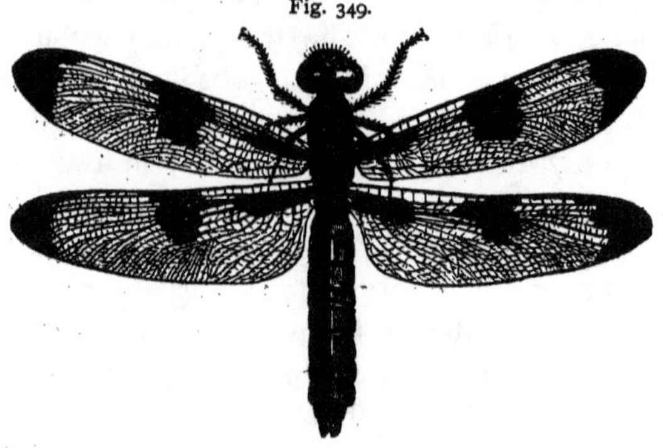

Dragon-Fly, *L. trimaculata*, DeGeer.

other insects, which constitute their food. In the larva and pupa states they live in the water, and when the time comes for the last change, they crawl up the stems of plants, and, having withdrawn from the pupa-skin, which remains clinging to the plant, and dried themselves a little, they spread their wings and dart swiftly away. Though they bite quite fiercely with their jaws, they are without any sort of sting, and are perfectly harmless to man.

The Genus *Agrion* embraces small slender species, of a

blue, green, bronze, or red color. *Æschna* is very large, and has the body cylindrical. The Hero-Dragon-Fly, *Æ. heros*, Fabr., is over three inches and a half long, and expands about five inches. *Libellula* has the body rather flat, as seen in Fig. 348. *Diplax* contains mainly yellowish, yellowish-red, or reddish species, with long and slender feet; some, however, are black or dark brown.

SIALIDÆ, *Leach*, OR CORYDALIS FAMILY.—This Family comprises sluggish neuroptera of moderate or very great size, with large heads, large jaws, and square thoracic rings. They frequent the neighborhood of water, in which they pass the larva state.

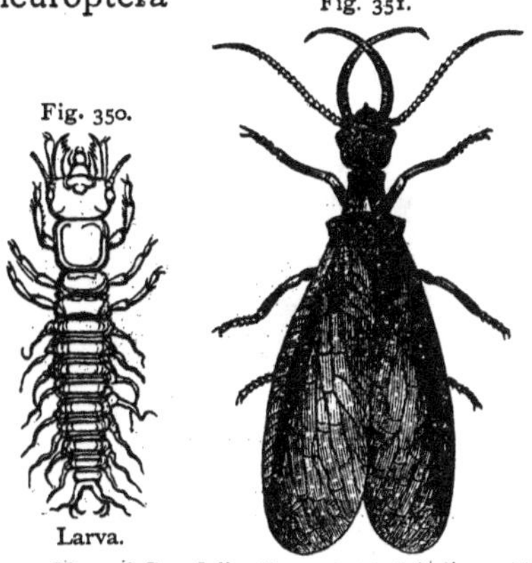
Fig. 350. Larva.
Fig. 351. Horned Corydalis, *C. cornuta* of authors. Reduced one half.

The Genus *Corydalis* contains the Horned Corydalis, *C. cornuta*, Linn. which expands five or six inches. *Sialis* and *Chauliodes* are genera containing mainly black, ferruginous, or dark brown species.

HEMEROBINI, *Latr.*, OR LACE-WINGED FAMILY.—This Family contains very delicate-winged neuroptera, whose larvæ feed upon other insects, especially plant-lice.

The Genus *Hemerobius* contains small brown species. *Polystœchotes* resembles the preceding, but is very much larger, with more acute wings. *Chrysopa* contains species which are about half an inch long, and which expand about one inch and a quarter; wings gauze-like, greenish. eyes golden. They give a fetid smell when disturbed. *Mantispa* has the prothorax much elongated, and the forward feet adapted to seizing prey.

The Genus *Myrmeleon* contains the Ant-Lions, which in

the larva state feed upon ants and other insects, securing them by making pitfalls, at the bottom of which the larva conceals itself, except its jaws, and awaits its prey.

Fig. 352.

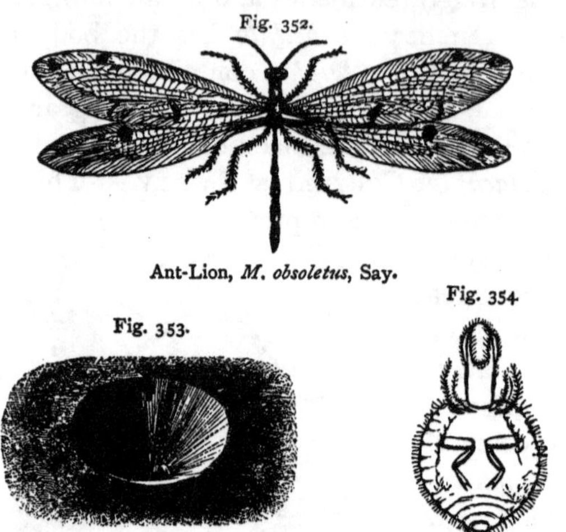

Ant-Lion, *M. obsoletus*, Say.

Fig. 353. Pitfall of Ant-Lion; tips of jaws exposed.

Fig. 354. Larva of Ant-Lion; enlarged.

PANORPATÆ, *Latr.*—This Family comprises small neuroptera which have the head long and narrow, wings narrow and banded, and the tail armed with a forceps-like apparatus. They are found in woods, and feed upon other insects.

PHRYGANIDÆ, *Latr.*, OR CADDICE-FLY FAMILY.—This Family embraces neuroptera which have the wings broad and parallel-veined, and long antennæ. The larvæ are found abundantly at the bottom of ponds and streams in cases composed of bits of wood, or grass, or of shells, or grains of sand, and lined with silk. They carry the case about with them, crawling along the bottom, and even rising to

Fig. 355.

Caddice-Fly, *N. fasciata*, Say.

the surface of the water. Many of them load one side of the case with heavier pieces, so as to keep that side downward.

The Genus *Neuronia* contains the Half-banded Caddice-Fly, *N. fasciata*, Say, which is about an inch long, and expands more than an inch and a half, with the general color tawny.

SUB-SECTION VIII.

THE ORDER OF ARACHNIDA, OR SPIDERS.

THIS Order embraces insects which have the body divided into only two well-marked regions, the head and the hind body; the head and thorax being closely united into one piece. They are wholly destitute of wings, have simple eyes, eight legs, and are subject to no changes in form in coming to maturity, which they reach after moulting their skins six times. Their legs are attached to the forward region, and the hind body is soft or little protected, and with some exceptions is very large in comparison with the head. Most of them feed on insects proper. Some are parasitic on vertebrated animals. They naturally divide into two groups, — Tracheary and Pulmonary Arachnida.

The Tracheary Arachnida perform their respiration by means of tracheæ, which divide near their origin into various branches, — not, as in insects proper, forming two trunks which run parallel to each other through the whole length of the body, receiving air through numerous stigmata, — and receive air through only two stigmata, and these situated near the base of the abdomen; the number of their simple eyes is never more than four. Such are the Pseudo-Scorpions and their allies.

The Pulmonary Arachnida perform their respiration by means of pulmonary sacs, situated in the under part of the abdomen, into which the air is admitted by means of transverse fissures or stigmata varying in number from two to eight. Such are the true Spiders and Scorpions.

ARANEIDÆ, *Latr.*, OR TRUE SPIDER FAMILY. — This Family contains pulmonary arachnida which have palpi resembling feet, with no forceps at the end, or at most terminated in the females by a little hook. The first joint in the males gives origin to more or less complicated sexual appendages. Their frontal cheliceræ, or forceps-antennæ, or mandibles, — for all these names are applied to the same things, — are terminated by a movable hook, flexed inferiorly, underneath which, and near its pointed extremity, is a little opening for the passage of a venomous fluid contained in a gland of the preceding joint. Their jaws are never more than two, and the abdomen is always furnished with from four to six closely approximated cylindrical or conical jointed protuberances with fleshy extremities, which are perforated with numberless small holes for the passage of silky filaments, that have their origin in internal reservoirs. The newly spun filaments are adhesive, and a certain amount of drying is required to fit them for their destined purposes. When the temperature is favorable, a single instant is sufficient to dry and harden them. All the spiders that do not roam about in search of prey weave webs of more or less compact tissue, and varying in form according to the species, where insects upon which they feed may become entangled. As soon as an insect is caught in the web, the spider, hitherto at the centre of his web, or at the bottom, or in a covert at one corner, rushes towards his victim, and endeavors to

Fig. 356.

Spider, *Lycosa lenta*, Hentz.

pierce him, distilling into the wound a prompt and fatal poison. Should the victim be too large or too powerful for the spider, the latter retires till the former becomes more entangled and exhausted. But as soon as practicable, the spider binds his victim firmly with silken bands, and proceeds to feast upon it. The females enclose their eggs in sacs made of the same kind of silk as that of which they make their webs. Some species tear open the egg-sac when the young are ready to hatch; others carry their egg-sacs under their abdomen, or stay near to watch them. The bite of an ordinary spider will kill a fly in a few minutes; the bite of some of the large kinds in South America kills humming-birds; and that of the larger species is poisonous to man, and in some cases fatal. Our numerous species are well described by Hentz.

PEDIPALPI, *Latr.*, OR SCORPION FAMILY.—This Family comprises pulmonary arachnida which have a long body, terminated by a long tail ending in an arcuated point or sting, which discharges a venomous fluid contained in an internal reservoir; they have the palpi very large, with forceps at the extremity. Scorpions inhabit warm countries of both hemispheres, live on the ground under rubbish, among ruins, and sometimes in houses. They run quite rapidly, curving the tail over the back; and they can throw the tail in any direction, and use it both for attack and defence.

Fig. 357.

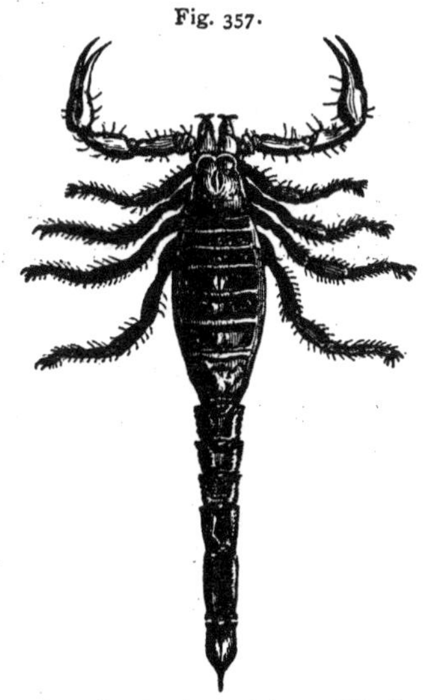

Scorpion, *Buthus spinigerus*, Wood, Texas.

Pseudo-Scorpiones, *Latr.* — This Family comprises small tracheary, scorpion-like animals represented by *Chelifer*, which has a flattened abdomen, the palpi enlarged and bearing a claw at the extremity much like that of a lobster. One minute species is common in books and neglected drawers.

Phalangita, or Long-legs Family. — This Family embraces tracheary arachnids which are popularly known as Daddy-long-legs, or Harvest-men, and which are at once distinguished by the round oval body and long slender legs, which are very easily detached.

Acarina, *Nitsch.*, or Mite Family. — This Family comprises very small, and in many cases microscopic, tracheary arachnida, some of which are found almost everywhere, and which have the forward region joined in a mass with the abdomen, and not divided apparently into rings. The majority are parasitic on other animals.

The Genus *Trombidium* includes the little, square, velvet-red mite seen in spring in flower-beds. *Gamasus* is found on beetles. *Acarus* causes the loathsome disease known as the itch, by burrowing in the skin and flesh of the unfortunate victim. *Ixodes* lives in the woods, and attaches itself to animals, and is known as the Tick.

SUB-SECTION IX.

THE ORDER OF MYRIAPODA, MYRIAPODS, OR CENTIPEDES.

This Order comprises insects which have a very long body, made up of numerous, and generally equal, segments, each of which generally bears two pairs of feet, mostly terminated with a single hook. Their organs of sight consist of a few ocelli. The larvæ when hatched generally have nine

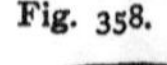

Fig. 358.

American Myriapod, or Galley-Worm, *Iulus.*

rings, but the number increases with age till there are in some cases many times that number. They may be divided into Chilognatha and Chilopoda; the former embracing those which have a large number of rings, each of which bears two pairs of legs, and short and few-jointed antennæ, and feed mainly on decomposing vegetable substances; the latter, those which have the body flattened, with a smaller number of rings, each of which has a single pair of legs, of which the last pair is largest and extended behind, the antennæ long and with numerous joints, and the jaws strong, and which are carnivorous in their habits.

GLOMERIDÆ, *Leach.* — This Family contains chilognaths which have an oval form and few segments, and which have the habit of rolling themselves into a ball.

IULIDÆ, *Leach.* — This Family contains chilognaths which are long, cylindrical, and hard, with numerous short and weak feet, and which crawl rather slowly, and coil the body when at rest. The Genus *Iulus* contains the species. The species of temperate regions are seldom more than one or two inches long; but tropical species are six or seven inches in length.

POLYDESMIDÆ, *Leach.* — This Family embraces chilognaths which are closely related to Iulidæ in structure and habits, but have the body much flattened.

Fig. 359.

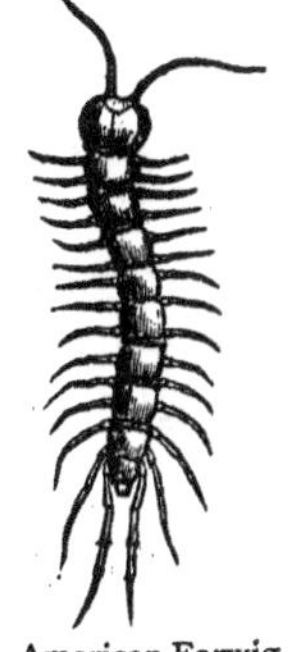

American Earwig, *Lithobius.*

LITHOBIIDÆ, *Newport.* — This Family embraces chilopods, which are well represented by the Genus *Lithobius*, found everywhere under rubbish, and distinguished by the orbicular head, long forty-jointed antennæ, and sixteen rings. They feed upon insects and worms, and run rapidly. They are called Earwigs in this country, but the writer is not aware that they ever enter people's ears.

SCOLOPENDRIDÆ, *Leach*, OR CENTIPEDE FAMILY. — This Family comprises chilopods represented by the Genus *Scolopendra*, which has twenty rings besides those that form the head. Our species are only two or three inches long; but in the warm and tropical regions they are much larger, even a foot in length in some cases. The bite of the tropical species is very poisonous.

GEOPHILIDÆ, *Leach*. — This Family comprises chilopods which are greatly elongated and slender, and with one or two hundred rings in some cases.

Fossil Insects are found in the rocks as low as the Carboniferous inclusive.

SECTION II.

THE CLASS OF CRUSTACEA, OR CRUSTACEANS.

THE Class of Crustacea includes all articulated animals which have the head and thorax essentially in one piece, called cephalo-thorax, and which respire by means of gills, being thus aquatic in their mode of respiration, and which have a straight alimentary canal, and shed, at more or less regular intervals, their usually solid calcareous covering or external skeleton. The body of a Crustacean consists normally of twenty-one segments, fourteen belonging to the cephalo-thorax, and seven to the abdomen; but in the adult these are not generally apparent. These animals are carnivorous, mainly aquatic, and mostly marine; but some live on the land, others in fresh water, and all can remain out of water for a considerable time without perishing. The locomotive organs of Crustaceans are very numerous; for in many cases every segment has its pair of appendages; but notwithstanding all these appendages have the same fundamental structure, they are specialized so as to perform very various functions as those of antennæ, eyes, jaws, claws, feet, paddles, and tail; that is, these extremely different organs are only modified locomotive appendages. The voluntary muscles of these animals are composed of transversely striated and perfectly colorless fibres, and are always inserted on the interior of the skeleton, either directly or by means of its processes. Isolated muscles have a ribbon-like form. The nervous system, in its central mass, consists of an abdominal cord connecting with the cerebral ganglia by a ring enclosing the œsophagus. In the long-bodied crustaceans, the abdominal cord is composed of numerous ganglia, arranged in successive pairs, from before backwards, and connected by longitudinal commissures. When the body is shortened by the fusion of segments, the number of ganglia diminishes by the coalescence or disappearance of several. The sense of touch is highly developed; the sense of sight is present in nearly or quite all; but the organs of hearing have been detected only in the highest. The mouth is generally situated underneath and somewhat back from the anterior border of the head. The heart is situated in the axis of the body, directly under the shell, at the fore part of the back, and is often attached to the internal surface of the skeleton by muscular fibres. The blood is generally colorless. Crustaceans have a wonderful power of repairing injuries to themselves; if a leg or other appendage be broken off, another like it soon grows in its place.

Crustaceans may be divided, according to Dana, into three orders,—Decapods, Tetradecapods, and Entomostracans.

SUB-SECTION I.

THE ORDER OF DECAPODS, OR TEN-FOOTED CRUSTACEANS.

THE Order of Decapods comprises crustaceans which normally have nine cephalic segments, and but five foot segments, each of the latter bearing a pair of so-called feet. It is the highest of the Crustaceans, and embraces Brachyurans, including Anomurans; Macrurans; and Gastrurans.

Fig. 360.

Brachyuran: Fiddler Crab, *Gelasimus vocans*, Milne-Edw.

Fig. 361.

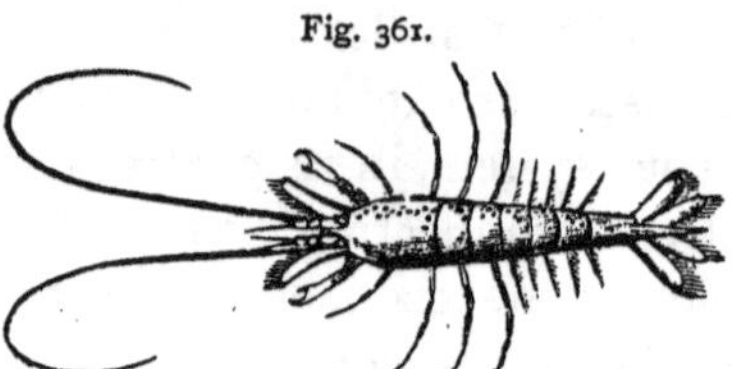

Macruran: Bait Shrimp, *Crangon septemspinosus*, Say.

The Brachyurans include crustaceans which have the hind body — popularly called the tail — shorter than the cephalo-thorax, and, in a state of rest, doubled under the

Fig. 362.

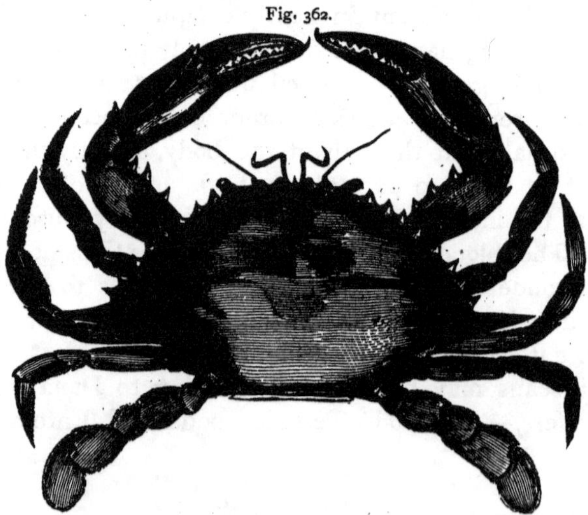

American Edible Crab, *Lupa dicantha*, Milne-Edw. Reduced one half.

latter, where it fits into a groove. In the males it is triangular, and furnished at the base with two or four horn-like appendages; in the females it is wider, and has beneath it four pairs of double hairy appendages to support the eggs. Such are the Crabs, of which there is a large number of genera and species, varying from a very small size to those which, with all their appendages, cover an area of two or three feet square; their forms also are almost endlessly varied. They walk with equal facility forward, backward, sidewise, and oblique. The Anomurans are represented by the Hermit Crabs, *Pagurus*, which inhabit the spiral shells of dead Gasteropods.

The Macrurans comprise long-tailed decapods, those which have the hind body as long as the cephalo-thorax, and generally curved downwards and forwards, as Lob-

Fig. 363.

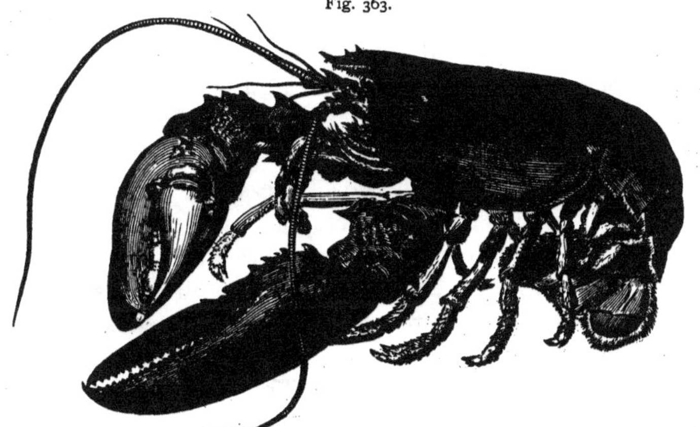

American Lobster, *Homarus americanus*, DeKay.

sters, Cray-Fishes or Fresh-water Lobsters, Shrimps, and Prawns. *Homarus* contains the American Lobster, *H. americanus*, DeKay, which is from one to two feet long.

The Gastrurans — Stomapods, *Latr.* — contain those decapods which have the viscera extending into the abdomen, and the feet mainly approximating the mouth. Their general form bears considerable resemblance to that of Shrimps. They are all marine. *Squilla*, containing the Sea Mantes, is a characteristic genus.

SUB-SECTION II.

THE ORDER OF TETRADECAPODS, OR FOURTEEN-FOOTED CRUSTACEANS.

THE Order of Tetradecapods comprises the crustacea which normally have seven cephalic segments and pairs of appendages, and seven foot-rings or pairs of feet. It contains three groups, — Isopods, including Anisopods; Amphipods, including Læmodipods; and Trilobites. The living species are mainly very small.

The Isopods have the four posterior pairs of thoracic legs in one series, and the three anterior in another; the branchiæ abdominal; and the abdominal members in two sets, the five anterior pairs branchial, and the sixth more or less styliform. Many of the marine species are parasitic on other animals. The land genera inhabit dark, damp situations. Of the latter, *Oniscus*, containing the Sow-Bug, and *Armadillo*, the Pill-Bug, are examples.

Fig. 364.

Sand-Flea, *Orchestia longicornis*, Gould.

The Amphipods have the three posterior pairs of thoracic legs in one series, the four anterior pairs in two other series of two pairs each, and the branchiæ thoracic; the abdominal members in two sets, the three anterior pairs subnatatory, the three posterior styliform. They are known as Sand- and Beach-Fleas.

Fig. 365.

Trilobite.

Trilobites may perhaps be placed here, according to the classification presented by Dana in his learned papers on Cephalization. These very curious animals are all fossil, and abound in the Silurian and Devonian rocks. They also occur in the Carboniferous, where they at last disappear.

SUB-SECTION III.

THE ORDER OF ENTOMOSTRACA, OR ENTOMOSTRACANS.

The Order of Entomostraca comprises crustacea which are defective both in segments and feet as compared with the preceding orders, and rank lower. They have normally six or five cephalic rings, the eight or nine posterior ones belonging to the foot series, but three or more hind pairs of these are usually obsolete. The abdomen is also without appendages. This group embraces Carcinoids; Ostracoids, including Cirripeds; Limuloids; and Rotifers.

The Carcinoids are very small or minute, and are represented by *Cyclops*, *Caligus*, and *Argulus*; the last two genera containing little crustaceans often found adhering to the surfaces respectively of marine and fresh-water fishes.

The Ostracoids have a bivalve carapax, and a short incurved abdomen without terminal appendages. Ex-

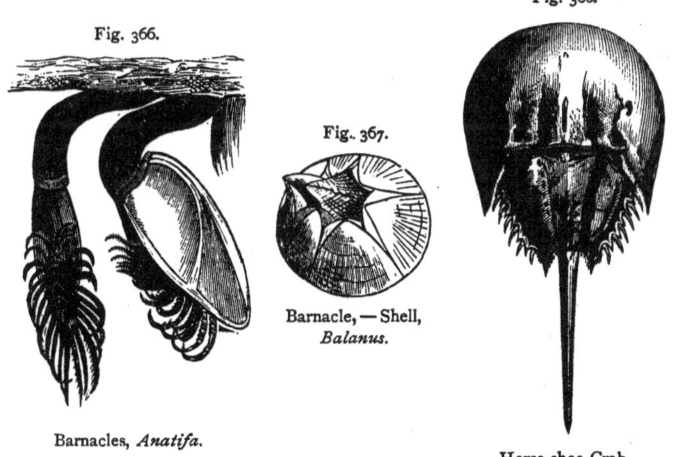

Fig. 366. Barnacles, *Anatifa*.

Fig. 367. Barnacle,— Shell, *Balanus*.

Fig. 368. Horse-shoe Crab, *Limulus polyphemus*.

cepting the Cirripeds or Barnacles, which are from half an inch to several inches long, and marine, they are mainly very small or minute inhabitants of fresh water. *Daphnia*, *Cypris*, and *Limnadia* are characteristic genera.

The Limuloids have the abdomen reduced to a spine. The Genus *Limulus* is represented by the Horse-shoe Crab, Fig. 368, which attains the length of nearly two feet in some cases. This curious crustacean uses the same organs both for walking and eating, — the haunches of the first six pairs of legs performing the functions of jaws.

The Rotifers are minute and mainly microscopic crustaceans, varying from one sixteenth of a line to a line in length, and the organs of locomotion are merely cilia arranged around the head. They are radiate in general appearance, but crustacean in structure.

SECTION III.

THE CLASS OF WORMS.

The Class of Worms includes the lowest articulates, those that present the typical structure of the branch in the most simple and uniform manner. The body is long, and composed of numerous similar rings or segments, and the first, though scarcely differing from the others in appearance, is the head. The nervous system is distributed equally throughout the whole length of the body, and hence these animals are not destroyed when cut asunder, as is the case in the higher animals, where there is a great centre of the nervous system and nervous force. When severed, worms not only do not immediately die, but in many cases the head part at length produces a tail, and the tail part a head, so that one individual in this way becomes two. Division and self-repair, as above, are in some a normal mode of multiplication. Worms have been divided into three orders; — Annelides; Nematoids, including Gordiacei and Acanthocephala; and Trematods, including Leeches, Planariæ, and Cestoids.

SUB-SECTION I.

THE ORDER OF ANNELIDES.

THIS Order comprises worms which have red blood that circulates in a double system of complicated vessels.

SERPULADÆ, OR SERPULA FAMILY. — This Family embraces marine worms whose organs of respiration are in tufts attached to the head and anterior part of the body. In most cases they live in tubes, and hence are often called Tubicolæ. In some the tubes are calcareous, in others horny, the result of transudation; others still are formed of grains of sand, or other particles bound together by a membrane also transuded.

Fig. 369.

Marine Worm, *Serpula*.

The Genus *Serpula* has the anterior portion spread out in the form of a disk armed on each side with bundles of coarse hairs, and on each side of the mouth is a tuft of branchiæ shaped like a fan, and generally tinged with bright colors. At the base of each tuft is a fleshy filament, one of which is always elongated, and expanded at its extremity into a disk which serves as an operculum, and seals up the opening to the tube when the animal is withdrawn into it. The calcareous tubes of the serpulæ cover submarine bodies.

ARENICOLADÆ, OR SAND-WORM FAMILY. — This Family comprises worms which have the organs of respiration in the form of trees, tufts, laminæ, or tubercles, placed on the middle of the body. They are marine, and known as Dorsibranchiatæ, and live free in sand, mud, or water.

LUMBRICIDÆ, OR EARTH-WORM FAMILY. — This Family embraces worms which have no visible external organs of respiration, but appear to respire by the entire surface.

The Genus *Lumbricus* contains the common Earth-

Fig. 370.

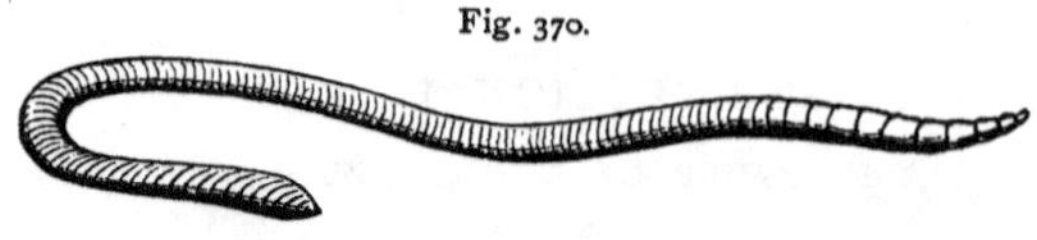

Earth-Worm, *L. terrestris*, Linn.

Worm, or Angle-Worm, *L. terrestris*, Linn., which, when fully grown, is six inches long, reddish, and is composed of more than a hundred rings. It inhabits moist, rich soils.

SUB-SECTION II.

THE ORDER OF NEMATOIDS.

THIS Order comprises worms known as Helminthes, Entozoa, or Intestinal Worms. They live and multiply in the interior of other animals. There is scarcely an animal that is not inhabited by one or more species belonging either to this order or the next.

GORDIACEIDÆ, OR GORDIUS FAMILY. — This Family contains worms which in their larva state inhabit other animals, but not in the adult. They are long, thread-like or hair-like in appearance, and live in fresh water and mud. They are often called hair-worms, and persons ignorant of their history suppose them to be horse-hairs transformed into worms!

SUB-SECTION III.

THE ORDER OF TREMATODS.

THIS Order comprises worms which are provided with organs either at one or both extremities, by which they are able to fix themselves firmly to the walls of internal cavities, to the flesh, or to the external surfaces of animals.

HIRUDINIDÆ, OR LEECH FAMILY. — This Family comprises worms which are oblong, and generally depressed. The mouth is encircled with a lip, and the hind extremity with a flattened disk, both of which are well adapted to adhere to other bodies, and are the principal organs of locomotion. Leeches abound in stagnant waters.

CESTOIDÆ, OR TAPE-WORM FAMILY. — This Family embraces tape-like worms, narrow towards the head and widening behind, which in their mature state live only in the intestines of vertebrated animals. They occur in all the classes of vertebrates; and generally different species are inhabited by different species of cestoids; and sometimes two or three species of cestoids inhabit the same species of vertebrate at the same time, and in some cases the same intestine. Some are scarcely visible; others, the largest, attain in some cases the length of one hundred feet! The width is nearly an inch in some of the widest. The eggs of a cestoid never hatch in the same intestine in which the cestoid lives, but only after they have been taken into the stomach of another and suitable animal. Thence the embryos pierce their way into the blood-vessels, and are carried by the circulation of the blood into various parts of the body, where they develop into larvæ called hydatids. The so-called "measly pork" is pork containing these hydatids, — that is, measly hogs are such as have their muscles more or less filled with the larvæ of cestoids or tape-worms; and if the flesh of such hogs be eaten before cooking, which kills the hydatids, the man or animal eating it takes these hydatids into his intestines, where they are sure to develop into tape-worms. And so in regard to all animals which have tape-worms; they get them by eating other animals in whose tissues there exist hydatids; and the way those animals afflicted with the hydatids get the latter is, as stated above, by swallowing with their food or drink some of the infinitesimal small eggs of the tape-worm. Two hundred species of cestoids have already been described, quite a number of which inhabit man.

CHAPTER IV.

THE BRANCH OF MOLLUSCA, OR MOLLUSKS.

THE Branch of Mollusca embraces animals which have the body soft and enveloped in a muscular skin, and in most cases protected by a shell. They have a distinct nervous system, consisting of ganglia, some of which surround the œsophagus, and others, connected by nervous filaments, are scattered throughout the body; and they

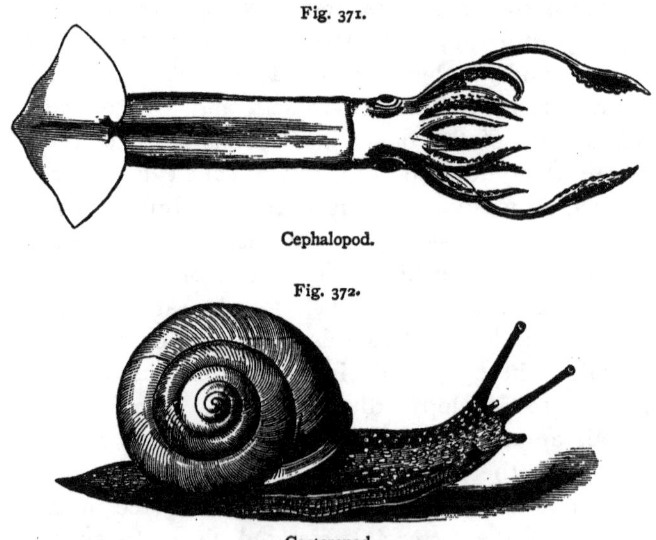

Fig. 371.

Cephalopod.

Fig. 372.

Gasteropod.

are furnished with organs more or less developed appropriate to the five senses of higher animals, and with organs by which food is secured and digested, a heart

Fig. 373. Fig. 374.

Acephals.

with arteries and veins, and organs of respiration and reproduction.* They are divided into three classes, — Cephalopoda or Cephalopods, Gasteropoda or Gasteropods, and Acephala or Acephals.

SECTION I.

THE CLASS OF CEPHALOPODA, OR CEPHALOPODS.

THE Class of Cephalopoda comprises mollusks whose head is distinctly marked, and furnished with a large and prominent eye on either side, and crowned with longer or shorter fleshy flexible appendages, or arms covered with cups, suckers, or hooks. These arms serve both as organs of locomotion and prehension, and the cups or suckers enable these animals to adhere with the greatest tenacity to whatever body they embrace. They swim with the head backwards, and crawl with the head beneath and the body above. Surrounded by the arms or fleshy appendages mentioned above is the mouth, armed with two stout horny jaws resembling the beak of a parrot; the tongue bristles with horny points; the œsophagus swells into a crop, and then communicates with a gizzard as fleshy as that of a bird, to which succeeds a third membranous and spiral stomach, which receives the

* Professor J. D. Dana defines Mollusca as follows: "The structure essentially a soft, fleshy bag, containing the stomach and viscera, without a radiate structure, and without articulations."

bile from the two ducts of a very large liver. A fleshy funnel before the neck affords a passage to the water which aerates the gills, and also an exit for the excretions. The eye consists of several membranes, and is covered by the skin, which becomes diaphanous in that particular spot. The ear is a slight cavity on each side near the brain, where a membranous sac containing a little stone is suspended. Cephalopods are marine, and are remarkable for a peculiar and intensely black fluid which they secrete, and which, when they apprehend danger, they eject into the water, thus discoloring it, and enabling the animals to conceal themselves. They are quick in their movements, predaceous, and very voracious.

SUB-SECTION I.

THE ORDER OF DIBRANCHIATA, OR TWO-GILLED CEPHALOPODS.

THIS Order comprises cephalopods which have two branchiæ, an ink-gland always present, and, with few exceptions, a rudimentary internal shell. Representatives are found in all latitudes, and in open ocean as well as near the shores. The skin of the naked cephalopods contains variously colored pigment-cells; and these animals have the power of effecting such changes in these cells, that the hues of the skin differ from one moment to another.

The Dibranchiata embrace Argonautidæ, Octopodidæ, Teuthidæ, Belemnitidæ, Sepiadæ, and Spirulidæ. The first two families have eight arms, and fixed eyes; the remaining ones have eight arms and two elongated tentacles with expanded ends, and movable eyes.

ARGONAUTIDÆ, OR PAPER-SAILOR FAMILY. — This Family contains cephalopods which have the dorsal arms webbed at the extremity, secreting a symmetrical convoluted shell, which is thin and translucent. The argonaut sits in its shell, with its siphon turned towards the keel, and its dorsal, sail-shaped arms closely applied to the shell, and swims by ejecting water from the funnel.

Fig. 375.

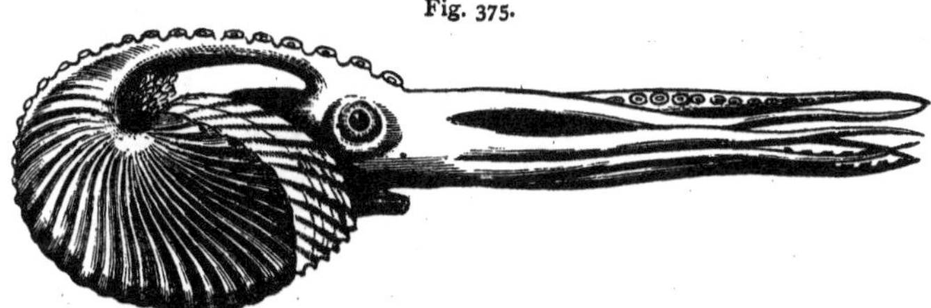

Argonaut, or Paper-Sailor, *Argonauta argo*, Linn. One fourth.* Warm Seas.

This is the Nautilus of Aristotle, who described it as sailing on the surface of the sea, with its sail-shaped arms spread to the breeze; a mere fable.

OCTOPODIDÆ. — This Family comprises cephalopods with eight arms which are similar, elongated, and united

Fig. 376.

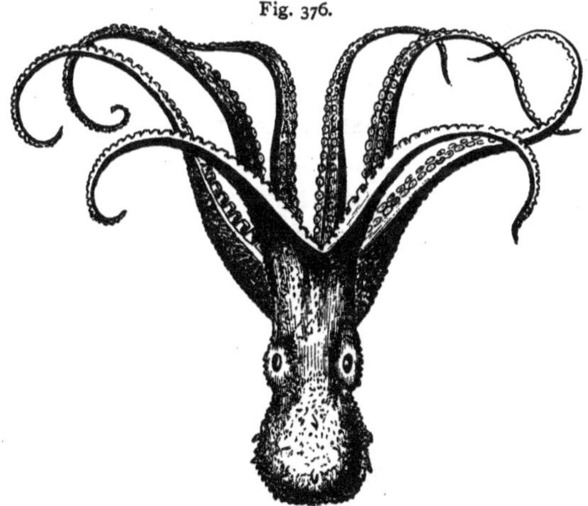

Octopus tuberculatus, Bl. One fifth. Mediterranean.

at the base by a web; and whose shell is represented by two short styles imbedded in the substance of the mantle.

* One fourth of the linear dimensions. So in all similar cases.

They frequent rocky shores. They are exposed for sale in the markets of Naples and Smyrna, and in the bazaars of India. They vary from an inch to two feet in length.

TEUTHIDÆ, LOLIGO, OR SQUID FAMILY.—This Family comprises cephalopods with an elongated body, fins short,

Fig. 377.

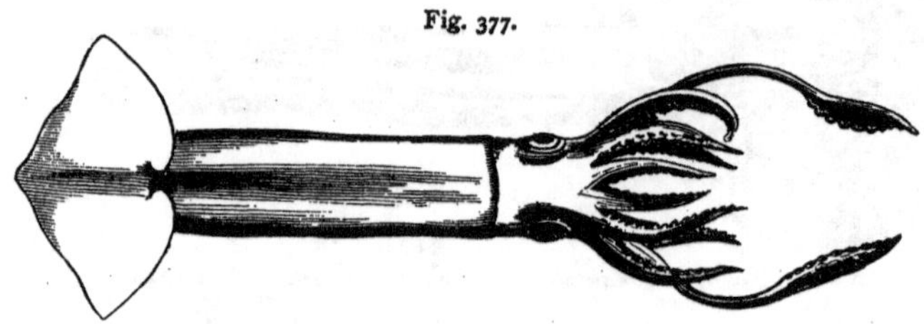

Squid, *Loligo Bartramii*, LeS. One half. Atlantic Coast of the United States.

broad, and mostly terminal. The internal shell, or pen, consists of three parts, a shaft and two lateral expansions.

BELEMNITIDÆ, OR BELEMNITE FAMILY.—This Family embraces cephalopods which have a shell consisting of a pen terminating at the hind extremity in a chambered cone. A hundred species have been found imbedded in the rocks, but there are no living representatives.

SEPIADÆ, OR CUTTLE-FISH FAMILY.—This Family comprises cephalopods whose calcareous internal shell or cuttle-bone consists of a broad laminated plate, terminating behind in a hollow imperfectly chambered apex. They are distributed world-wide, and are from three inches to three feet long.

SPIRULIDÆ.—This Family comprises cephalopods whose shell is wholly nacreous, discoidal, whorls separate,

Fig. 378.

Spirula lævis, Gray. One half. New Zealand.

chambered, and with a ventral siphuncle. Three species are known, which inhabit warm seas.

SUB-SECTION II.

THE ORDER OF TETRABRANCHIATA, OR FOUR-GILLED CEPHALOPODS.

THIS Order comprises cephalopods which have four branchiæ, an external chambered shell, eyes pedunculated, mandibles calcareous, and arms very numerous. The shell is an extremely elongated cone, and is straight or variously folded or coiled, and is divided into chambers by partitions called septa, the animal as it grows forming a wall behind itself at regular intervals, and always living in the outer chamber, communicating, however, by a tube or siphuncle with all the others. The Tetrabranchiates are best known under the name of Chambered Shells. Although but one or two species are now living, more than fourteen hundred species have been found imbedded in the rocks.

Fig. 379.

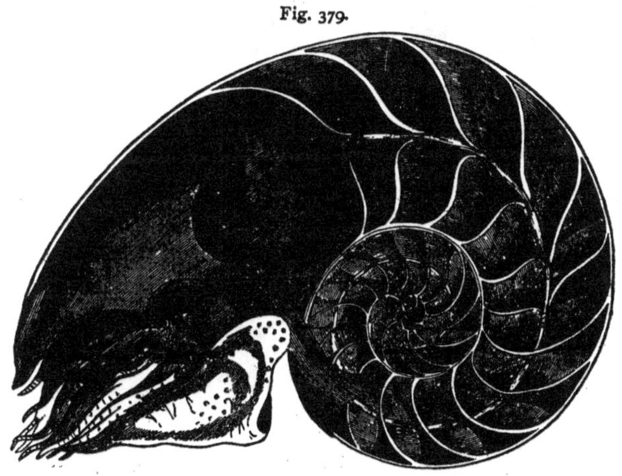

Pearly Nautilus, *Nautilus pompilius*, Linn. Cut open to show the chambers and siphuncle. One half. Pacific and Indian Oceans.

NAUTILIDÆ, OR NAUTILUS FAMILY. — This Family contains a single living representative species, or perhaps two, and its fossil allies. They have the sutures of the shell simple, and siphuncle central.

ORTHOCERATIDÆ. — This Family contains cephalopods which have the shell straight, curved, or discoidal. They are all fossil, and abound in the Paleozoic Rocks.

AMMONITIDÆ, OR AMMONITE FAMILY. — This Family embraces cephalopods which have a shell bearing a close resemblance to that of the Nautilidæ, but with the sutures angulated, or lobed and foliated, and the siphuncle external, that is dorsal, as regards the shell. They are all fossil, and abound in the rocks below the Tertiary. The species are numerous, and vary from an inch to two or three feet in diameter.

Fig. 380.

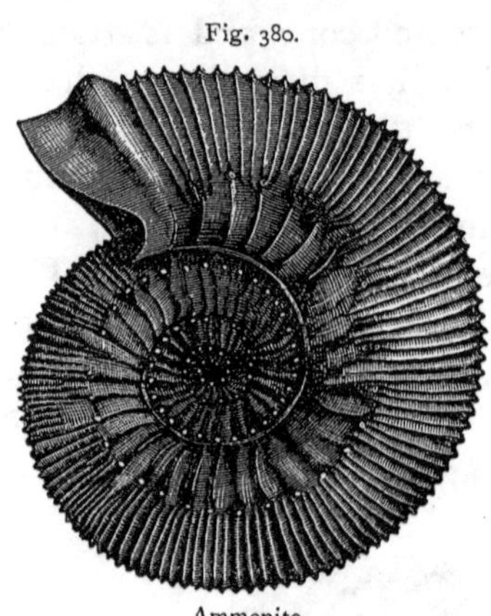

Ammonite.

SECTION II.

THE CLASS OF GASTEROPODA, OR GASTEROPODS.

THE Class of Gasteropoda comprises mollusks which effect their locomotion by means of a broad muscular expansion called a foot. Some of them are destitute of a shell, but most are protected by a single shell, and are often called Univalves. When first hatched, they are always provided with a shell, but in many families this is soon concealed in the mantle, or wholly disappears.

Gasteropods, considered in regard to their manner of breathing, may be divided into two groups; — Pulmonifera, or Air-Breathers; and Branchifera, or Water-Breathers. The former undergo no apparent metamorphosis; the latter at first have a shell capable of entirely concealing them, and closed by an operculum, and, instead of creeping, they swim with a pair of ciliated appendages growing from the sides of the head. The shells of Gasteropods, though always single, and usually spiral, vary almost endlessly in form and color. The names applied to the different parts of a gasteropod shell are shown by Fig. 381. Shells which are always concealed by the mantle are colorless; those which are covered when the animals expand are glazed or enamelled on the

Fig. 381.

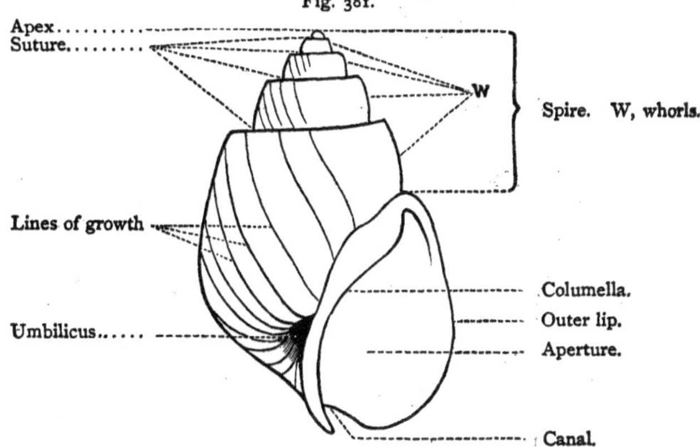

surface; and all other shells are more or less covered with an epidermis. The Class is composed of three orders, — Gasteropoda proper, Heteropoda, and Pteropoda.

SUB-SECTION I.

THE ORDER OF GASTEROPODA PROPER.

THIS Order embraces about forty families, or, according to the more recent writers, a much greater number, as several of the old families have been much subdivided.

STROMBIDÆ, OR STROMBUS FAMILY. — This Family

contains gasteropods which have a shell with an expanded lip deeply notched near the canal, and a claw-shaped operculum. They feed on dead animals. Their shells

Fig. 382.

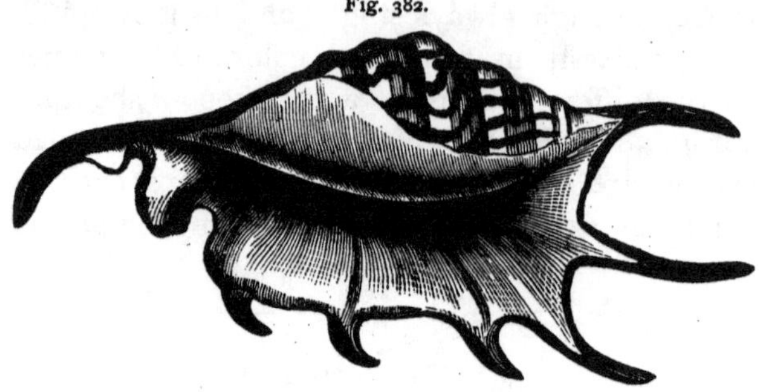

Pteroceras lambis, L. One half. Chinese Seas.*

Fig. 383.

Fig. 384.

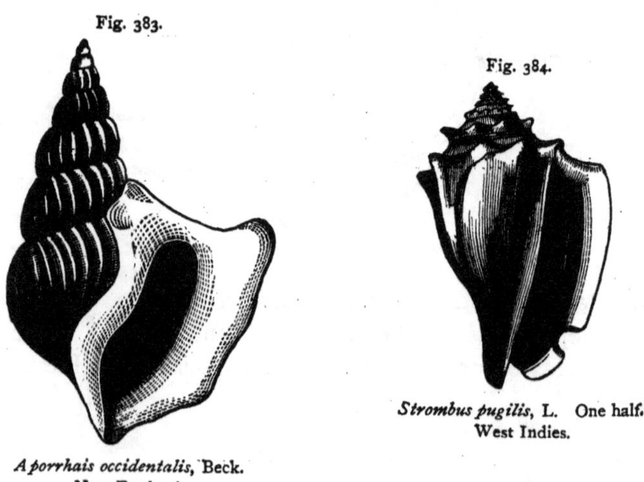

Aporrhais occidentalis, Beck. New England.

Strombus pugilis, L. One half. West Indies.

are extensively used in the manufacture of shell cameos. The living species are estimated at about eighty, and the fossil at two hundred. Marine.

* Foreign shells are quite as likely to fall into the hands of American students as those of our own coast.

MURICIDÆ, OR MUREX FAMILY. — This Family is represented by about nine hundred living, and about seven hundred fossil species of gasteropods. They are predatory on other mollusks. Marine.

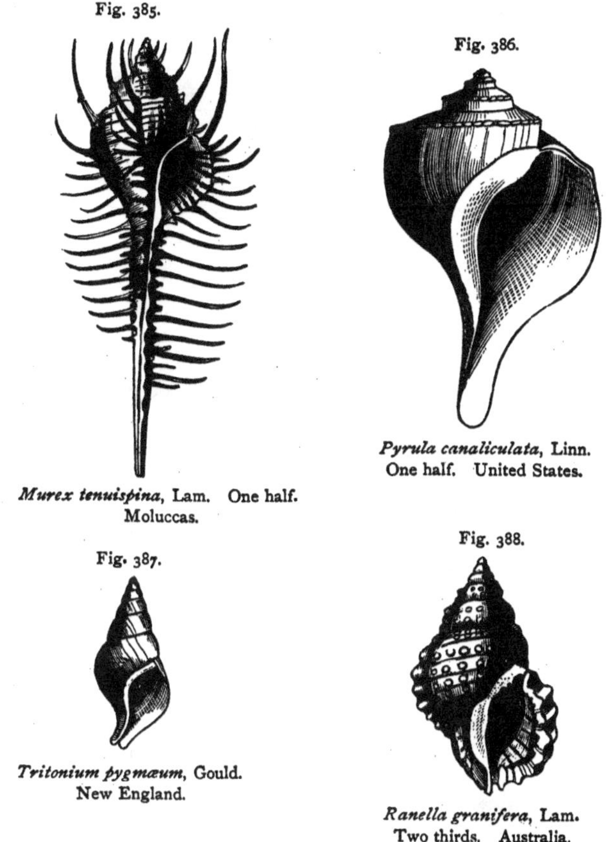

Fig. 385.

Murex tenuispina, Lam. One half. Moluccas.

Fig. 386.

Pyrula canaliculata, Linn. One half. United States.

Fig. 387.

Tritonium pygmæum, Gould. New England.

Fig. 388.

Ranella granifera, Lam. Two thirds. Australia.

BUCCINIDÆ, OR WHELK FAMILY. — This Family contains gasteropods which have the shell notched in front, or the canal abruptly reflected. There are about one thousand living, and three hundred and fifty fossil species. They are marine and carnivorous.

Fig. 389.

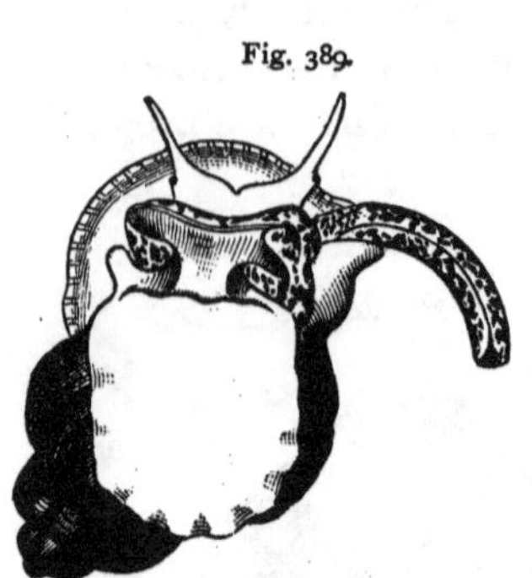

Buccinum undatum, L.
North Atlantic.

Fig. 390.

Harpa ventricosa, Lam.
One half. Mauritius.

Fig. 391.

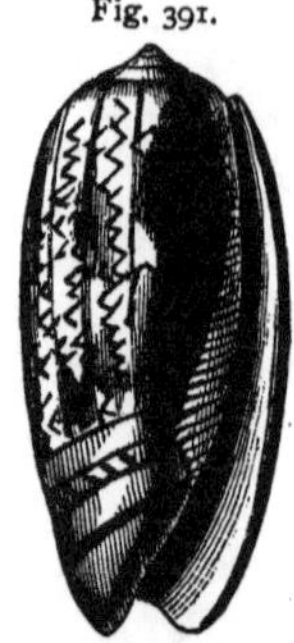

Oliva porphyria, L.
One half. Panama.

Fig. 392.

Ricinula arachnoides, Lam.
China.

Fig. 393.

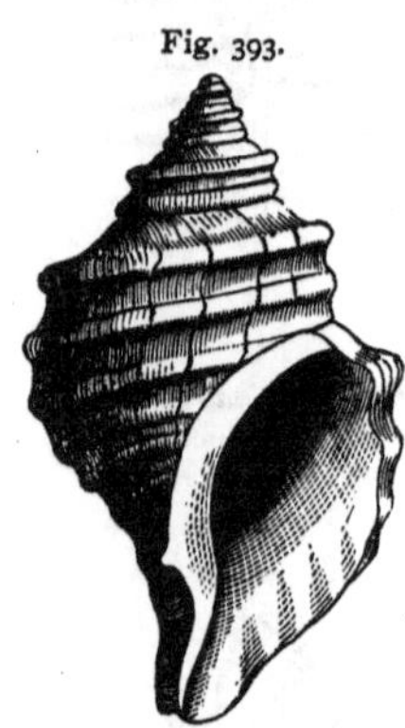

Fusus decemcostatus, Say.
United States.

CONIDÆ, OR CONE FAMILY. — This Family embraces gasteropods which have the shell inversely conical, aperture large and narrow, outer lip notched, and the operculum minute. There are eight hundred and fifty living, and four hundred fossil species.

Fig. 394.

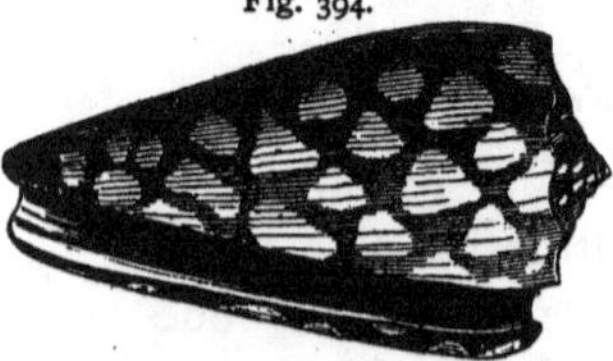

Conus marmoreus, Gm. Two thirds.
China.

VOLUTIDÆ, OR VOLUTE FAMILY.—This Family contains gasteropods which have the shell turreted or convo-

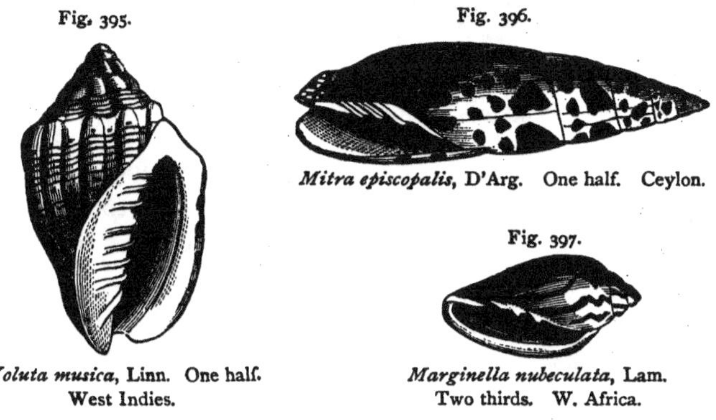

Fig. 395. *Voluta musica*, Linn. One half. West Indies.

Fig. 396. *Mitra episcopalis*, D'Arg. One half. Ceylon.

Fig. 397. *Marginella nubeculata*, Lam. Two thirds. W. Africa.

lute, and the aperture notched in front. There are seven hundred living, and two hundred fossil species. Marine.

CYPRÆIDÆ, OR COWRY FAMILY.—This Family comprises gasteropods which have the shell convolute, enam-

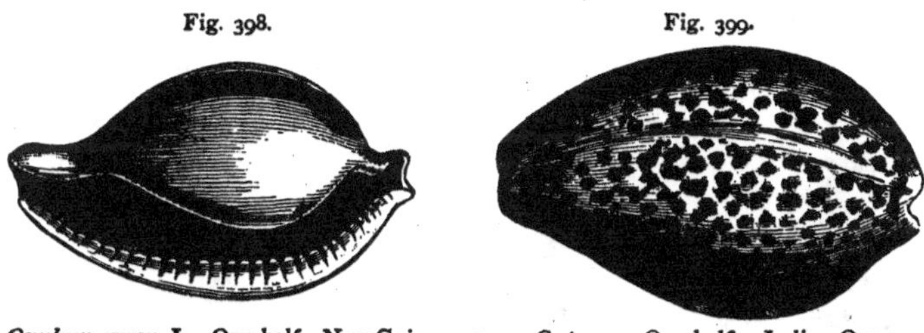

Fig. 398. *Ovulum ovum*, L. One half. New Guinea.

Fig. 399. *Cypræa*. One half. Indian Ocean.

elled, spire concealed, aperture narrow and channelled at each end, the outer lip inflected, and no operculum. There are two hundred living, and one hundred fossil species. Marine.

Figs. 400, 401. *Trivia europea*, Mont. Britain.

NATICIDÆ, OR NATICA FAMILY.—This Family contains gasteropods which have the shell globular, few-whorled, spire small, obtuse, aperture semi-lunar, and

Fig. 402.

Sigaretus haliotoides, L. West Indies.

Fig. 403.

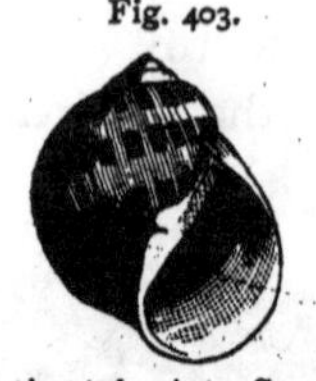

Natica triseriata, Say. Coast of New England.

the operculum horny. There are two hundred and fifty living, and three hundred and fifty fossil species. Marine.

PYRAMIDELLIDÆ. — This Family comprises gasteropods which have the shell spiral, turreted, and the aperture small. The living species are about two hundred, and the fossil over three hundred; all marine. Many of the fossil species are gigantic compared with those now existing.

Fig. 404.

Pyramidella dolobrata, Gmel. West Indies.

Fig. 405.

Pyramidella elegantissima, Mont. One half. Britain.

CERITHIADÆ. — This Family contains gasteropods which have the shell spiral, elongated, many-whorled, aperture channelled in front, the lip generally expanded in the adult, and the operculum horny and spiral. The species are marine, estuary, or freshwater. The living species are two hundred, and the fossil six hundred.

Fig. 406.

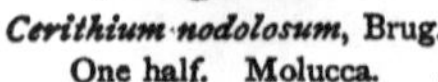

Cerithium nodolosum, Brug. One half. Molucca.

Fig. 407.

Melania. Western States.

Fig. 408.

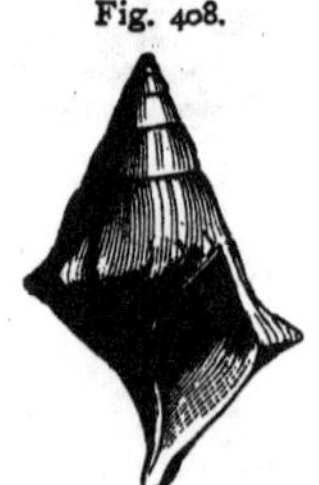

Io. Southern States.

MELANIADÆ. — This Family comprises gasteropods which have the shell spiral, turreted, aperture often channelled or notched in front, and a dark thick epidermis; operculum horny and spiral. They inhabit fresh waters. There are five hundred living species. Figs. 407, 408.

Turritellidæ, or Wentle-trap Family. — This Family embraces gasteropods which have the shell tubu-

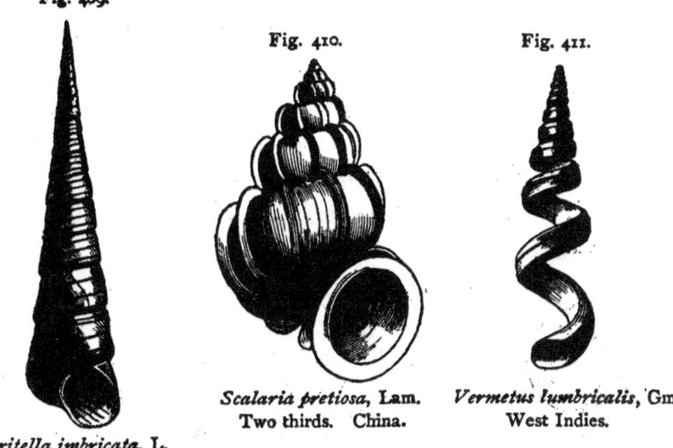

Fig. 409. *Turritella imbricata*, L. West Indies.

Fig. 410. *Scalaria pretiosa*, Lam. Two thirds. China.

Fig. 411. *Vermetus lumbricalis*, Gm. West Indies.

lar or spiral, and the upper part partitioned off; all marine. There are two hundred living and three hundred fossil species.

Litorinidæ, or Periwinkle Family. — This Family contains gasteropods which have the shell spiral, turbinated or depressed, the aperture rounded, and the operculum horny. They inhabit the sea near the shore, and feed on algæ. The living species are more than three hundred, and the fossil more than two hundred.

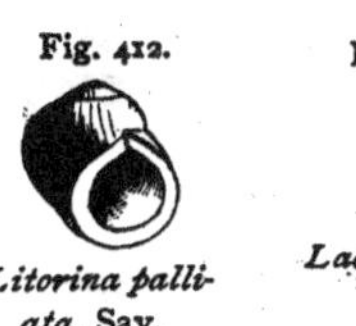

Fig. 412. *Litorina palliata*, Say.

Fig. 413. *Lacuna vincta*, Mont.

Paludinidæ, or River-Snail Family. — This Family embraces fresh-water gasteropods which have the shell conical or globular, and covered with an olive-green epidermis; tentacles long and slender, and eyes on short

Fig. 414.

Valvata tricarinata, Say.
United States.

Fig. 415.

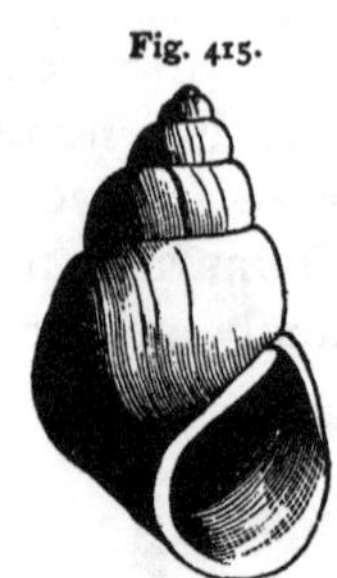

Paludina integra, Say.
Western and Middle States.

pedicels outside the tentacles. Operculum horny or shelly. There are one hundred and fifty living species.

Fig. 416.

Neritina zebra, Brug.
Pacific.

Fig. 417.

Nerita ustulata, L.
Scinde.

NERITIDÆ, OR NERITA FAMILY.—This Family contains gasteropods which have the shell thick, globular. Three hundred living species.

Fig. 418.

Trochus zizyphinus, L. Britain.

TURBINIDÆ, OR TOP-SHELL FAMILY.—This Family comprises gasteropods which have the shell turbinated and the operculum very small. The shell is brilliant pearly when the epidermis is removed. There are about nine hundred living, and as many fossil species. Marine.

Fig. 419.

Haliotis tuberculata, L.
Britain.

HALIOTIDÆ, OR EAR-SHELL FAMILY.—This Family contains gasteropods which have the shell spiral, ear-shaped, or trochiform, and no operculum. *Haliotis* is much used for ornamental work. There are one hundred living, and more than one hundred fossil species. Marine.

IANTHINIDÆ. — This Family is represented by the Violet-Snail and its half-dozen allies.

Fig. 420.

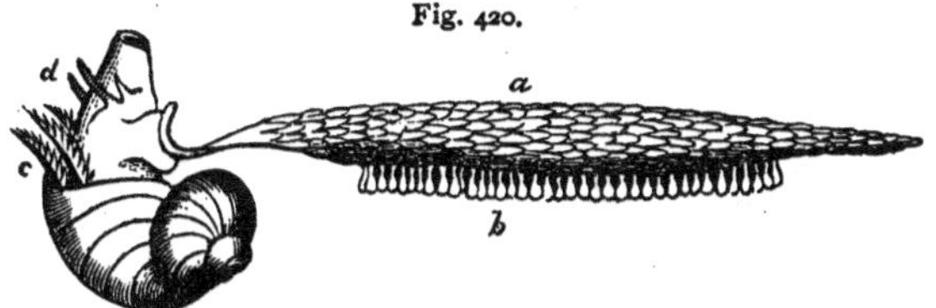

Ianthina fragilis, Lam. Atlantic.
a, raft; *b*, egg capsules; *c*, gills; *d*, tentacles and eye-stalks.

FISSURELLIDÆ, OR KEY-HOLE LIMPET FAMILY. — This Family embraces gasteropods which have the shell conical, and apex recurved. There are about two hundred living, and one hundred fossil species. Marine.

Fig. 421.

Fissurella Listeri, Orb.
West Indies.

CALYPTRÆIDÆ, OR BONNET-LIMPET FAMILY. — This Family contains gasteropods which are found adhering to stones and shells, most of them never quitting the spot where they first settle. There are one hundred and sixty living, and one hundred fossil species, all marine.

Fig. 422.

Calyptræa equestris, L.
Philippines.

Fig. 423.

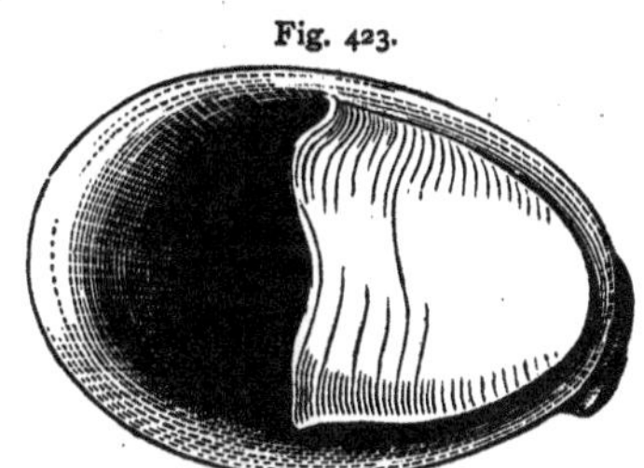

Crepidula fornicata, Say.
New England.

PATELLIDÆ, OR LIMPET FAMILY. — This Family comprises gasteropods which have the shell conical, with the apex turned forward. There are more than one hundred living, and one hundred fossil species.

Fig. 424.

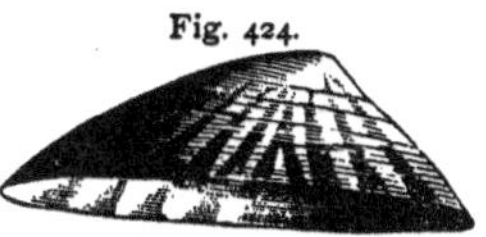

Patella testudinalis, Müll.
Coast of New England.

Fig. 425.

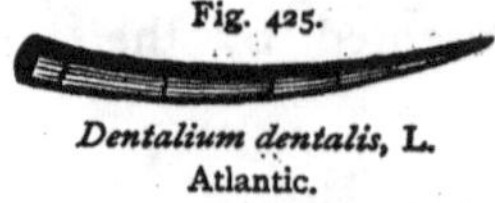

Dentalium dentalis, L. Atlantic.

DENTALIADÆ. — This Family contains the Tooth-shells, of which there are fifty living, and seventy fossil species; all marine.

Fig. 426.

Chiton ruber, L. New England.

CHITONIDÆ, OR CHITON FAMILY. — This Family comprises gasteropods which have a shell composed of eight transverse imbricating plates, lodged in a coriaceous mantle. There are more than two hundred living species, and about one eighth as many fossil; all marine.

Fig. 427. *Helix albolabris*, Say. United States.

Fig. 428. *Bulimus excelsus*, Gould. California.

Fig. 429. *Pupa incana*, Say. Florida.

Fig. 430. *Succinea obliqua*, Say. Western States.

Fig. 431. *Helix albolabris*, Say. Northern States.

HELICIDÆ, OR LAND-SNAIL FAMILY. — This Family contains terrestrial gasteropods which are distributed world-wide, and of which there are four thousand living species, and three hundred fossil.

LIMACIDÆ, OR SLUG FAMILY. — This Family comprises terrestrial gasteropods which have the shell small or rudimentary, and usually internal. Seventy living and a few fossil species.

Fig. 432.

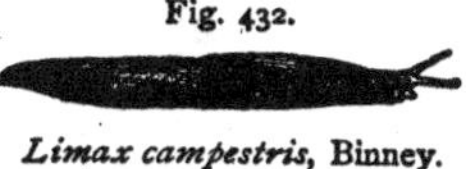

Limax campestris, Binney. New England States.

LIMNEIDÆ, OR POND-SNAIL FAMILY. — This Family contains gasteropods which have the shell thin and horn-

Fig. 433. *Physa heterostropha*, Say. United States.

Fig. 434. *Planorbis lentus*, Say. United States.

Fig. 435. *Limnæa desidiosa*, Say. United States.

colored, and capable of containing the whole animal when retracted. Limnæids inhabit fresh water, and deposit their spawn in oblong transparent masses on aquatic plants and stones. There are one hundred and sixty living and about as many fossil species.

AURICULIDÆ. — This Family comprises gasteropods which have the shell spiral, with a horny epidermis, and the body whorl large. Fifty or sixty species are known.

CYCLOSTOMIDÆ. — This Family embraces seven hundred species of gasteropods, whose general form is shown by Figs. 436, 437. Twenty fossil species. Terrestrial.

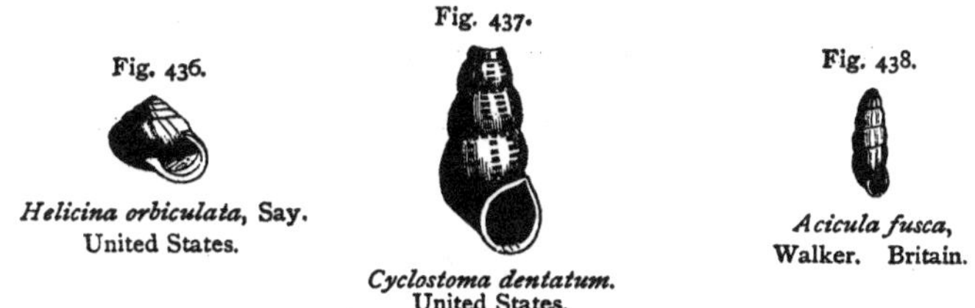

Fig. 436. *Helicina orbiculata*, Say. United States.

Fig. 437. *Cyclostoma dentatum*. United States.

Fig. 438. *Acicula fusca*, Walker. Britain.

ACICULIDÆ. — This Family contains gasteropods which have the shell long and cylindrical. Twenty-five species.

Fig. 439.

Tornatella tornatilis, L. Britain.

Fig. 440.

Bulla solitaria, Say. United States.

TORNATELLIDÆ.—This Family comprises gasteropods which have the shell solid, aperture long and narrow, and columella plaited. There are fifty living and one hundred and fifty fossil species.

BULLIDÆ. — This Family contains one hundred and fifty living and nearly one hundred fossil species. They have the shell globular or cylindrical, convoluted and thin. Marine.

DORIDÆ, TRITONIDÆ, ÆOLIDÆ, ELYSIADÆ, &c. — These Families contain sea-slugs which have no shell except in the embryo state. Three hundred species. Figs. 441–444.

Fig. 441.

Eolis coronata, Forbes. Britain.

Fig. 442.

Doris Johnstoni, Alder & Hancock. Britain.

Fig. 444.

Elysia viridis, Mont. Britain.

Fig. 443.

Tritonia plebeia, Johnston. Britain.

SUB-SECTION II.

THE ORDER OF HETEROPODA, OR HETEROPODS.

Figs. 445, 446.

Atlanta Peronii, LeS. South Atlantic.

THIS Order comprises mollusks whose general appearance differs considerably from the true gasteropods. They live in the open sea, and swim at the surface. They belong to two families, — Firolidæ and Atlantidæ.

SUB-SECTION III.

THE ORDER OF PTEROPODA, OR PTEROPODS.

THIS Order also embraces mollusks which inhabit the open sea, where they move in immense swarms, sometimes leagues in extent. They comprise three families, — Hyaleidæ, Limacinidæ, and Cliidæ.

Fig. 447.

Hyalea tridentata, Gmel. Atlantic and Mediterranean.

Fig. 448.

Limacina antarctica, Hooker. South Polar Seas.

Fig. 449.

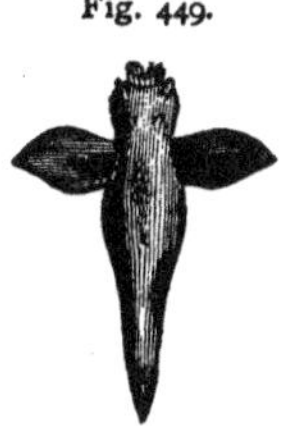

Clio borealis, Brug. Arctic Seas.

SECTION III.

THE CLASS OF ACEPHALA, OR ACEPHALS.

THIS Class comprises mollusks which have the head obscurely indicated. There are four orders, — Lamellibranchiata, Tunicata, Brachiopoda, and Bryozoa.

SUB-SECTION I.

THE ORDER OF LAMELLIBRANCHIATA, OR LAMELLIBRANCHIATES.

THIS Order contains acephalous mollusks which have their gills in lamellæ on the sides, and protected by a shell composed of two valves occupying a similar position, namely, right and left. A single valve with the names of its different parts is shown in Fig. 450, on the next page. There are more than twenty families.

OSTREIDÆ, OR OYSTER FAMILY. — This Family comprises bivalves, of which the common Oyster is a well-known representative. There are about three hundred living and one thousand fossil species. *Ostrea* and *Pecten*, Fig. 451, are the prominent genera.

Fig. 450.

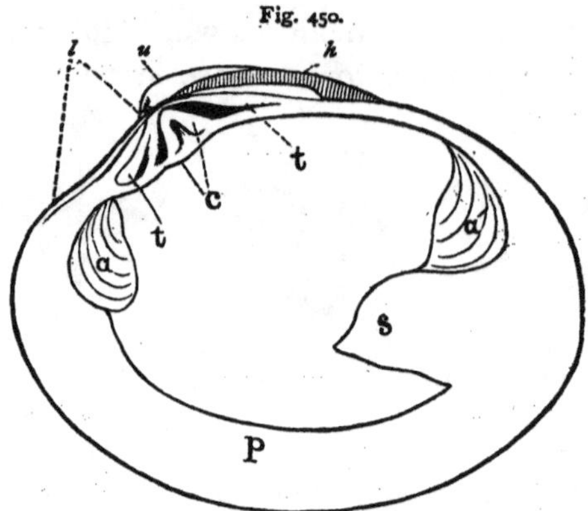

Names of the parts of a Bivalve Shell.

a, anterior retractor; *a'*, posterior retractor; *t*, lateral teeth; *c*, cardinal tooth; *l*, lunale; *u*, umbo; *h*, hinge ligament; *s*, sinus occupied by retractor of siphons; *p*, pallial impression.

AVICULIDÆ, OR PEARL-OYSTER FAMILY. — This Family embraces acephalous mollusks which have the valves unequal and very oblique. They inhabit tropical and temperate seas, and yield the Mother-of-pearl, and the Oriental pearls, so highly prized. They are found in

Fig. 451.

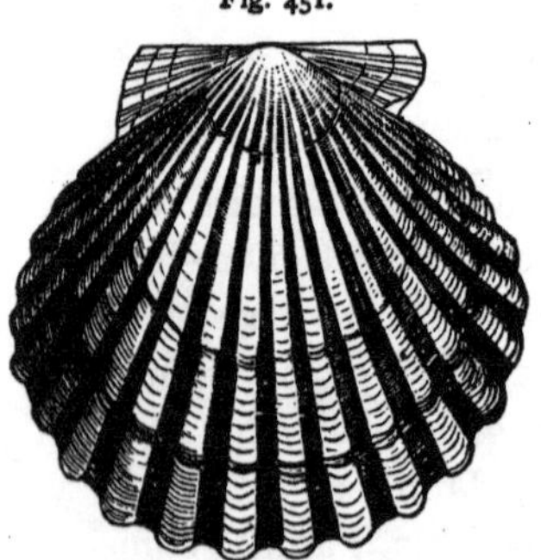

Pecten irradians, Lam. About one half. From Cape Ann southward.

Fig. 452.

Mytilus edulis, L. Grows larger. Both shores of the Atlantic.

about twelve fathoms of water. There are about one hundred living and six hundred fossil species. Fig. 453.

MYTILIDÆ, OR SEA-MUSSEL FAMILY. — This Family comprises acephala which have the shell equivalve, oval, or elongated, and the epidermis thick and dark. They seek concealment, and spin a nest of sand, or burrow in mud-banks. There are more than one hundred living, and two hundred and fifty fossil species. Fig. 452.

Fig. 453.

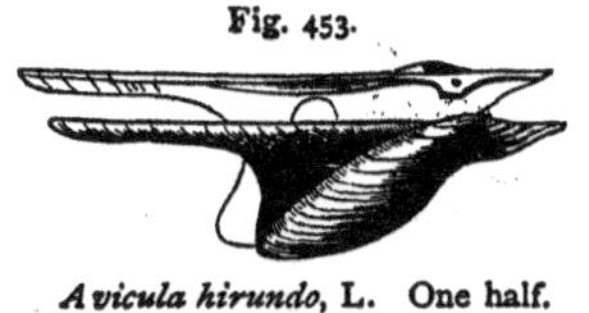

Avicula hirundo, L. One half. Mediterranean.

Fig. 454.

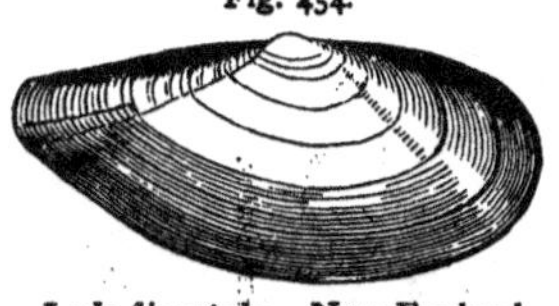

Leda limatula. New England.

ARCADÆ. — This Family embraces acephala which have the shell regular, with a strong epidermis, and the hinge with a row of comb-like teeth. There are three hundred living, and six hundred fossil species. Fig. 454.

TRIGONIADÆ, OR TRIGONIA FAMILY. — This Family contains acephala which have the shell trigonal, and the hinge teeth few and diverging. There are three living species, and one hundred and thirty or more fossil.

UNIONIDÆ, OR POND AND RIVER MUSSEL FAMILY. — This Family embraces a large number of bivalves which are found in ponds, brooks, rivers, and fresh-water lakes. They are the Naïdes of authors. Figs. 456–460.

CHAMIDÆ.— This Family comprises acephala which have the valves unequal, thick, two hinge teeth in one valve, and one in the other. Fifty living species.

TRIDACNIDÆ. — This Family comprises very large bivalves of the Indian and Pacific Oceans. The shell sometimes weighs five hundred pounds, the animal twenty.

CARDIADÆ, OR COCKLE FAMILY. — This Family embraces acephala which have the shell cordate and ornamented, with radiating ribs. There are two hundred living, and three hundred fossil species. Marine. Fig. 455.

Fig. 455.

Cardicum islandicum, Linn. One half. New England.

Fig. 456.

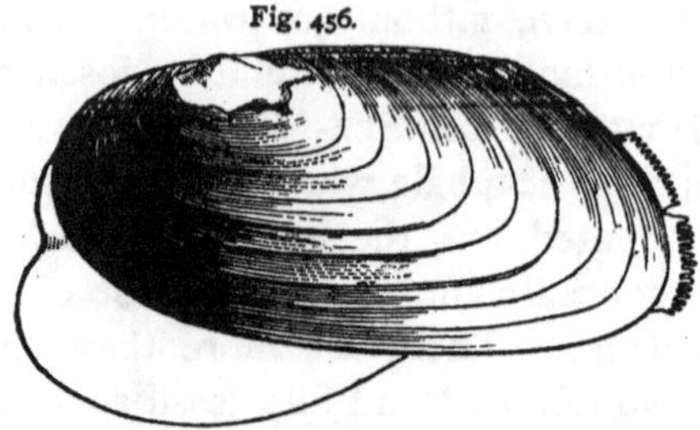

Unio complanatus, Lea. New England and westward.

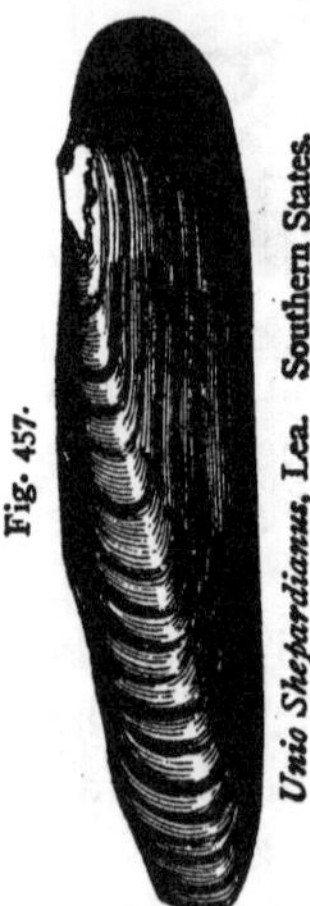

Fig. 457. *Unio Shepardianus*, Lea. Southern States.

Fig. 458.

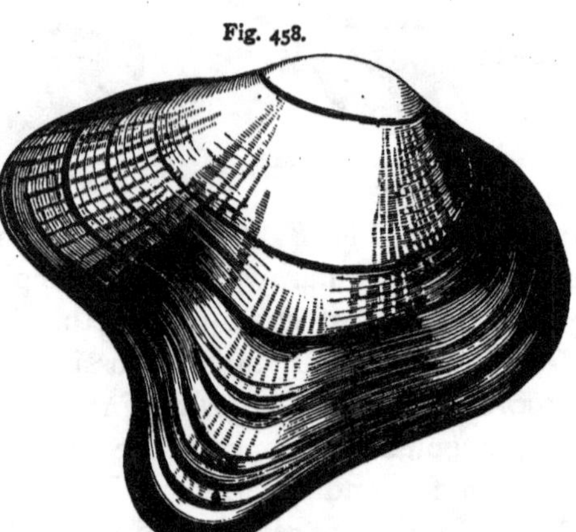

Unio flexuosus, Raf. Two thirds. Western States.

Fig. 459.

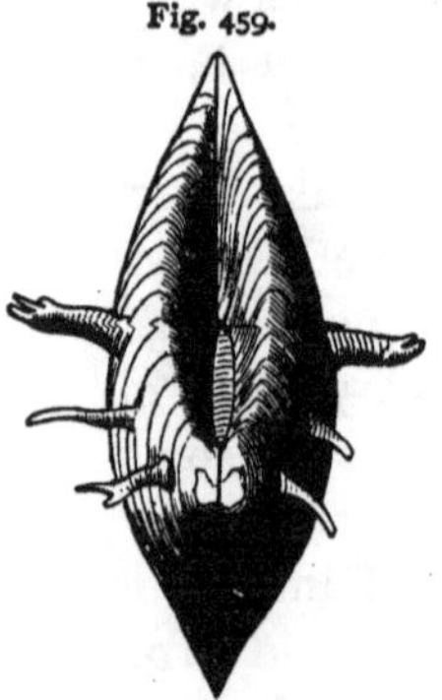

Unio spinosa, Lea. One half. Georgia.

Fig. 460.

Unio clava, Lam. Two thirds. Western States.

LUCINIDÆ. — This Family contains acephala which have the shell orbicular. There are one hundred and twenty living, and three hundred and fifty fossil species. Marine. Fig. 463.

CYCLADIDÆ. — This Family contains acephala which have the shell sub-orbicular, with a thick horny epidermis. There are two hundred living, and one hundred fossil species. Fresh water. Figs. 462, 464.

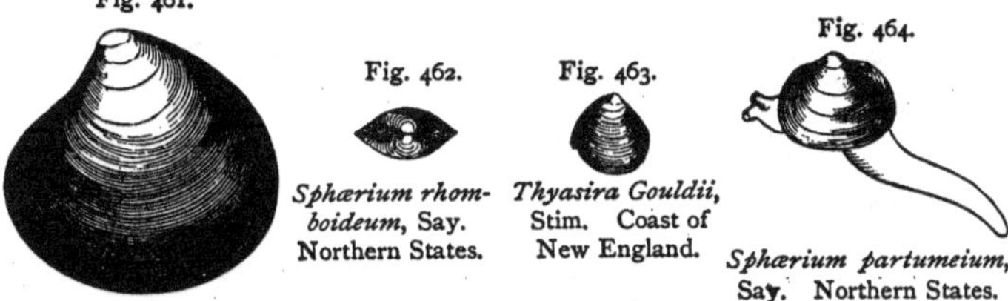

Fig. 461. *Astarte castanea*, Say. Casco Bay and southward.

Fig. 462. *Sphærium rhomboideum*, Say. Northern States.

Fig. 463. *Thyasira Gouldii*, Stim. Coast of New England.

Fig. 464. *Sphærium partumeium*, Say. Northern States.

CYPRINIDÆ. — This Family comprises acephala which have the shell oval or elongated, valves solid, and the epidermis thick and dark. There are more than one hundred living, and three hundred and fifty fossil species. Marine. Fig. 461.

VENERIDÆ. — This Family embraces acephala which have the shell sub-orbicular or oblong, and the hinge with

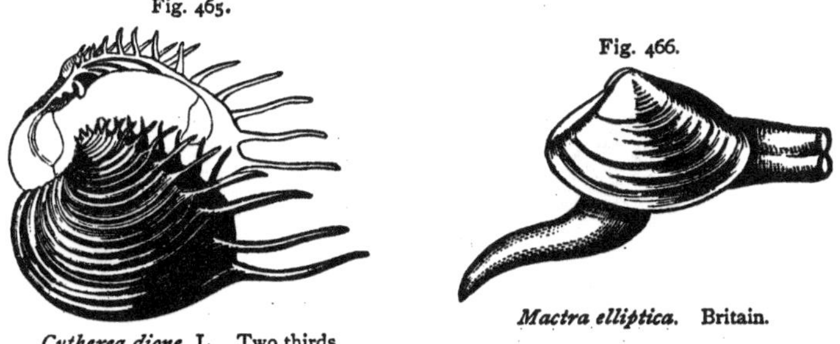

Fig. 465. *Cytherea dione*, L. Two thirds. West Indies.

Fig. 466. *Mactra elliptica*. Britain.

three diverging teeth in each valve. There are nearly six hundred living, and three hundred fossil species. Marine.

MACTRIDÆ. — This Family contains acephala which have the shell trigonal, and the hinge with two cardinal teeth, besides the lateral ones.

TELLINIDÆ. — This Family comprises acephala which have the shell compressed, and the muscular impressions rounded and polished. There are three hundred living and two hundred fossil species.

Fig. 467. *Tellina donacina.* Britain.

Fig. 468. *Tellina tenera*, Say. Our coast.

Fig. 469. *Tellina tenta*, Say. Our coast.

SOLENIDÆ, OR RAZOR-SHELL FAMILY. — This Family comprises acephala which have the shell much elongated

Fig. 470. *Solen ensis*, Linn. About one third. Both shores of the Atlantic.

and gaping at both ends. Fifty or more species are living, and as many more fossil.

MYACIDÆ, OR CLAM FAMILY. — This Family contains the common edible Clam and its allies, of which there

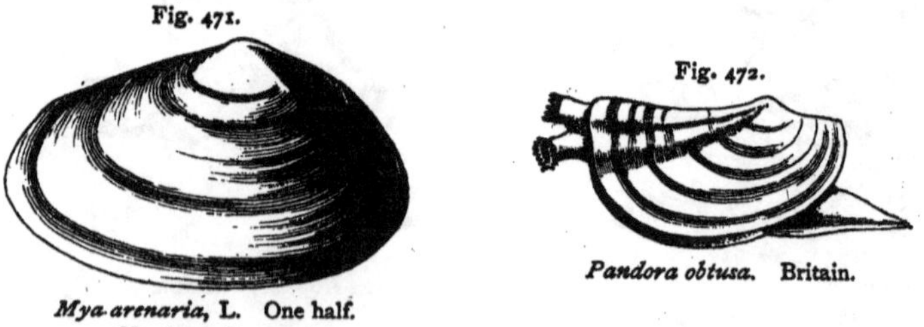

Fig. 471. *Mya arenaria*, L. One half. Northern States.

Fig. 472. *Pandora obtusa.* Britain.

are about one hundred living and two hundred and fifty fossil species. Marine.

ANATINIDÆ, OR LANTERN-SHELL FAMILY. — This Family embraces acephala which have the shell thin and often inequivalve. There are less than one hundred living species, and four hundred fossil. Fig. 472.

GASTROCHÆNIDÆ.— This Family comprises acephala which have the shell thin, and gaping; often cemented into a shelly tube when adult. There are twenty or thirty living, and as many fossil species.

Fig. 473.

Gastrochæna modiocena, Lam. Galway.

Fig. 474.

Aspergillum vaginiferum, Lam. One half. Red Sea.

PHOLADIDÆ, OR PHOLAS AND SHIP-WORM FAMILY. — This Family embraces acephala which have the shell open at both ends, thin, white, exceedingly hard, and armed with rasp-like imbrications. They burrow in almost all substances. Fifty or sixty species are living, and as many more are fossil. Marine.

Fig. 475.

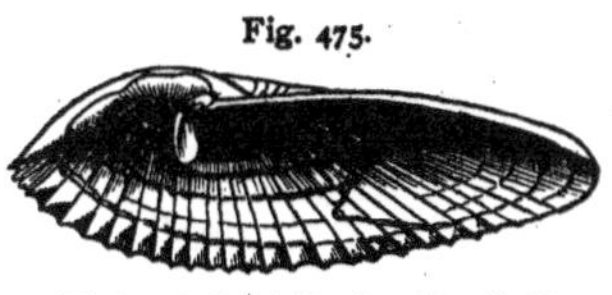

Pholas Bakeri, Desh. One half. India.

Fig. 476.

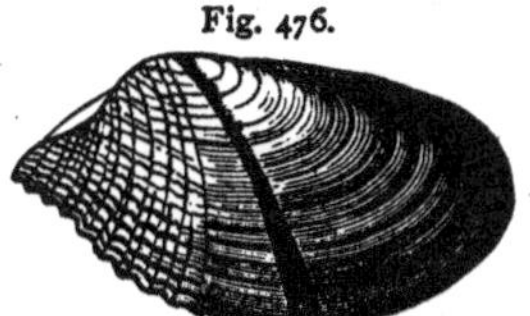

Pholas crispata, Linn. New England and eastward.

SUB-SECTION II.

THE ORDER OF TUNICATA, OR ASCIDIANS.

THIS Order comprises acephalous mollusks which have no hard parts, but which are protected by an elastic tunic instead of a shell. Some are transparent, so that their whole internal structure may be easily seen. Several families, and quite a large number of species, are known.

Fig. 477.

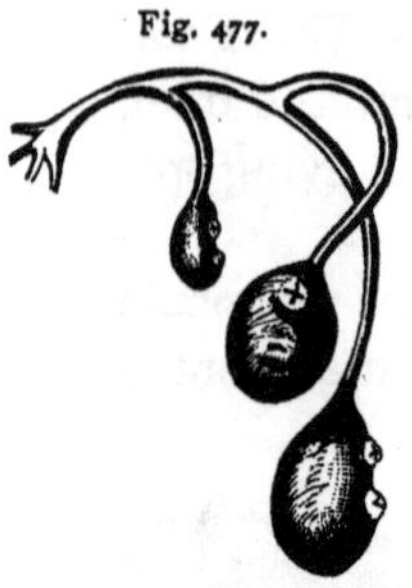

Boltenia pedunculata, M.-Edw. One eighteenth. New Zealand.

They are not uncommon on the coast of the United States.

Ascidiadæ. — This Family contains Simple Ascidians which are fixed, and solitary or gregarious.

The Genus *Ascidia* contains *A. rustica*, Linn., found in clusters adhering to stones and floating timbers in Boston Harbor. It is from the size of a pea to a half or three quarters of an inch in diameter.

SUB-SECTION III.

THE ORDER OF BRACHIOPODA, OR BRACHIOPODS.

This Order comprises mollusks whose shells, composed of two valves, occupy a dorsal and ventral position in relation to the animal. These valves, though unequal in size, are symmetrical in shape. The dorsal valve is the smaller, and is always free and imperforate. The larger valve is the ventral, and has a prominent beak through which the organ of adhesion passes, by which the animal is attached to submarine bodies. The two valves are articulated by two curved teeth developed from the margin of the ventral valve, and received by sockets in the dorsal, and this makes a hinge so complete that the two valves cannot be separated without injury.

Fig. 478.

Brachiopods take their name from two long ciliated arms, shown in Fig. 478, growing from the sides of the mouth, by which they create currents in the water, and thus secure their food. These animals are found hanging from the under sides of shelving rocks, from coral branches, and other submarine bodies. They inhabit all seas. There are seventy-five living, and more than twelve

Fig. 479.

Terebratula septentrionalis, Couth.

Fig. 480.

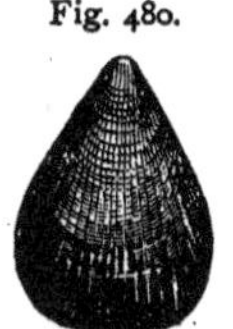

Terebratula septentrionalis, Couthouy. Small specimen. New England and northward.

Fig. 481.

Lingula anatina, Lam. One half. Philippines.

hundred fossil species. The principal families are Terrebratulidæ, Spirulidæ, Rhynchonellidæ, Orthidæ, Productidæ, and Lingulidæ.

Mr. Edward S. Morse, in a paper published in the Essex Institute Proceedings, entitled "Classification of the Mollusca based on the Principle of Cephalization," claims to have shown that what has been regarded as the dorsal valve in Brachiopods is the ventral valve, and what has heretofore been considered the posterior pole of the same is the anterior.

SUB-SECTION IV.

THE ORDER OF BRYOZOA, OR BRYOZOANS.

THE Bryozoa — also called Polyzoa — are very small or minute mollusks growing in clusters upon rocks, shells, and sea-weeds, which they ornament with their delicate ramifications. Some kinds, however, inhabit only fresh waters. All are polyp-like in general appearance, but molluscan in structure. The aggregated cells of some genera are coral or coral-like.

CHAPTER V.

THE BRANCH OF RADIATA, OR RADIATES.

THIS Branch includes all animals whose parts radiate from a vertical axis; or, in other words, whose structure clearly exhibits the idea of radiation. The Radiates are all aquatic, mainly marine, and constitute the lowest Branch of the Animal Kingdom. There are at least ten thousand living species distributed among three classes, — Echinodermata or Echinoderms, Acalephs or Jelly-Fishes, and Polypi or Polyps.

SECTION I.

THE CLASS OF ECHINODERMATA, OR ECHINODERMS.

THIS Class comprises radiates which have a tough covering containing more or less calcareous particles, or a shell composed of pieces which are movable, or bound together and covered with tubercles or spines, the body divided into two well-marked regions, the oral and ab-oral, — or actinal and abactinal,* — and all the parts radiating from the oral opening and meeting in the ab-oral region, and along certain of the rays regular rows of tubular suckers used in locomotion, and called ambulacra; and

* Professor Agassiz proposes the following new names for the different parts of Radiates, viz.: — *actinostome* for the so-called mouth; *spherosome* for the body wall; *actinal* for the pole where the actinostome is situated; *abactinal* for the opposite pole; and *spheromeres* for the homological segments of the body.

the internal organs are contained within walls of their own. The muscular system is well developed. The nervous system consists of a ring around the commencement of the œsophagus, which sends off branches along the rays. Respiration is performed by means of branchiæ, by organs performing other functions, and by water passing into the cavity of the body, and thus aerating the blood through the capillary vessels of the viscera. Echinoderms increase by means of eggs, are marine, and are abundant on almost every coast; and the remains of extinct species fill the rocks in many regions. It is an interesting fact, that in Radiates there is generally a definite or reigning number to which the parts conform. In Echinoderms this number is *five;* that is, the parts of a given kind are generally five or a multiple of five. They embrace five orders,* — Holothurioids, Echinoids, Asterioids, Ophiurioids, and Crinoids; ranking in the order named.

SUB-SECTION I.

THE ORDER OF HOLOTHURIOIDS, OR HOLOTHURIANS.

THE Holothurioids comprise echinoderms which have the body long, cylindrical, somewhat worm-like in general appearance, with a row of appendages around the oral opening, and without a calcareous shell, but with a

Fig. 482.

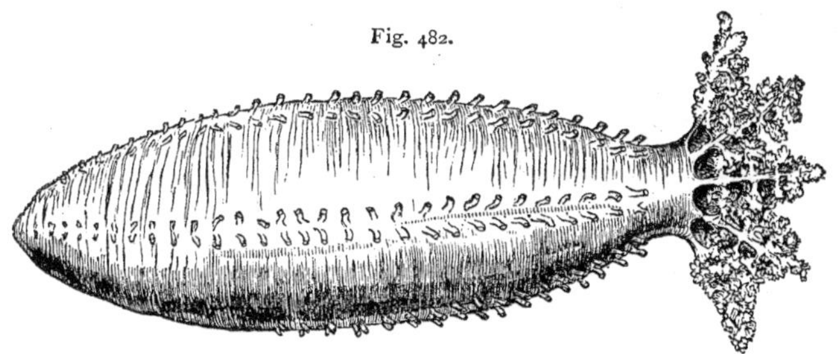

Holothurian, or "Sea Cucumber," *Pentacta frondosa*, North Atlantic.

* Six, according to those who regard Sipunculoids as Echinoderms. Some class them with Worms. Forbes calls them Annelidous Radiates.

tough, leathery envelope, capable of great dilation and contraction, and generally containing more or less of calcareous particles. There are at least four families:— Pentactidæ, with the locomotive suckers in five regular rows; Thyonidæ, with suckers scattered over the whole body; Psolidæ, with suckers in three rows on an oblong disk; and Synaptidæ, destitute of suckers. The different species vary from an inch to a foot or more in length.

SUB-SECTION II.

THE ORDER OF ECHINOIDS, OR SEA-URCHINS.

THE Order of Echinoids contains echinoderms which have a more or less spherical or discoidal shell composed of definitely formed and symmetrically arranged plates, which are firmly bound together, and which bear tubercles crowned with spines. These plates are so arranged as to divide the shell into more or less distinctly marked zones radiating from the oral opening. In every alternate zone, the plates are perforated for the passage of the locomotive suckers or ambulacra, and are called ambulacral plates;

Fig. 483.

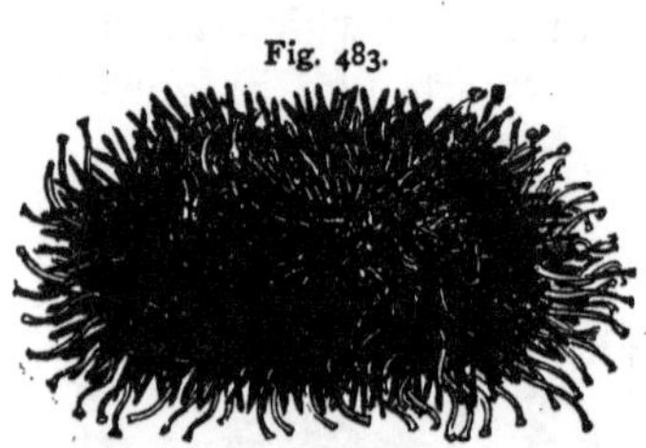

Sea-Urchin, *Toxopneustes drobachiensis*, Ag. Both coasts of the United States, at the North.

Fig. 484.

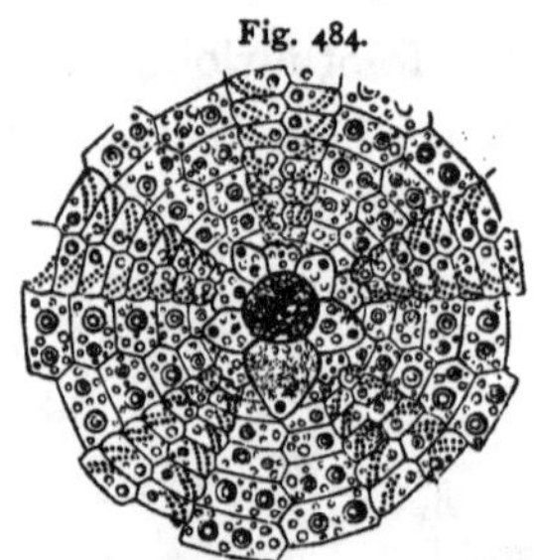

Top view of Sea-Urchin, spines removed. Shows ambulacral and interambulacral plates.

and the plates of the other zones are not perforated, and are called interambulacral plates.

Echinoids may be divided into two great groups,— Regular and Irregular Echinoids, and these, according to Desor's Synopsis, into families and genera, as given below.

Regular Echinoids have the mouth below, vent above, both central, and the ambulacra in five pairs continuous from vent to mouth ; ovaries five, and the mouth furnished with a complicated dental apparatus. They are from an inch to three or four inches in diameter, and different species have spines from a fraction of an inch to three or four inches in length.

CIDARIDÆ, OR CIDARIS FAMILY. — This Family embraces echinoids which have the interambulacral areas composed of two rows of plates. The Genera *Cidaris*, *Leiocidaris*, *Goniocidaris*, *Diadema*, *Savignya*, *Astropyga*, *Echinocidaris*, *Psammechinus*, *Echinus*, *Tripneustes*, *Boletia*, *Sphærechinus*, *Toxopneustes*, *Heliocidaris*, *Loxechinus*, *Echinometra*, *Acroladia*, and *Podophora*, are among the principal ones. *Cidaris*, *Echinometra*, and *Tripneustes* are found on the coast of Florida ; *Echinocidaris*, coast of South Carolina and island of Naushon ; *Toxocidaris*, of A. Agassiz, San Francisco ; *Toxopneustes*, both coasts of the United States, North ; *Loxechinus*, San Francisco ; and *Psammechinus* or *Lytechinus*, South Carolina and Florida.

The Irregular Echinoids have the mouth below, the vent sometimes below, sometimes at one side, and the ambulacra not continuous.

Fig. 485.

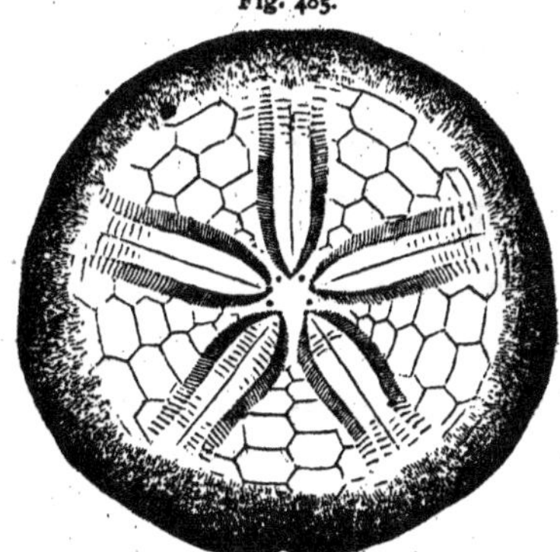

Echinarachnius parma, Gray. Northeast coast of the United States.

ECHINONIDÆ. — This Family has the ambulacral areas simple, and a mascatory apparatus. *Echinoneus* is the principal genus.

CLYPEASTRIDÆ. — This Family contains those which have the ambulacra petaloid, and peristome central. The Genera *Echinocyamus*, *Moulinia*, *La-*

Fig. 486.

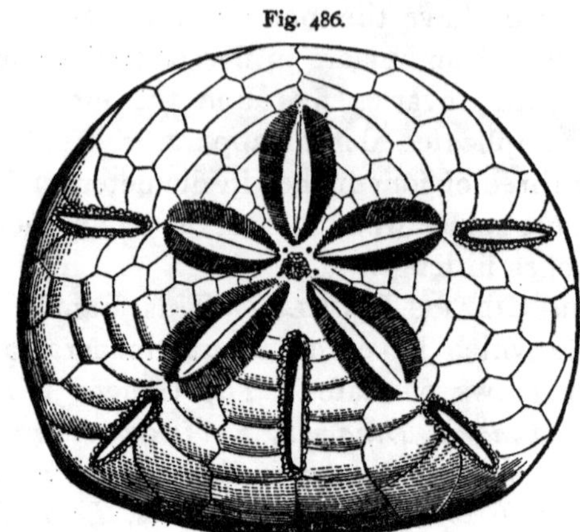

Mellita quinquefora, Ag. Tropical America.

Fig. 487.

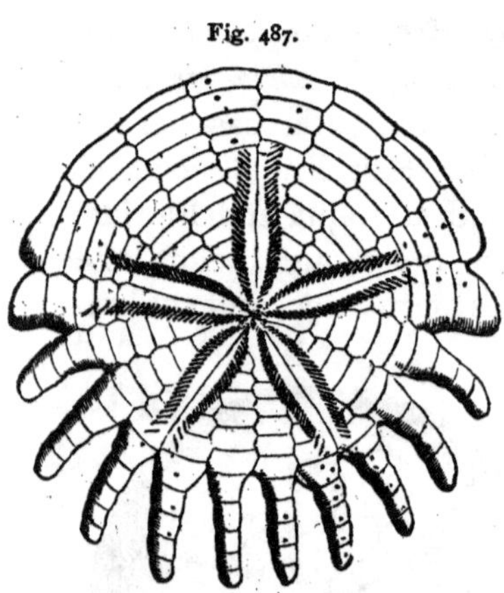

Rotula Rumphii, Klein. Cape Palmas.

ganum, Rumphia, Arachnoides, Echinarachnius, Dendraster, Lobophora, Mellita, Encope, Rotula, Echinodiscus, and *Clypeaster*, are the principal ones. *Clypeaster, Encope*, and *Mellita* are found on the southeastern coast of the United States; *Dendraster*, at San Francisco; *Echinarachnius*, coast of New England and northward.

CASSIDULIDÆ. — This Family has the ambulacra petaloid, no jaws, peristome angular, central or subcentral. Genera: *Cassidulus, Pygorhynchus, Echinolampus*, etc.

SPATANGIDÆ. — This Family contains echinoids which are more or less depressed, heart-shaped, with the ambulacra petaloid, and the peristome eccentric. They are known as Heart-Urchins. The Genera *Spatangus, Brissus, Amphidotus,* and others belong here. Most of the Spatangidæ bury themselves in mud or sand. *Brissus* is found on the coast of Florida.

SUB-SECTION III.

THE ORDER OF ASTERIOIDS, OR STAR-FISHES.

THIS Order comprises echinoderms which are more or less star-shaped, the disk or central portion gradually merging into the rays, beneath which the locomotive suckers extend the whole length; and the calcareous skeleton is composed of movable pieces, so that these animals can bend themselves in any direction. There is al-

Fig. 488.

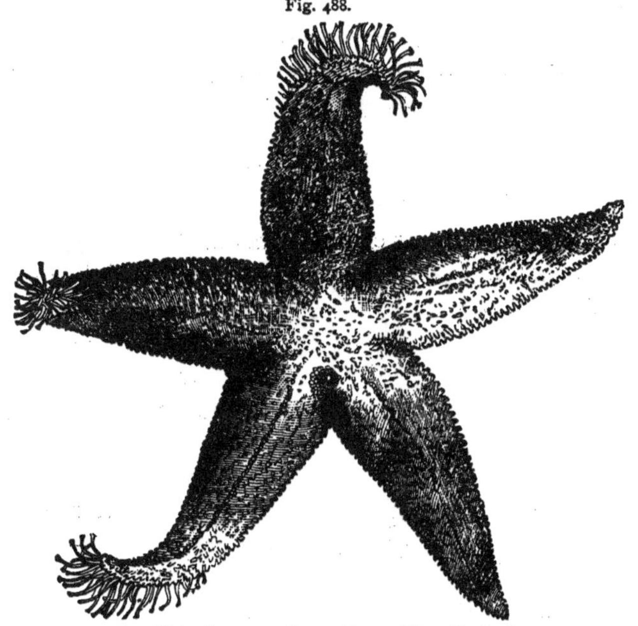

Star-Fish, *Asteracanthion.* Coast of New England.

most every possible form, from those which have the arms long and graceful to those in which the arms and body are merged in one; and they vary in size from an inch or two to a foot or more in diameter. Star-fishes have a wonderful power of reproducing lost parts; if an arm be broken off, another soon grows in its place. Even after all the arms but one have been destroyed, a star-fish has lived, and new arms have sprouted out in the places of the lost ones. They are common on almost all coasts.

Müller and Troschel divide the Star-fishes into three groups. The first contains those which have four rows of suckers, as *Asteracanthion* including *Uraster;* the second, those which have two rows of suckers, as *Echinaster* including *Asterias* and *Cribella, Solaster, Chætaster, Ophidiaster, Dactylosaster, Tamaria, Cistina, Scytaster, Culcita, Astericus* including *Asterina* and *Palmipes, Pteraster, Oreaster, Astrogonium, Goniodiscus, Stellaster, Asteropsis,* and *Archaster;* the third, those which have two rows of suckers, and no vent, as *Astropecten, Ænodiscus,* and *Luidia.*

SUB-SECTION IV.

THE ORDER OF OPHIURIOIDS, OR OPHIURANS.

THIS Order embraces echinoderms which have the central disk very small in comparison to the size of the arms, and circular; and the arms start off abruptly from its circumference. Locomotion is effected by means of spines.

Fig. 489.

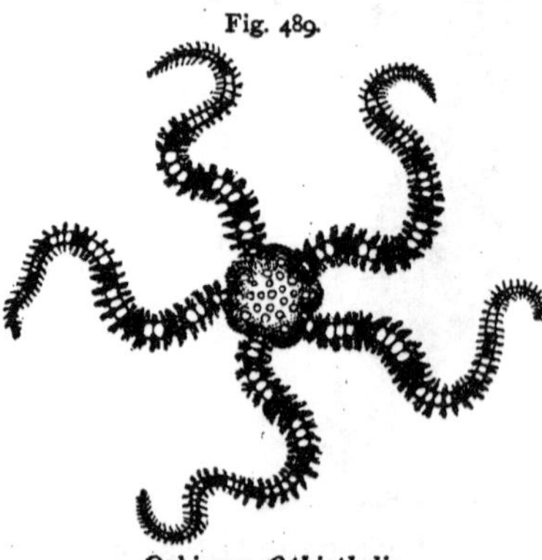

Ophiuran, *Ophiopholis.*

Müller and Troschel divide the Ophiurans into two groups, — Ophiuræ and Euryalæ.

The Ophiuræ are characterized by simple arms, and contain the Genera *Ophioderma* including *Ophiura, Ophiocnemis, Ophiolepis,*

Ophiocoma, *Ophiarachna*, *Ophiacantha*, *Ophiopholis*, *Ophiomastix*, *Ophiomyxa*, *Ophioscolex*, and *Ophiothrix*.

The Euryalæ are characterized by branched arms, and

Fig. 490.

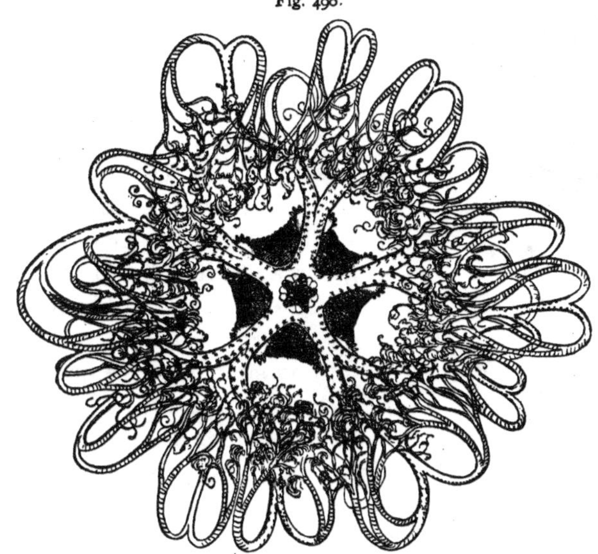

Astrophyton Agassizii, Stimpson. Less than half the natural size. Northeast coast of North America.

comprise the Genera *Asteronyx*, *Trichaster*, and *Astrophyton*.

Since the above was in type, we have seen the excellent Catalogue of "Ophiuridæ and Astrophytidæ" by Theodore Lyman, in which the list of genera is much greater than that here given, and from which we learn that more than a dozen genera of Ophiurans are found on the coasts of the United States.

SUB-SECTION V.

THE ORDER OF CRINOIDS.

The Order of Crinoids embraces echinoderms which are characterized by the great preponderance of the

Fig. 491.

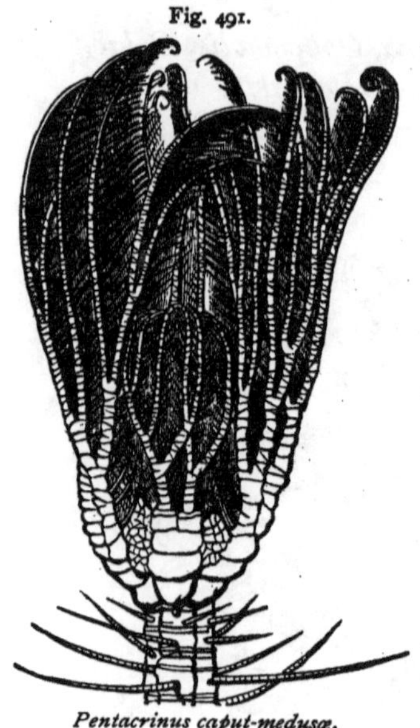

Pentacrinus caput-medusæ.

ab-oral region over the oral, the former being developed into a cup or calyx-like projection, and composed of solid, immovable plates. With one exception, the living representatives bear a great resemblance to Star-Fishes and Ophiurans; but the fossil species, which are exceedingly numerous, and literally fill the rocks in many regions, and whose beauty and ornamentation are beyond description, have a long, jointed stem, and are popularly known as Stone Lilies. These stemmed crinoids, so abundant in the past, have but a single representative in the present seas, — the *Pentacrinus caput-medusæ* of the West Indies. *Comatula* has a stem in its early stages, thus for a time appearing like the old crinoids; but at length it becomes detached, and spends the remainder of its life as a free crinoid.

SECTION II.

THE CLASS OF ACALEPHS, OR JELLY-FISHES.

THE Class of Acalephs comprises jelly-like radiated animals, with a central cavity hollowed out of the mass of the body, — which is generally built of four, eight, or twelve spheromeres, — and this cavity with a central oral opening; and instead of radiating plates, as in the next order, there are radiating tubes, which pass from the central portion and unite with a circular tube which follows the outline of the periphery of the animal. The external edges of the central or oral opening are turned outward, and more or less prolonged into fringe-like appendages. Tentacular appendages wanting, or present in almost every degree of development both as regards number and extent. Jelly-Fishes, in their most common forms, are known as Medusæ, and have long been objects of great interest to voyagers and residents by the sea, as well as to the most learned naturalists. Their jelly-like bodies, scarcely denser than the water in which they move; their curious and beautiful forms, and often beautiful colors; their complicated structure, delighting all who study it; their movements, varied and graceful as those of the butterfly or the bird of the air; their phosphorescence by night, making the track of the vessel, and the breakers upon the shore, glow with light, and gaining for them the appellation of "Lamps of the Sea"; and, above all, their wonderful modes of reproduction and development, — conspire to excite our interest, our wonder, and, as we study them carefully, our admiration.

Fig. 492.

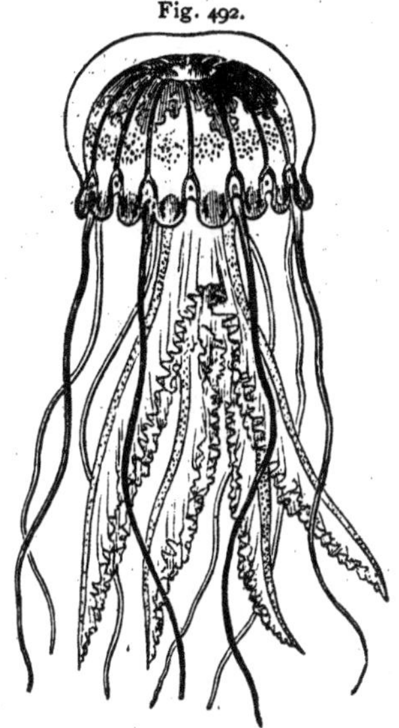

Jelly-Fish, *P. cyanella*, Per. & LeS.

In treating this group, we follow Agassiz, in his "Contributions to the Natural History of the United States."

Jelly-Fishes embrace three orders, — Ctenophoræ, Discophoræ, and Hydroids, ranking in the order named, Ctenophoræ being highest.

SUB-SECTION I.

THE ORDER OF CTENOPHORÆ, OR BEROID MEDUSÆ.

THIS Order contains more or less spherical or ovate jelly-fishes, which have the body built of eight homologous segments bearing eight rows of locomotive appendages more or less distinctly indicated. Agassiz enumerates about seventy species, which he distributes among four sub-orders, twelve families, and over thirty genera.

CYDIPPIDÆ. — This Family embraces *Pleurobrachia* and allied genera.

The Genus *Pleurobrachia* has the body nearly spherical or slightly elongated and compressed, the locomotive appendages extending from near the margin of the mouth, in eight rows, towards the opposite centre.

Fig. 493.

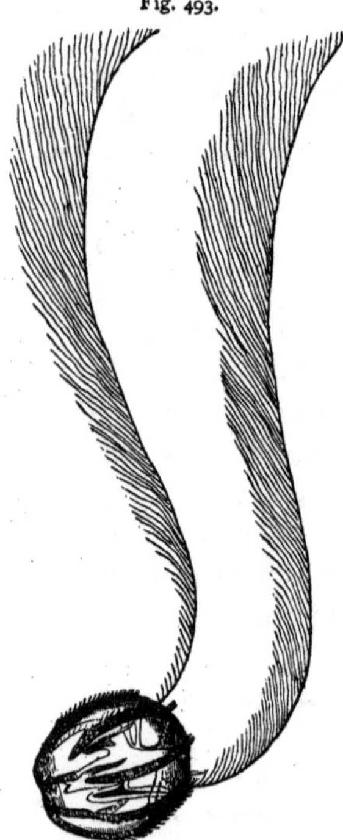

Pleurobrachia, *P. rhododactyla*, Ag.

Pleurobrachia rhododactyla, Ag., is common on the north-east coast of North America, and has received very special attention from Agassiz, who says, that, "when active, it hangs out a pair of most remarkable appendages, the structure and length and contractility of which are equally surprising, and exceed in wonderful adaptation all I have ever known among animal structures. Two apparently simple, irregular, and unequal threads hang out from opposite sides of the sphere. Presently, these appendages may elongate, and equal in length the diameter of the sphere, or surpass it, and increase to two, three, five, ten, and twenty times the diameter of the body, and more and more; so much so, that it would seem as if these threads had the power of endless extension and development. But, as they lengthen, they appear more complicated: from one of their sides other delicate threads shoot out like fringes, forming a row of beards like those of the most elegant ostrich feather, and each of these threads itself elongates till it equals in length the diameter of the whole body, and bends in the most graceful curves."

BEROIDÆ. — This Family contains *Beroe, Idyia,* and their allies.

The Genus *Idyia* contains *I. roseola,* Ag., of the coast of New England and northward. This species attains the length of three or four inches or more, and is of a beautiful rose-color. It sometimes appears in such numbers during the summer months as to tinge large patches of the sea with a delicate rosy hue. It is very voracious, and feeds chiefly on other ctenophoræ.

SUB-SECTION II.

THE ORDER OF DISCOPHORÆ, OR MEDUSÆ PROPER.

THE Discophoræ comprise jelly-fishes which have the form of a hemispheric disk spreading uniformly in all directions. They abound in all seas, and the species are numerous. Agassiz recognizes three sub-orders, fifteen families, and about seventy genera.

The Sub-Order Rhizostomeæ includes those discophoræ which are generally regarded as destitute of the so-called mouth, and which absorb their food through innumerable tubes traversing the arms, and reaching the digestive cavity through narrow channels; and they are wholly destitute of marginal tentacles. As to the mouth, however, Agassiz holds, if we understand him rightly, that their essential structure, in this respect, is the same as that of other discophoræ, differing only in degree.

The Sub-Order Semæostomæ comprises discophoræ which have the so-called mouth plainly represented, though surrounded by more or less extensive appendages, and marginal tentacles more or less developed.

AURELIDÆ, OR AURELIA FAMILY. — This Family embraces discophoræ which are characterized by the even curve of the outer surface of the disk, while the lower surface is excavated in its central portion by four large genital pouches, between which hang four stout arms, closing upon one another in the centre, so as to form a rectilinear opening, prolonged in the undulating curves or folds between the lower margins of the arms. The margin of the disk has small tentacles, except where the eight eyes occupy slight indentations.

The Genus *Aurelia* contains the common "Sun-Fish," *A. flavidula*, Per. & LeS., of the northeast coast of North America. It attains eight or ten inches or more in diameter, and lives but a single year. When first seen in the spring, it is hardly a quarter of an inch in diameter; and when the sky is clear and the sea smooth, it floats in immense numbers near the surface of the water. They grow rapidly, reaching their average size in early summer. As they increase in size, they separate more and more, but reassemble towards the close of summer, which is the spawning season. Later, they are broken into fragments, and destroyed by the autumnal winds; but the planulæ — as the newly-hatched jelly-fishes are called — soon appear moving freely about by means of vibratile cilia. After a little time, each becomes attached to the rocks, sea-weed, or shells, and is then known as *Scyphostoma*, Fig. 494. Then the body begins to be divided into rings by transverse constrictions, and the rings or segments become more and more numerous and more distinct, and in this form is called *Strobila*, Fig. 495. At length, by deeper and deeper constriction, the segments become more and more isolated, and the uppermost segment drops off, then the next one, and so on, till each

Fig. 494.

Fig. 495.

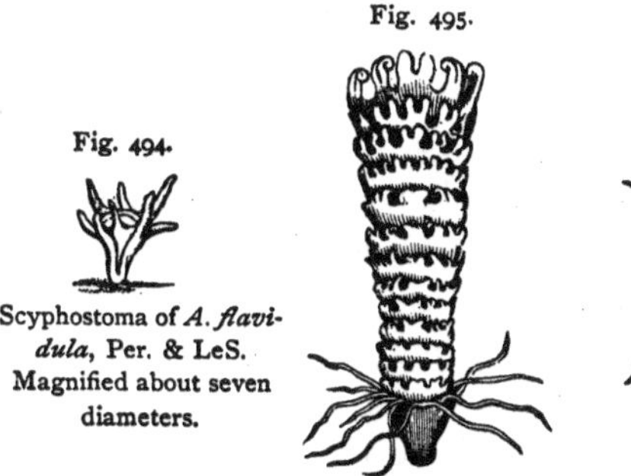

Scyphostoma of *A. flavidula*, Per. & LeS. Magnified about seven diameters.

Strobila of *A. flavidula*, Per. & LeS. Magnified about seven diameters.

Fig. 496.

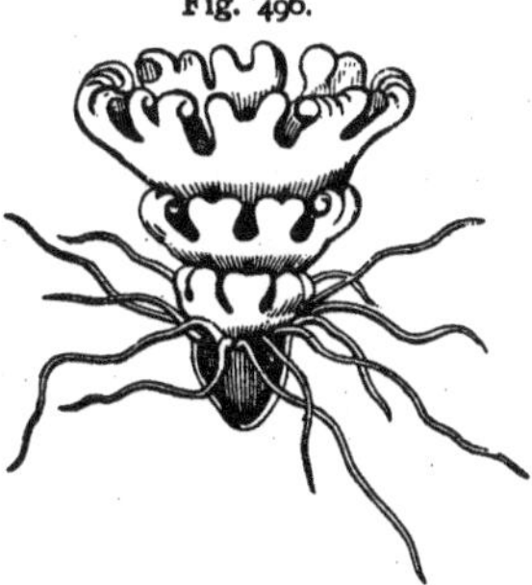

Strobila of *A. flavidula*, Per. & LeS. Magnified fifteen diameters.

Fig. 497.

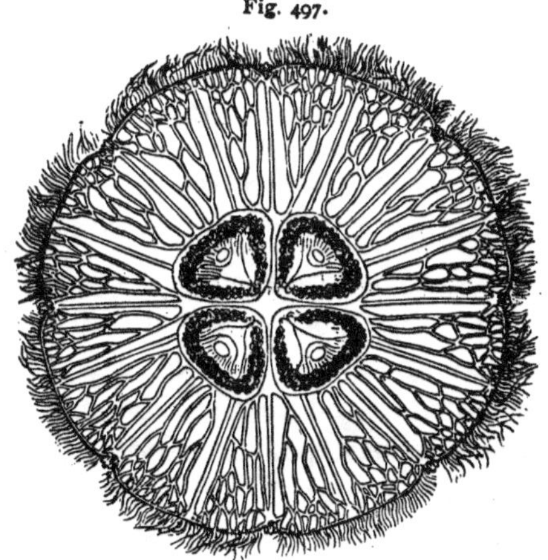

Aurelia flavidula, Per. & LeS. Offspring of Figs. 494 - 496.

segment or disk has separated from the one below itself; and the base, having reproduced tentacles, remains still alive. Each segment or disk, as it separates, turns over, and floats away, and in this form is known as *Ephyra*.

Soon each ephyra assumes the form of the perfect jelly-fish. Thus a single scyphostoma, the product of a single egg, becomes a strobila, which at length divides into numerous parts, each of which becomes a perfect jelly-fish.

CYANEIDÆ.—This Family contains *Cyanea* and its allies.

The Genus *Cyanea* contains *C. arctica*, Per. & LeS., which is one of the most magnificent of all jelly-fishes, attaining a diameter of two or three feet, and with tentacles which extend to the enormous length of twenty or thirty feet. The color of the disk is bright purplish red, the margin whitish with a tinge of grayish blue; and the tentacles vary from orange to deep purple. Its mode of development is essentially the same as that of Aurelia. It inhabits the northeast coast of North America.

PELAGIDÆ.—This Family comprises *Pelagia*, and allied genera. In the mode of development, they differ essentially from the two preceding families, inasmuch as the young hatched from the egg passes directly into the ephyra form. *P. cyanella*, Per. & LeS., Fig. 492 about half the natural size, is found on the coast of Florida.

SUB-SECTION III.

THE ORDER OF HYDROIDÆ, OR HYDROIDS.

THIS Order includes the lowest acalephs, and embraces, according to Agassiz, two more or less distinct forms, one of which, though having the structure of acalephs, reminds us of polyps, described in the next section, and the other closely resembling medusæ proper; and between these there is every possible gradation. All the so-called hydroid polyps, and the naked-eyed medusæ, belong to this order, which is divided by the distinguished author just named into eight sub-orders, forty-six families, and about one hundred and thirty genera. Occurring, as they do in many cases, in one stage of their existence at least, as mere discolored lichen-like patches on sea-weed, stone, or shell, or in appearance like little tufts of moss, or miniature shrubs, the careless observer might well mistake the fact of their animal character.* But, thanks to learned, persevering, and patient investigators, we know much of their curious and wonderful history. We now know that these little vegetable-like, but acalephian forms produce medusæ buds which expand into genuine medusæ, and in some kinds sever their connection, and float away and lead an independent life, and in other kinds remain attached to the hydroid stalk, and in both cases produce eggs which serve to establish new communities of hydroids, like the ones from which they themselves were developed. Thus the hydroid communities, and certain medusæ, are alternate generations of the same beings.

The Sub-Order Tubulariæ comprises hydroid acalephs in which the medusa is free or persistent, deep bell-shaped, the hydra pedunculated, and the head club-shaped.

SARSIDÆ. — This Family, as restricted by Agassiz, embraces those in which the medusæ are deep bell-shaped, and have four long tentacles, and a long simple proboscis upon which the eggs are developed.

The Genus *Coryne*, formerly *Sarsia*, is a well-known representative of this family. Fig. 501 shows the form of one species of Coryne in the adult state. Nothing can excel the delicacy of the structure of these animals. Soft as jelly, transparent as the dew-drop, almost as perish-

* Of the sub-orders alluded to above, one is Tabulatæ, whose representatives produce solid parts known as Coral, and formerly referred to the next class, which is the principal coral-producing group. The Genera *Millepora*, *Seriatopora*, *Favosites*, *Heliopora*, and *Pocillopora* are now included in this sub-order. In the corals referred to Tabulatæ, the cells have a horizontal partition or floor extending from wall to wall; and these floors are formed one above another as the animal grows; and the radiating partitions never extend vertically through the successive floors.

Fig. 498.

Acalephian Coral, *Pocillopora cæspitosa*, Dana.

able as a bubble, yet they perform varied and rapid movements, contract and expand their tentacles, catch and vo-

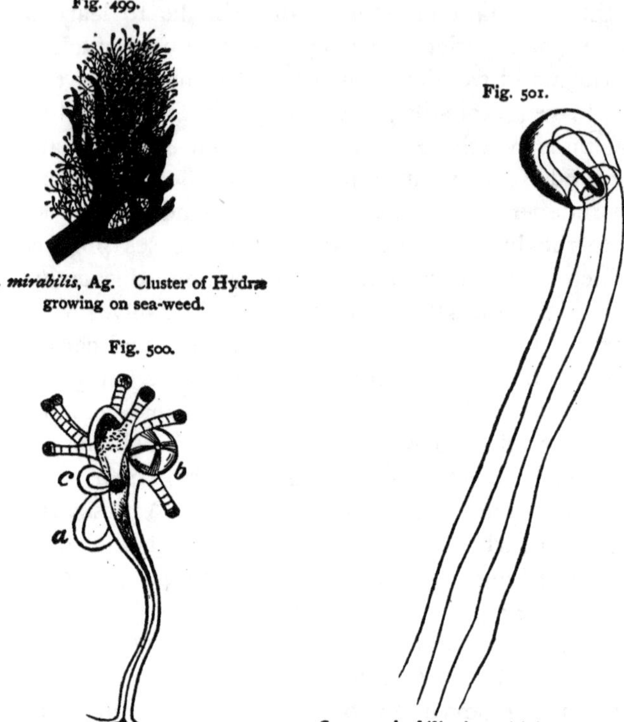

Fig. 499. *C. mirabilis*, Ag. Cluster of Hydræ growing on sea-weed.

Fig. 500. Single individual of Fig. 499 enlarged, showing *a*, *b* just ready to become free medusæ, Fig. 501; *c*, young bud.

Fig. 501. *Coryne mirabilis*, Ag. Adult. Massachusetts Bay.

raciously devour medusæ of their own kind or those of other species, and other marine animals. They may be seen in the spring, and in great numbers; about the middle of summer they lay their eggs and perish. But these eggs, as already noticed, do not hatch medusæ, like the parents; but each hatches a little hydroid, which is first free, then afterwards becomes attached to a shell, or sea-weed, or stone, and from this hydroid other hydroids bud and branch until a community of hydroids resembling a tuft of moss has grown up, Fig. 499. From these hydroids, in turn, bud the free medusæ, like Fig. 501.

TUBULARIDÆ. — This Family, as restricted by Agassiz, embraces only hydroids whose head is furnished with a wreath of simple coronal tentacles, and a proboscis with simple tentacles around the mouth; and which produce either persistent or free medusæ, more or less one-sided, budding from the floor between the coronal tentacles and the proboscis. The Genera *Tubularia*, *Thamnocnidia*, and *Parypha* have the medusæ persistent; *Hybocodon* free.

Fig. 502.

Tubularia Couthouyi, Ag. Massachusetts Bay. *m*, medusæ; *ct*, coronal tentacles; *p*, proboscis.

The Sub-Order of Sertulariæ embraces those in which the hydra is always pedunculated and attached, and protected by a horny sheath forming a cup around the head; and the medusæ are either free or persistent, generally flat, but in some cases bell-shaped, and furnished with numerous tentacles. This

Fig. 503.

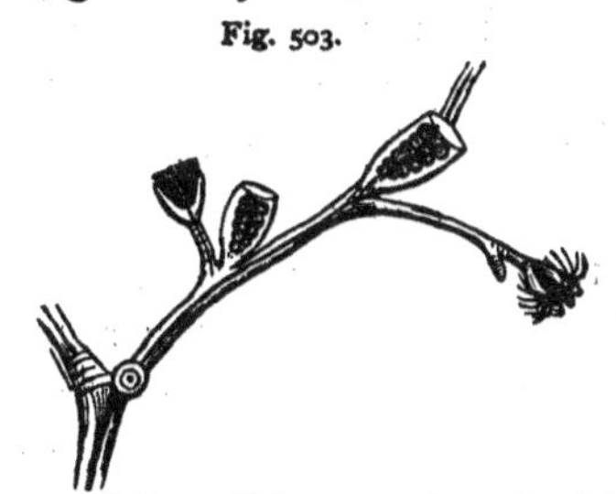

Campanularian, *Obelia commissuralis*, McCr. The hydro-medusæ in the cups drop out and become free medusæ, similar to Fig. 504. Atlantic coast of North America.

Fig. 504.

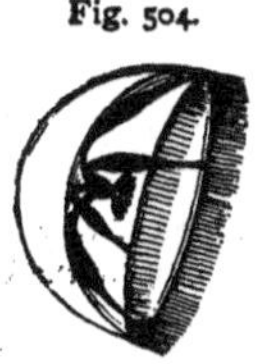

Campanularian, *Tiaropsis diademata*, Ag. Northeast coast of North America.

group includes the hydroids known as Sertularians and Campanularians.

The Sub-Order of Siphonophoræ comprises hydroid acalephs which exist as free moving communities, each community being made up of individuals of different kinds, yet all conspiring to give the appearance of one complicated animal.

Fig. 505.

Portuguese Man-of-War, *P. arethusa*, Til. Southern coast of the United States.

The Genus *Physalia* is a prominent one. It contains the Portuguese Man-of-War, *P. arethusa*, Til., one of the most remarkable and best known of this group. It consists of a pear-shaped and elegantly crested air-sac, floating lightly upon the surface of the water, and giving off from its under surface numerous long and varied appendages. These appendages are the different members of the community, and perform different functions, some of them eating for the whole, others producing medusa buds, and others being the locomotive members,—the latter having tentacles that stretch out behind the floating community even to the length of thirty feet. The air-sac is three or four inches long, or more.

On the tentacles of both Acalephs and Polyps there are numerous microscopic lasso-cells — Cnidæ — each containing a spirally coiled lasso which can be darted forth at will, and which aids in securing prey.

SECTION III.

THE CLASS OF POLYPI, OR POLYPS.

The Class of Polyps embraces radiates which have a tubular or sack-like body, with a circular summit or disk, in the centre of which is an opening popularly called the mouth, surrounded by one or more rows of tentacles; and which have the interior of the body, except the main central cavity, divided into vertical chambers by vertical plates or partitions, extending from the inner wall to the main cavity, for the whole height of the body. The mouth leads directly into an interior central sac, which is the stomach, and which opens at the bottom into the main cavity. From the main cavity there is free communication with all the radiating chambers, and from the latter free communication through the hollow tentacles which crown the summit. The chambers also communicate with each other by circular openings near the top. Polyps are all marine, and, according to the kinds, are free or attached, single or associated, often in numbers that defy computation. They increase by means of eggs, by budding in a manner analogous to that of trees and shrubs, and by division and subdivision; so that the largest communities arise from a single parent. Polyps readily reproduce a lost part. Even if cut in pieces, each considerable fragment will, in some cases, become a new animal. With few exceptions, Polyps do not flourish at depths greater than twenty or thirty fathoms, and they abound in comparatively shallow wa-

Fig. 506.

Polyp, *Bunodes stella*, Verrill.

ters. They vary in size from microscopic forms to several inches, and even a foot or more, in diameter. Some are wholly soft, others secrete more or less solid parts. In this respect there is every grade, from those wholly fleshy to those which secrete a solid framework. This framework is called *Coral.** The too common notion that coral is built by an *insect*, or that the coral animals build coral at will, as the bee builds comb, or as workmen masonry, is wholly erroneous. Coral is simply the framework or skeleton, or aggregate skeletons, of polyps, — or, in some cases, as we have seen, of acalephs, — and is a necessary result of their existence, and is entirely independent of the volition of the animals themselves. In fact, polyps form coral in a manner not different in kind from that in which the higher animals form bones; and the coral is wholly inside the polyps, and is in no sense a house, as is too commonly supposed, in which the latter live; and it is only when the polyps die, wither, and disappear that we see the solid coral itself. From their resemblance to plants, the animals of this class were regarded by the early naturalists as vegetable forms; and later they have been regarded as partaking of the nature of both plants and animals; but now their strictly animal character is established beyond any question. Still, they are often called Zoöphytes, as well as Polyps. The forms and hues exhibited by them are almost endless. Some parts of the tropical seas, where polyps especially flourish, rival in graceful and varied forms, and in beauty and splendor of colors, the most beautiful flower-gardens of the land. There is scarcely a form of vegetation, either trunk or branch, leaf or flower, fern, moss, lichen, or fungus, that is not imitated with striking exactness by these wonderful animals of the sea.

* *Corallum* is the term used by Dana in his most excellent works on Zoöphytes.

According to Professor Verrill, Polyps embrace three orders, — Alcyonaria, Actinaria, and Madreporaria.

SUB-SECTION I.

THE ORDER OF ALCYONARIA.

This Order contains polyps which have well-developed actinal, mural, and abactinal regions; eight long, pinnately-lobed tentacles around a narrow disk; and which are compound by budding. It comprises three sub-orders, — Pennatulacea, Gorgonacea, and Alcyonacea.

The Sub-Order Pennatulacea, or Sea-Pens, includes polyps which form free, moving communities, and contains four families, — Pennatulidæ, Pavonaridæ, Veretillidæ, and Renillidæ.

RENILLIDÆ. — This Family contains polyps which are arranged symmetrically on the upper surface of a more

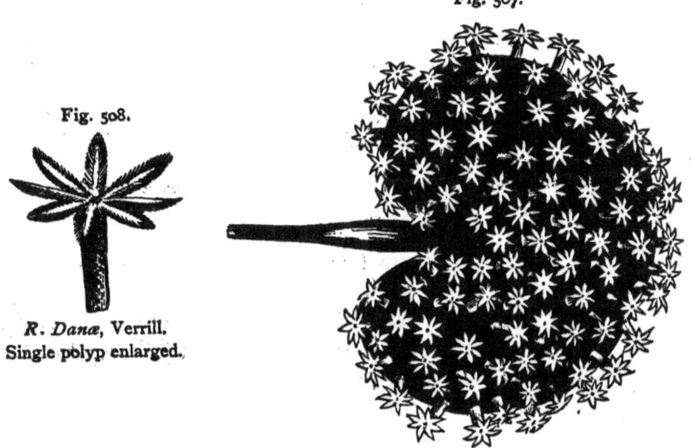

Fig. 508. *R. Danæ*, Verrill. Single polyp enlarged.

Fig. 507. Renilla, *R. Danæ*, Verrill.

or less flattened disk, to the lower surface of which there is attached a hollow locomotive organ in the form of a peduncle. The Genus *Renilla* is the principal one.

R. reniformis, Cuv., is found on the coast of North and

South Carolina and Georgia; *R. Danæ*, Verrill, on the Atlantic coast of South America; and *R. peltata*, Verrill, near the mouth of the Mississippi.

The Sub-Order Gorgonacea embraces polyps which are cylindrical, short, connected laterally, and which secrete a solid central axis. It contains seven families, — Gorgonidæ, Plexauridæ, Primnoidæ, Gorgonellidæ, Isidæ, Corallidæ, and Briaridæ. The forms are exceedingly varied, and often extremely delicate and beautiful. They abound in tropical seas; some species are found in temperate zones. Fig. 509 belongs to Gorgonellidæ.

Fig. 509.

Verrucella gemmacea, Val.

Fig. 510.

Red Coral, *C. rubrum*, Lamk. Single polyp enlarged.

Fig. 511.

Rhipidogorgia flabellum, Val. Portion of a large frond.

Fig. 512.

Red Coral, *Corallium rubrum*, Lamarck.

Fig. 513.

Primnoa myura, M.-Edw.

Fig. 514.

Organ-pipe Coral, *Tubipora syringa*, Dana.

GORGONIDÆ, OR SEA-FAN FAMILY. — This Family comprises those which are usually much branched, and have a tendency to spread in a plane, forming a flattened or fan-shaped, and often reticulated frond; axis horn-like. Fig. 511.

The Genus *Gorgonia* has the corallum much branched, the branchlets slender, and the cells prominent and arranged in two or more rows on the edges of the branches.

G. humilis, Dana, is found on the coast of South Carolina attached to stones and shells, and is four or five inches high; color reddish brown.

The Genus *Leptogorgia* has the corallum branching, the branches slender, and with a space on each side destitute of cells.

L. virgulata, Milne-Edw., is slender, fasciculate, and the color variable, but frequently lemon or reddish purple. It is abundant a few feet below low-water mark, from North Carolina to Florida.

L. tenuis, Verrill, of Long Island Sound, is closely allied to the preceding.

PLEXAURIDÆ. — This Family has the corallum branching, arborescent, with a horn-like axis, often calcareous at the base, and the cells arranged equally on all sides of the branches.

The Genus *Muricea* has the axis horn-like in the branches, often very solid at the base; cells prominent, spiculose.

M. pendula, Verrill, of the coast of South Carolina, has the trunk large, erect, and giving off, pinnately, numerous irregular branches, many of which are also irregularly pinnate. It is about two feet high, and the trunk three quarters of an inch in diameter.

PRIMNOIDÆ.— This Family is represented on the coast of the United States by the Genus *Primnoa*, which contains *P. Reseda*, Verrill, found at St. George's Bank and in the Bay of Fundy. The axis is calcareous, and the cells bell-shaped. Fig. 513 shows *P. myura*, M.-Edw.

Corallidæ. — This Family includes the Red Coral of commerce, *Corallium rubrum*, Lamarck, which has the axis calcareous, and very hard. Figs. 510 and 512 represent it alive. It lives in the Mediterranean.

The Sub-Order of Alcyonacea comprises polyps which are turbinate at the base, and which are found encrusting foreign bodies. It embraces four families, — Alcyonidæ, Xenidæ, Cornularidæ, and Tubiporidæ.

Alcyonidæ. — This Family contains those in which the polyps are united, forming lobed or arborescent clusters of fleshy or coriaceous texture, filled with calcareous particles. The Genus *Alcyonium* is the principal one.

A. carneum, Ag., is found from Cape Cod northward, and is attached to shells and stones in from eight to twenty fathoms of water; usually delicate flesh-color.

Tubiporidæ. — This Family contains those which have the coral tubular, calcareous or semi-calcareous, and the tubes not striate within; color red. Fig. 514.

SUB-SECTION II.

THE ORDER OF ACTINARIA.

This Order embraces polyps which have a well-developed abactinal region, conical or cylindrical tentacles around the mouth, and the ambulacral spaces always open. It divides into three sub-orders, — Actinacea, Antipathacea, and Zoanthacea.

The Sub-Order of Actinacea comprises those which are free, capable of locomotion, and which have from ten to hundreds of tentacles, and the mouth with special lobes or folds. Most are simple, a few are compound, and a few secrete from the base a horn-like deposit. There are five families, — Actinidæ, Thalassianthidæ, Minyidæ, Ilyanthidæ, and Cerianthidæ.

Actinidæ, or Sea-Anemone Family. — This Family contains polyps which are more or less cylindrical, rising from a broadly expanded disk used in locomotion; tentacles simple and in several rows near the margin.

Figs. 515, 516, 517.

Sea-Anemone, *M. marginatum*, Milne-Edw. Expanded, closed, and just opening.

The Genus *Bunodes* has the column elongated, subcylindrical in expansion, walls firm, with numerous prominent papillæ arranged in vertical lines corresponding to the chambers within; tentacles large, not numerous.

B. stella, Verrill, of the coast of Maine and northward, is about two inches high in the largest specimens, and pale olive-green; sometimes flesh-color. Fig. 506. It is found among rocks.

B. cavernata, Verrill, of the coast of South Carolina, is about one inch and a half in diameter.

The Genus *Rhodactinia* has the column low, shorter than broad, the mouth large and often everted, and the tentacles large with distinct openings at the ends.

R. Davisii, Ag., is two inches or more in diameter, and is found from Nantucket Shoals northward.

The Genus *Aulactinia* has the base adherent, column elongated, and the mouth with a fold at each angle, one of which is much larger than the others. Below each tentacle there is a three-lobed appendage.

A. capitata, Ag., of the coast of North and South Carolina, is six inches high in some cases; lives in the mud.

The Genus *Metridium* has the column very contractile and changeable in form; often much elongated.

The "Sea-Anemone," or Fringed Actinia, *M. marginatum*, Milne-Edw., is the most common of all the polyps on the northeast coast of North America. The larger specimens are about four inches high, and three inches across the disk, in expansion. Figs. 515–517.

The Sub-Order of Antipathacea comprises polyps which have from six to twenty-four simple tentacles, are connected, and secrete a solid axis. It contains two families, — Antipathidæ and Gerardidæ.

The Sub-Order Zoanthacea contains compound, fixed polyps, which have the tentacles simple, short, and at the edge of the disk. Families: Zoanthidæ and Bergidæ.

SUB-SECTION III.

THE ORDER OF MADREPORARIA.

This Order embraces polyps which are simple or compound, with a broadly expanded form, simple conical tentacles, and whose dermal tissues and usually the radiating lamellæ deposit solid coral. They abound in the warm seas, to which they are mainly confined. There are four sub-orders, — Madreporacea, Astræacea, Fungacea, and Stauracea.

The Sub-Order of Madreporacea contains those which have the tentacles in definite numbers, twelve or more, well developed, and encircling the narrow disk. The coral is porous, dermal, and septal. The polyps are mainly compound by budding, and the growth chiefly vertical. There are four families, — Eupsammidæ, Figs. 523, 524, and 526, Gemmiporidæ, Poritidæ, and Madreporidæ.

Poritidæ, or Porites Family. — This Family is characterized by shallow cells, often hardly traceable within the coral.

Fig. 518.

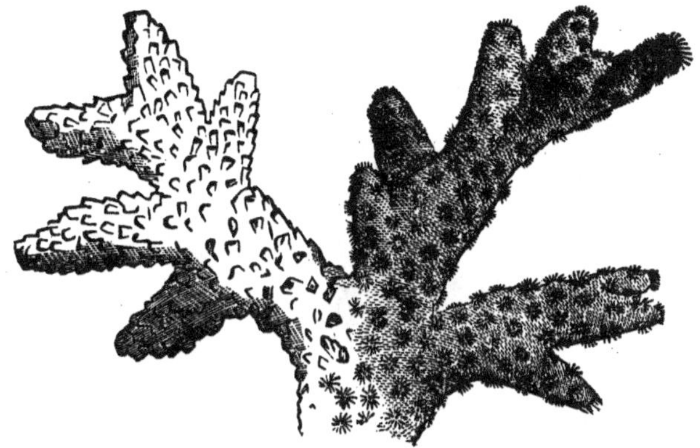

Madrepore, *Madrepora aspera*, Dana. Right-hand branches alive.

Fig. 519.

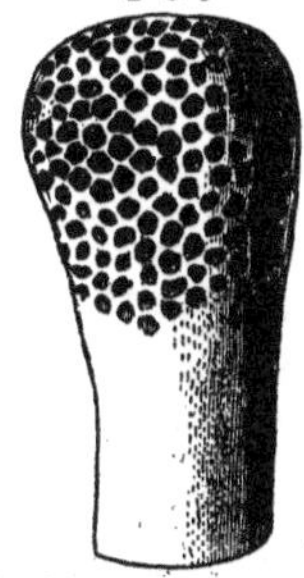

Porites flexuosa, Dana.

Fig. 520.

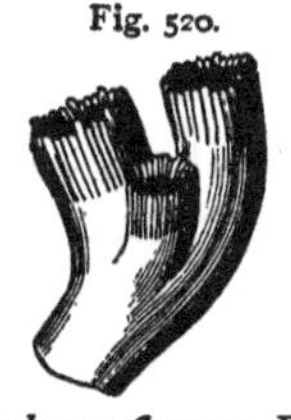

Cladocora flexuosa, Ehr.

Fig. 521.

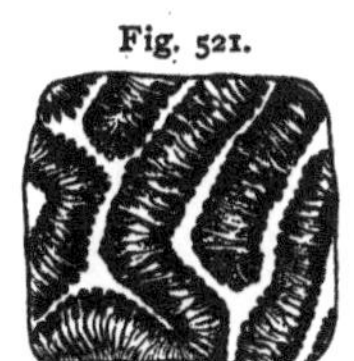

Mæandrina gracilis, Dana. Small piece of a large mass.

Fig. 522.

Merulina speciosa, Dana.

Fig. 523.

Astroides calycularis, Milne-Edw. Coral Polyps in various stages of expansion.

Fig. 524.

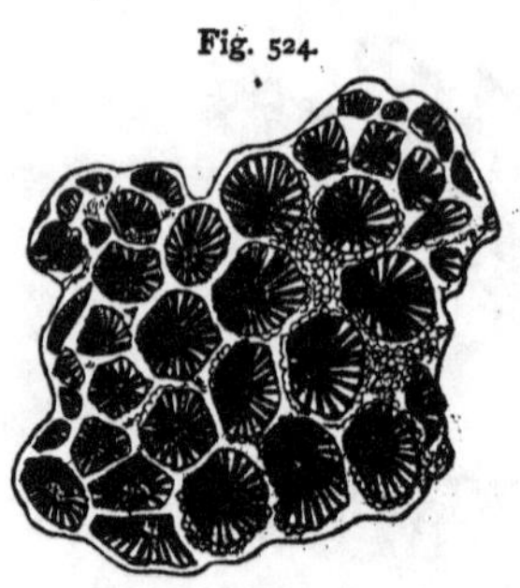

Astroides calycularis, Milne-Edw. Coral of Fig. 523.

Fig. 525.

Oculina horrescens, Dana.

Fig. 526.

Cœnopsammia nigrescens, Milne-Edw.

The Genus *Porites* has the coral massive or branching, and the cells not over a line in diameter; polyps with twelve short tentacles. Fig. 519. The massive specimens are sometimes fifteen feet in diameter.

MADREPORIDÆ, OR MADREPORE FAMILY.—This Family contains polyps which are not coral-producing at the base, and hence the cells of the corallum are very deep, and not crossed by septa within. The species are very numerous, and the forms extensively varied. Fig. 518.

The Sub-Order of Astræacea contains polyps which are mostly compound by budding or fissiparity, with well-developed tentacles in multiples of six. It contains Lithophyllidæ, Mæandrinidæ, Eusmillidæ, Caryophyllidæ, Stylinidæ, Astræidæ, Oculinidæ, and Stylophoridæ.

MÆANDRINIDÆ, OR BRAIN-CORAL FAMILY.—This Family is well represented by the Genus *Mæandrina*, which has the disks trench-like, sinuous, and the tentacles forming a series along either margin. The form is generally hemispherical, and from six inches to twelve feet in diameter. Fig. 521.

CARYOPHYLLIDÆ.—This Family has the cells of the coral with margin thin, and the coral within not transversely septate. Fig. 520.

ASTRÆIDÆ, OR STAR-CORAL FAMILY. — This Family has the coralla with concave radiate cells; septa in aggregate species not continuous from one centre to another, but generally interrupted half-way. The prevailing forms are hemispherical or dome-shaped, and some of the large domes of the Astræas are even twenty feet in diameter. The Genus *Astræa* of Lamarck is the principal one. The polyps are often an inch in diameter.

Fig. 527.

Astræa pallida, Dana. Portion of a large dome, alive

The Genus *Astrangia* is represented on the coast of North and South Carolina by *A. astræiformis*, M.-Edw. & Haime; and in Long Island Sound by *A. Danæ*, Ag. The latter is found incrusting rocks from just below low-water mark to ten fathoms. It thrives well in the aquarium, eating mollusks with avidity.

OCULINIDÆ. — This Family contains the Genus *Oculina*, which has the corallum, while young, spreading laterally by basal budding, and forming an incrusting base from which branches arise in tufts or arborescent forms; cells rather deep, and edges of the septa entire. Fig. 525. *O. arbuscula*, Ag., occurs off Charleston, South Carolina.

The Sub-Order of Fungacea contains those in which the polyps are simple, or compound by marginal or disk budding; tentacles numerous in multiples of six, usually short, lobe-like, and scattered on the actinal surface; and the coral broad and low, and not transversely septate. There are four families, — Cycloclitidæ, Lophoseridæ, Fungidæ, and Merulinidæ.

Fig. 528.

Fungia. Some species a foot in diameter.

The Sub-Order of Stauracea has the coral simple or compound by budding, and the septa apparently in multiples of four. All the species are fossil. It contains Stauridæ, Cyathophyllidæ, Cyathoxonidæ, and Cystiphyllidæ.

Wherever circumstances have favored their growth, Coral Reefs and Islands abound in all the hot regions. When a reef is near the shore, it is called a Fringing Reef; when at a distance, a Barrier Reef; and when it surrounds a body of water, an Atoll or Coral Island. The body of water thus surrounded is called a Lagoon. Coral Reefs are in all stages of formation, from those which have just begun to form to those which have grown to the surface of the water, and, having received the *débris* thrown upon them by the waves, have become dry land, and even the home of man. There are scores of islands in the Pacific which are thus made up of the skeletons of coral polyps; and the islands which skirt the coast of Florida — the Keys — are a reef which has reached and risen above the surface here and there. According to Agassiz, a large part of Florida itself is composed of old coral reefs. There are reefs in the Pacific which are several hundred miles long, and one on the northeast coast of Australia is a thousand miles in length. The reef-forming corals are mainly Astræas, Mæandrinas, Porites, and Madrepores. The frailer corals, those of the Order Alcyonaria, such as the Gorgonias and their allies, adorn the reef as it approaches the surface of the water, but contribute little to its growth. Only an inch or less of the surface of a growing coral mass is alive, — death going on below, growth above.

The reef corals are essentially of the same composition as limestone or marble, being composed of ninety to ninety-six per cent of carbonate of lime; the other parts are organic matter, phosphates, fluorides, magnesia, silica, and oxide of iron.

CHAPTER VI.

THE GEOGRAPHICAL DISTRIBUTION OF ANIMALS.

THE Geographical Distribution of Animals furnishes one of the most important and most interesting departments of study in zoölogy, and one which is receiving more and more attention from the ablest naturalists. The limits of this book allow, and our present purpose requires, only the most general statement of the facts and principles of the subject.

It is in the torrid zone that animal life appears in the most numerous and varied forms, and the species of animals diminish in number and brilliancy of color as we go towards the poles.

Again, each climatic zone of the earth's surface, each zone of altitude, each hemisphere, each grand division of the earth, has its own peculiar fauna. Nay, each of the different parts of every country generally has animals peculiar to itself. The same principles obtain in the seas as on the land. Each ocean and sea, each gulf and bay, each zone of depth, — so far as life obtains, — has its own peculiar animal forms.

A few facts will serve to illustrate the above principles. The White Bear, the Walrus, the Seal, the Whale, the Narwhal, the Auk, and the Jægar have their true home in or near the Arctic regions. The Bats and Moles; the Bears, the Wolves, the Foxes, the Lynxes, the Martens, and the Weasels; the Squirrels, the Beavers, the Woodchucks, the Rabbits, and the Porcupines; the Wild

Boar, and the Ass; the various kinds of Deer, the Sheep, the Goats, and Oxen; the Birds of Prey, the Perching and Singing Birds, the Pigeons, the Grouse, the Waders, and the Swimming Birds; the Fishes and Reptiles; the Insects and the shells, and other and lower forms of life of the North Temperate zone, — are unknown in all the Arctic regions. Not only so, but the animals which bear these names are not of the same species in North America that they are in Europe or Asia. The Grizzly Bear is confined to Western North America; the Brown Bear, to the northern parts of the Eastern hemisphere. The American Sable, Fisher, and Weasel inhabit Northern North America; the Russian Sable and true Ermine inhabit Siberia, and the Beach Marten is found in Europe. A species of Reindeer inhabits Lapland, but in Northern North America are two species of Reindeer, both of which are different from the European one. The Moose of Maine and Canada closely resembles the Elk of Europe, but is not identical with it; the Stag of Europe and the American Deer are two species; and the noble Wapiti, with antlers six feet in length, and the curious Musk-Ox and Bison, belong exclusively to North America; though in the forests of Lithuania the latter has an analogue in the European Buffalo. The Golden Eagle and Peregrine Falcon may be identical in the two hemispheres; but the White-headed Eagle, and the Great Vulture of California, are never found outside of North America, and the Lammergeyer never quits the limits of Europe. The European may justly boast of the sweet singing of the Nightingale; but the indescribable and ravishing notes of the Wood Thrush and of the Hermit Thrush are only heard in the deep groves of North America. The Box-Turtles, the Wood-Tortoise, the Painted Turtle, the curious Trionyx, the Snapper, and all the numerous species of Turtle of North America, are represented in Europe by only one species, the *Testudo græca* of Linnæus.

The tropical and sub-tropical regions of the earth are the home of the Monkeys; of the noble carnivorous animals, like the Lion, Tiger, Leopard, and the like; of the gigantic Pachyderms, such as the Elephant and Rhinoceros; of numerous Ruminants; of Sloths, Ant-Eaters, and Armadillos; of Birds with gorgeous plumage; of gigantic and powerful Reptiles; and of Insects of the most varied forms and most splendid hues. But the ninety American species of Monkeys are all different from a nearly equal number found in the Eastern hemisphere. The latter have the nostrils near together, only thirty-two teeth, cheek-pouches, and the tail non-prehensile; while the American Monkeys have the nostrils widely separated, thirty-six teeth, no cheek-pouches, and in many cases a prehensile tail. The Lion, Tiger, and Leopard, the Elephant and Rhinoceros, are confined to Africa and Asia, and their only representatives in America are the Puma, Jaguar, and Tapir. The Camel belongs to Africa and countries adjacent, although it is true it has an analogue in the Llama of South America. The Sloths of South America, and the Armadillos of North and South America, have not even analogues in any other quarter of the globe. Four hundred species of Humming-birds, many of which vie with the rainbow in the colors of their plumage, belong mainly to Tropical America, but not one was ever found in all the vast realms of the Eastern Continent.

If we compare the animals of South America, of Africa, and of Australia, we find in most cases not even family resemblances. Australia, though lying with a large part of its northern half within the tropics, has no Monkeys, no Pachyderms, no Edentates, and no Ruminants. The numerous mammals of this island are all Marsupials; and it is a remarkable fact, that, outside of Australia and vicinity, no marsupials exist on the globe, except one group—the Opossum Family—which is found in America.

If we look at North America alone, we find ample illustrations of some of the principles first stated. As a general rule, the animals are specifically different on the two sides of the continent, and in many cases the Rocky Mountains are the line of separation. Some animals, as the Pronghorn, Rocky Mountain Goat, and Rocky Mountain Sheep, have not even an analogue in the eastern portion of the continent. Even those animals that resemble each other so much that they were formerly regarded as the same in both halves of the continent, are now known to differ specifically from one another. This is true of Mammals and Birds, Reptiles and Fishes, Insects, and the lower forms.

From the facts stated above, it would seem that climate has no power to mould or shape the species of animals,—and the same is true in regard to plants,—or to change one species into another. Were it so, any given climate would produce, in the course of time, the same species of animals in all the countries within its limits. But so far from this being the case, we find that, in spite of the influence of climate, animals of the different countries even of the same climatic zone are specifically, if not generically, distinct, and in many cases even family resemblances are wanting.

Although, through the agency of man, and in many other ways, animals of one region or country have been introduced into another, we are probably not to look to any such accidental operations for an explanation of the distribution of animals into many well-marked zoölogical provinces. On the contrary, careful observers have been led to believe that animals as well as plants have been created by an Omniscient Being, in the places, and for the places, which they now occupy.

CHAPTER VII.

CONCLUDING REMARKS.

In the preceding chapters we have obtained a glimpse — and only a glimpse — of the Animal Kingdom as it now appears on the surface of our globe. But the animals of the present, vast as are their numbers, are but a handful compared to those that have occupied the surface of the earth in past geologic ages, and that are now known only by their petrified remains, which fill the rocks in many countries to the depth of six or eight miles or more. Nature has embalmed these races, and handed them down to us so perfectly preserved, that we are able to get at least a faint view of the phases of life during all the past ages of the world. And it is a fact of the highest significance, that all the animals of the past, and all those of the present, are created according to the same great plan. Radiates, Mollusks, Articulates, and Vertebrates are the four * Types under which animal life has been exhibited

* We must not omit to mention, that there is a vast number of beings — regarded by some as animals and others as plants — which are believed

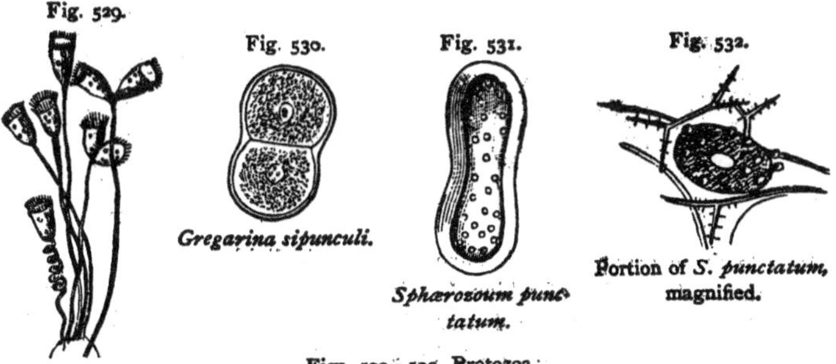

Fig. 529. *Vorticella.*

Fig. 530. *Gregarina sipunculi.*

Fig. 531. *Sphærozoum punctatum.*

Fig. 532. Portion of *S. punctatum*, magnified.

Figs. 529 – 539, Protozoa.

during all the ages since its first appearance upon the earth. When we consider the classes, the orders, the families, the genera, the vast number of living and perhaps the much greater number of extinct species, and then consider that each species is represented in many cases by millions of individuals, and that probably no two

by some eminent naturalists to constitute a fifth branch of the Animal Kingdom, called PROTOZOA. These organisms are almost wholly aquatic, and, excepting the Sponges, are mainly exceedingly minute. The Protozoa

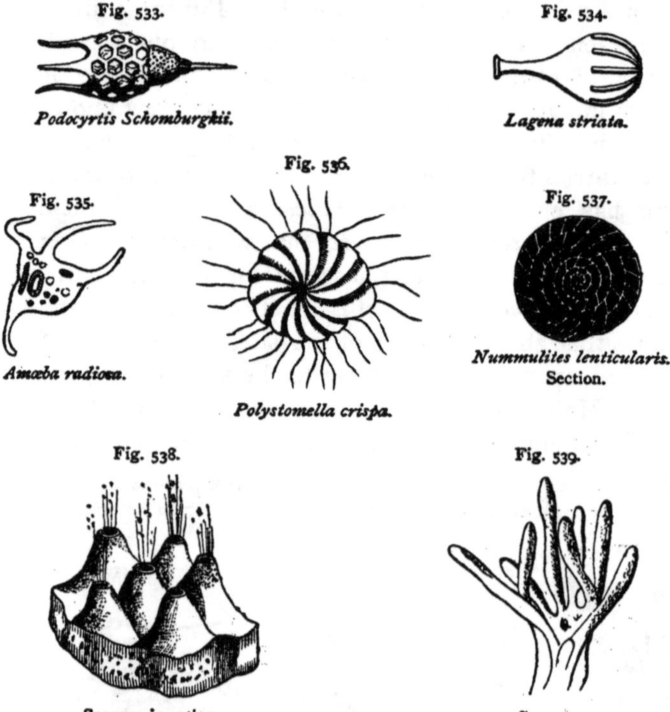

Fig. 533. *Podocyrtis Schomburgkii.*

Fig. 534. *Lagena striata.*

Fig. 535. *Amœba radiosa.*

Fig. 536. *Polystomella crispa.*

Fig. 537. *Nummulites lenticularis.* Section.

Fig. 538. Sponge, in action.

Fig. 539. Sponge.

include three groups, which have been much divided and subdivided by authors: — Infusoria, of which *Vorticella*, Fig. 529, is a prominent example; Rhizopoda, including Foraminifera, Figs. 535 - 537; and Spongidæ, or Sponges, well known to all, Figs. 538, 539.

individuals even of the same species are exactly alike in every particular, and yet that each one of all these uncounted millions bears within itself the stamp of a Radiate, or a Mollusk, or an Articulate, or a Vertebrate, so clearly, that, by patient study, the student of nature is able to refer every one of to-day, and of bygone ages, to its appropriate type, we are impressed with the great truth that in the Animal Kingdom — and it is so in all nature — there is the greatest possible diversity, and that in all this diversity there is perfect unity; and hence we are forced to believe that all the animals of the past, and all those of the present, have been created according to a plan wrought out in the Divine Mind before the foundation of the world.

INDEX.

U.

V.

W.

Y.

Z.

APPENDIX.

Those who desire to learn to mount and preserve Mammals, Birds, and other animals will do well to procure "The Taxidermist's Manual," by S. H. Sylvester, Middleboro, Mass.

Artificial eyes of all sizes and colors can be obtained of C. F. A. Hinrichs, Broadway, New York.

CLASSIFICATION OF ANIMALS.

THE BRANCH OF VERTEBRATA, OR VERTEBRATES.

THE CLASS OF MAMMALIA, OR MAMMALS.

ORDERS.	FAMILIES.	*Genera.*		
BIMANA or MAN.	HUMAN FAMILY.	*Homo*	or	Man.
QUADRUMANA or MONKEYS.	SIMIADÆ or OLD WORLD MONKEYS.	*Troglodytes*	or	Chimpanzee & Gorilla.
		Simia	"	Orang-Outang.
		Hylobates	"	Gibbons.
		Semnopithecus	"	Solemn Apes.
		Presbytis	"	Tailed Gibbons.
		Cercopithecus	"	Guenons.
		Colobus	"	Thumbless Apes.
		Macacus	"	Macacos.
		Inuus	"	Barbary Ape.
		Cynocephalus	"	Baboons.
	CEBIDÆ or NEW WORLD MONKEYS.	*Mycetes*	or	Howlers.
		Ateles	"	Spider Monkeys.
		Lagothrix	"	Glutton "
		Cebus	"	Weepers.
		Pithecia	"	Fox-tailed Monkeys.
		Callithrix	"	Squirrel "
		Nyctipithecus	"	Night "
		Jacchus and Midas	"	Marmosets.
	LEMURIDÆ or LEMURS.	*Lemur*	or	true Lemurs or Makis.
		Indris	"	Indri.
		Loris	"	Lorises.
		Galago	"	Galagos.
		Tarsius	"	Tarsiers.
		Chiromys	"	Aye-Aye.
CARNIVORA or FLESH-EATERS.	FELIDÆ or CAT FAMILY.	*Felis*	or	Lions, Tigers, Panthers, Leopards, Puma, Cat.
		Lynx	"	Wild Cats, Lynx, &c.
	HYENADÆ	*Hyena*	or	Hyenas.
	CANIDÆ.	*Canis*	or	Wolves, Jackals, Dogs.
		Vulpes	"	Foxes.
	VIVERRIDÆ or CIVET FAMILY.	*Bassaris*	or	Civet Cat.
		Viverra	"	Civets and Genet.
		Herpestes	"	Ichneumons.
	MUSTELIDÆ or WEASEL FAMILY.	*Mustela*	or	Martens or Sable.
		Putorius	"	Weasels and Minks.
		Gulo	"	Wolverines.
		Lutra	"	Otters.
		Mephitis	"	Skunks.
		Taxidea	"	Badgers.
	URSIDÆ or BEAR FAMILY.	*Procyon*	or	Raccoons.
		Ailurus	"	Panda.
		Ursus	"	Bears.
	PHOCIDÆ or SEAL FAMILY.*	*Phoca, &c.*	or	Seals.
		Rosmarus	"	Walrus or Morse.

* According to Professor Gill, the old family Phocidæ really comprises three families: PHOCIDÆ proper, including *Phoca*, *Pagomys*, *Pagophilus*, *Erignathus*, *Halichærus*, *Monachus*, *Cystophora*, *Macrorhinus*, *Lobodon*, *Stenorhynchus*, *Leptonyx* and *Ommatophoca*: OTARIIDÆ, including *Otaria*, *Callorhinus*, *Eumetopias*, *Zalophus*, and *Arctocephalus*; and ROSMARIDÆ, including *Rosmarus*.

VERTEBRATES: MAMMALS — *Continued.*

ORDERS.	FAMILIES.	Genera	
HERBIVORA or PLANT-EATERS.	ELEPHANTIDÆ or ELEPHANT FAMILY.	*Elephas* *Mastodon*	or Elephants. (Fossil.)
	RHINOCEROTIDÆ.	*Rhinoceros*	or Rhinoceroses.
	TAPIRIDÆ or TAPIR FAMILY.	*Tapirus*	or Tapirs.
	HYRACIDÆ.	*Hyrax*	or Damans.
	SUIDÆ or HOG FAMILY.	*Sus* *Phacochœrus* *Dicotyles*	or Wild Boar, &c. " Wart-bearing Hogs. " Peccaries.
	HIPPOPOTAMIDÆ	*Hippopotamus*	or "River Horse."
	EQUIDÆ.	*Equus*	or Horse, Ass, Zebra, &c.
	CERVIDÆ or DEER FAMILY.	*Alce* *Rangifer* *Cervus*	or Moose & European Elk. " Reindeer & Caribou. " Common Deer, Wapiti, Stag, &c.
	ANTILOPIDÆ or HOLLOW-HORNED RUMINANT FAMILY.	*Antilope* *Antilocapra* *Aplocerus* *Capra* *Ovis* *Ovibos* *Bos*	or Antelopes. " Pronghorn. " Rocky Mountain Goat. " Goats. " Sheep. " Musk Ox. " Oxen, Buffaloes, &c.
	CAMELOPARDALIDÆ.	*Camelopardalis*	or Giraffe.
	CAMELIDÆ.	*Camelus* *Auchenia*	or Camels. " Llamas.
	MOSCHIDÆ.	*Moschus*	or Musk Deer.
	SIRENIDÆ.	*Manatus* *Halicore* *Rytina*	or Sea-Cows. " Dugong. " Stellers.

[VERTEBRATES: MAMMALS — *Continued.*]

ORDERS.	FAMILIES.	*Genera.*	
MUTILATA or CETACEA or WHALES, &c.	BALÆNIDÆ or RIGHT WHALE FAMILY.	*Balæna* *Balænoptera*	or Greenland Whale, or Right Whale. " Rorquals or Finners.
	PHYSETERIDÆ or CATODONTIDÆ.	*Physter*, or *Catodon*	or Sperm Whales.
	DELPHINIDÆ or DOLPHIN FAMILY.	*Beluga* *Globicephalus* *Phocæna* *Delphinus* *Delphinorhynchus* *Delphinapterus* *Soosoo* *Inia*	or White Whale or White Grampus. " Bottleheads. " Porpoises & Grampuses. " Dolphins. " Beaked Dolphins. " Peron's Dolphin. " Soosoos of the Ganges.
	MONODONTIDÆ.	*Monodon*	or Narwhal.
CHEIROPTERA or BATS.	PTEROPODIDÆ or FRUGIVOROUS BAT FAMILY.	*Pteropus*	or Rousettes.
	MEGADERMATIDÆ or HORSE-SHOE BAT FAMILY.	*Rhinolophus* *Megaderma* *Macrotus* *&c.*	or Horse-shoe Bats. " Megaderms.
	PHYLLOSTOMATIDÆ.	*Phyllostoma*	or Vampire Bats.
	NOCTILIONIDÆ.	*Nyctinomus, &c.*	or Noctilios, &c.
	VESPERTILIONIDÆ or COMMON BAT FAMILY.	*Nycticejus* *Lasiurus* *Scotophilus* *Vespertilio* *Synotus* *Antrozous*	 or Red and Hoary Bats. " Carolina Bat, &c. " Little Brown Bat, &c. " Big-eared Bats. " Pale Bat.
INSECTIVORA or INSECT-EATERS.	DERMOPTERA.	*Galeopithecus*	or Galeopithecus.
	SCANDENTIA.	*Cladobates, &c.*	or Banxrings.
	SORICIDÆ or SHREW FAMILY.	*Neosorex* & *Sorex* *Blarina*	or Shrews. " Mole Shrews.
	TALPIDÆ or MOLE FAMILY.	*Scalops* *Condylura* *Talpa* *Urotrichus* *Chrysochloris*	or Shrew Moles. " Star-nosed Moles. " European Mole. " Bulb-nosed Mole. " Golden-green Moles.
	ACULEATA or HEDGEHOG FAMILY.	*Erinaceus* *Centetes*	or Hedgehogs. " Tenrecs.

[VERTEBRATES: MAMMALS — *Continued.*]

ORDERS.	FAMILIES.	*Genera.*		
RODENTIA or GNAWERS.	SCIURIDÆ or SQUIRREL FAMILY.	*Sciurus*	or	True Squirrels.
		Pteromys	"	Flying Squirrels.
		Tamias	"	Striped Squirrels.
		Spermophilus	"	Spermophiles or Gophers.
		Cynomys	"	Prairie Dogs.
		Arctomys	"	Woodchucks & Marmots.
		Myoxus	"	Dormice.
		Castor	"	Beavers.
		Aplodontia	"	Sewellel.
	SACCOMYIDÆ or POUCHED GOPHER FAMILY.	*Geomys* and *Thomomys*	or	Pouched Gophers.
		Dipodomys	"	Kangaroo Rats.
		Perognathus	"	Tuft-tailed Mice.
	MURIDÆ or RAT FAMILY.	*Dipus*	or	Jerboas.
		Jaculus	"	Amer. Jumping Mouse.
		Gerbillus	"	Gerbils.
		Mus	"	Rats.
		Cricetus	"	Hamsters.
		Reithrodon	"	Harvest Mice.
		Hesperomys	"	White-footed Mice.
		Neotoma	"	Wood Rats.
		Sigmodon	"	Cotton Rats.
		Arvicola	"	Field Mice.
		Myodes	"	Lemmings.
		Fiber	"	Muskrat.
	HYSTRICIDÆ or PORCUPINE FAMILY.	*Erethizon* & *Hystrix*	or	Porcupines.
		Dasyprocta	"	Agoutis.
		Dolichotis	"	Patagonian Cavies.
		Chinchilla	"	Chinchillas.
		Cavia	"	Guinea Pigs.
		Myopotamos	"	Couia.
		Hydrochœrus	"	Capybara.
	LEPORIDÆ.	*Lepus*	or	Hares and Rabbits.
		Lagomys	"	Pikas.
EDENTATA or EDENTATES.	BRADYPODA.	*Bradypus*	or	Sloths.
		Megatherium, Megalonyx, & Mylodon.		Extinct.
	EFFODIENTA.	*Dasypus, &c.*	or	Armadillos, &c.
		Glyptodon		(Extinct.)
		Myrmecophaga, &c.	"	Ant-eaters, &c.
MARSUPIALIA or MARSUPIALS.	PHALANGISTIDÆ.	*Phalangista, &c.*	or	Phalangers, &c.
	DASYURIDÆ.	*Dasyurus, &c.*	or	Bear Opossums, &c.
	MACROPODIDÆ.	*Macropus, &c.*	or	Kangaroos, &c.
	PERAMELIDÆ.	*Perameleso*	or	Bandicots, &c.
	DIDELPHIDÆ.	*Didelphys, &c.*	or	Opossums, &c.
	PHASCOLOMYIDÆ.	*Phascolomys*	or	Wombat.
MONOTREMATA.	PLATYPUS or DUCKBILL FAMILY.	*Ornithorhynchus* or *Platypus*	or	Duckbill.
		Echidna	"	Porcupine Ant-eater.

THE CLASS OF BIRDS.

ORDERS.	FAMILIES.	*Genera.*	
RAPTORES or BIRDS OF PREY.	VULTURIDÆ or VULTURE FAMILY.	*Vultur*	or Condor, &c
		Gypætos	" Læmmergyer.
		Cathartes	" North Amer. Vultures.
	FALCONIDÆ or FALCON FAMILY.	*Falco*	or Falcons.
		Astur	" Goshawk, &c.
		Accipiter	" Cooper's Hawk, &c.
		Buteo	" Buzzards.
		Archibuteo	" Rough-legged Hawk.
		Asturina	
		Nauclerus	" Swallow-tailed Hawk.
		Elanus	" White-tailed Hawk.
		Ictinia	" Mississippi Kite.
		Rostrhamus	" Black Kite.
		Circus	" Marsh Hawk.
		Aquila	" Golden Eagle.
		Halietus	" White-headed and Sea Eagles.
		Pandion	" Fish Hawks.
		Polyborus	" Caracara Eagle.
		Craxirex	
	STRIGIDÆ or OWL FAMILY.	*Strix*	or Barn Owls.
		Bubo	" Great Horned Owls.
		Scops	" Screech Owls.
		Otus	" Long-eared Owls.
		Brachyotus	" Short-eared Owls.
		Syrnium	" Gray Owls.
		Nyctale	" Sparrow Owls.
		Athene	" Burrowing Owls.
		Glaucidium	" Pigmy Owls.
		Surnia	" Day Owls.
SCANSORES or CLIMBERS.	PSITTACIDÆ.	*Conurus*, &c.	or Parrots.
	RHAMPHASTIDÆ.	*Rhamphastos*	or Toucans.
	TROGONIDÆ.	*Trogon*	or Trogons.
	CUCULIDÆ or CUCKOO FAMILY.	*Crotophaga*	or Black Parrot and Ani.
		Geococcyx	" Road Runner.
		Coccygus	" Cuckoos.
	PICIDÆ or WOODPECKER FAMILY.	*Campephilus*	or Ivory-billed Woodpeckers.
		Picus	" Hairy and downy Woodpeckers.
		Picoides	" Three-toed Woodpeckers.
		Sphyrapicus	" Yellow-bellied Woodpecker, &c.
		Hylatomus	" Black Woodcock.
		Centurus	" Red-billed Woodp., &c.
		Melanerpes	" Red-headed Woodpeckers.
		Colaptes	" Golden-winged Woodp.

[VERTEBRATES: BIRDS — *Continued.*]

ORDERS.		FAMILIES.	Genera.		
INSESSORES OR PERCHERS.	STRISORES.	TROCHILIDÆ.	*Trochilus, &c.*	or	Humming-Birds.
		CYPSELIDÆ.	*Chœtura, &c.*	or	Chimney Swallows and Swifts.
		CAPRIMULGIDÆ.	*Antrostomus*	or	Chuckwill's Widow and Whippoorwills.
			Chordeiles	"	Night-Hawks.
	CLAMATORES.	ALCEDINIDÆ.	*Ceryle*	or	King-fishers.
		PRIONITIDÆ.	*Momotus*	or	Saw-Bills.
		COLOPTERIDÆ or FLYCATCHER FAMILY.	*Pachyrhamphus*	or	Rose-throated Flycatchers.
			Milvulus	"	Forked-tailed Flycatchers.
			Tyrannus	"	King-birds, &c.
			Myiarchus	"	Great-crested Flycatchers, &c.
			Sayornis	"	Phœbe Bird or Pewee, &c
			Contopus	"	Wood Pewee, &c.
			Empidonax	"	Least Flycatcher, &c.
			Pyrocephalus	"	Red Flycatcher.
	OSCINES.	TURDIDÆ or THRUSH FAMILY.	*Turdus*	or	Wood & Hermit Thrushes, Robin, &c.
			Saxicola	"	Stone Chats.
			Erythaca	"	Robin Redbreast.
			Sialia	"	Blue-Birds.
			Regulus	"	Ruby-crowned Wren, &c.
			Hydrobata	"	Water Ouzels.
		SYLVICOLIDÆ or WARBLER FAMILY.	*Philomela*	or	Nightingales.
			Anthus	"	Tit Larks.
			Neocorys	"	Missouri Skylark.
			Mniotilta	"	Black & White Creepers.
			Parula	"	Blue Yellow-backed Warblers.
			Protonotaria	"	Prothonotary Warblers.
			Geothlypis	"	Maryland Yellow-throat, &c.
			Oporornis	"	Connecticut Warbler, &c.
			Icteria	"	Chats.
			Helmitherus	"	Worm-eating Warbler, &c.
			Helminthophaga	"	Golden-winged Warbler.
			Seiurus	"	Golden-crowned Thrush, &c.
			Dendroica	"	Yellow-rumped Warbler, Blackburnian, &c.
			Myiodioctes	"	Hooded Warbler, &c.
			Cardellina	"	Vermilion Flycatcher.
			Setophaga	"	Redstarts.
			Pyranga	"	Tanagers.
		HIRUNDINIDÆ or SWALLOW FAMILY.	*Hirundo*	or	Swallows.
			Cotyle	"	Bank Swallows.
			Progne	"	Purple Martins.
		BOMBYCILLIDÆ.	*Ampelis*	or	Wax-wings.
			Myiadestes	"	Townsend's Flycatcher.

[VERTEBRATES: BIRDS — *Continued.*]

ORDERS		FAMILIES.	*Genera.*		
INSESSORES or PERCHERS — *Continued.*	OSCINES — *Continued.*	LANIDÆ.	*Collyrio*	"	Shrikes.
			Vireo	"	Vireos.
		LIOTRICHIDÆ or MOCKING-BIRD FAMILY.	*Mimus, &c.*	or	Mocking and Cat Birds.
			Harporhynchus	"	Brown Thrushes.
			Catherpes	"	White-throated Wren.
			Salpinctes	"	Rock Wren.
			Thryothorus	"	Gt. Carolina & Bewick's.
			Cistothorus	"	Long-billed Marsh Wren.
			Troglodytes	"	House Wren, &c.
		CERTHIADÆ.	*Certhia*	or	Creepers.
			Sitta	"	Nuthatches.
		PARIDÆ.	*Polioptila*	or	Blue Gray Flycatcher.
			Lophophanes	"	Tufted Titmice.
			Parus, &c.	"	Titmice.
		DACNIDIDÆ.	*Certhiola*	or	Yellow-rumped Creeper.
		ALAUDIDÆ.	*Eremophila*	or	Skylarks.
		FRINGILLIDÆ or FINCH and SPARROW FAMILY.	*Hesperiphona*	or	Evening Grosbeaks.
			Pinicola	"	Pine Grosbeaks.
			Carpodacus	"	Purple Finches.
			Chrysomitris	"	Goldfinches.
			Curvirostra	"	Crossbills.
			Ægiothus	"	Red Polls.
			Leucosticte	"	Gray-crowned Finch.
			Plectrophanes, &c.	"	Snow Buntings, &c.
			Passerculus	"	Savannah Sparrow, &c.
			Pooecetes	"	Grass Finch.
			Coturniculus	"	Yellow-winged Sparrow.
			Ammodromus	"	Seaside Finch, &c.
			Chondestes	"	Lark Finch.
			Zonotrichia	"	White-crowned Sparrow.
			Junco	"	Snow Birds.
			Poospiza	"	Black-throated Sparrow.
			Spizella	"	Tree Sparrow, &c.
			Melospiza	"	Song Sparrow, &c.
			Peucaca	"	Bachman's Finch, &c.
			Embernagra	"	Texas Finch.
			Passerella	"	Fox-colored Sparrow,
			Calamospiza	"	Lark Bunting.
			Euspiza	"	Black-throated Bunting.
			Guiraca	"	Grosbeaks.
			Cyanospiza, &c.	"	Indigo Birds, &c.
			Cardinalis, &c.	"	Cardinal Birds.
			Pipilo	"	Ground Robins.
		ICTERIDÆ or BLACKBIRD FAMILY.	*Dolichonyx*	or	Bobolinks.
			Molothrus	"	Cow Birds.
			Agelaius, &c.	"	Blackbirds.
			Sturnella	"	Larks.
			Icterus	"	Orioles.
			Quiscalus	"	Grakles.
		STURNIDÆ.	*Sturnus*	or	Starlings.
		CORVIDÆ or CROW FAMILY.	*Corvus, &c.*	or	Ravens and Crows.
			Pica	"	Magpies.
			Cyanura, &c.	"	Jays.

[VERTEBRATES: BIRDS — *Continued.*]

ORDERS.		FAMILIES.	*Genera.*	
RASORES or SCRATCHERS.	COLUMBÆ.	COLUMBIDÆ or DOVE FAMILY.	*Columba*	or Doves.
			Ectopistes	" Wild Pigeon.
			Zenaida	" Zenaida Dove.
			Melopelia	" White-winged Dove.
			Zenaidura	" Carolina Dove.
			Scardafella	" Scaly Dove.
			Chamæpelia	" Ground Dove.
			Oreopeleia	" Key West Dove.
			Starnoenas	" Blue-headed Pigeon.
	GALLINÆ.	GOURIDÆ.	*Goura*	or Crown Pigeons.
		PENELOPIDÆ.	*Ortalida*	or Chiacalacca.
		MEGAPODIDÆ.		or Mound Birds.
		PHASIANIDÆ or PHEASANT FAMILY.	*Meleagris*	or Wild Turkeys.
			Pavo	" Peacocks.
			Numida	" Guinea Fowl.
			Gallus	" Domestic Cock, &c.
			Phasianus	" Pheasants.
		TETRAONIDÆ or GROUSE FAMILY.	*Tetrao*	or Spruce Partridge, &c.
			Centrocercus	" Cock of the Plains.
			Pediocetes	" Sharp-tailed Grouse.
			Cupidonia	" Prairie Chicken.
			Bonasa	" Ruffed Grouse.
			Lagopus	" Ptarmigans.
		PERDICIDÆ or PARTRIDGE FAMILY.	*Ortyx*	or Partridge or Quail.
			Oreortyx	" Plumed Partridge.
			Lophortyx	" California Quail.
			Callipepla	" Blue Partridge.
			Cyrtonyx	" Massena Partridge.
			Perdix	" Europ'n Gray Partridge.
			Coturnix	" European Quail.
CURSORES.		STRUTHIONIDÆ or OSTRICH FAMILY.	*Struthio*	or Ostriches.
			Rhea	" South American Ostrich.
			Casuarius	" Cassowaries.
			Apteryx	
		OTIDÆ.	*Otis*	or Bustards.
GRALLATORES or WADERS.	HERODIONES.	GRUIDÆ.	*Grus*	or Cranes.
		ARAMIDÆ.	*Aramus*	or Courlans.
		ARDEIDÆ or HERON FAMILY.	*Demigretta*	or Egrets.
			Garzetta	" Snowy Herons.
			Herodias	" White Herons.
			Ardea	" Great Blue Herons.
			Audubonia	" Great White Herons.
			Florida	" Blue Herons.
			Ardetta	" Least Bittern.
			Botaurus	" Stake Drivers.
			Butorides	" Green Heron.
			Nyctiardea	" Night Heron.
			Nyctherodius	" Yellow-crowned Heron.
		CANCROMIDÆ.	*Cancroma*	or Boat-Bills.

[VERTEBRATES: BIRDS — *Continued.*]

ORDERS.		FAMILIES.	*Genera.*	
GRALLATORES or WADERS — *Continued.*	HERODIONES — *Continued.*	CINCONIDÆ.	*Cinconia*	or Storks.
			Jabiru	
		TANTALIDÆ.	*Tantalus*	or Ibises.
			Ibis	" Scarlet Ibis, &c.
		PLATALEIDÆ.	*Platalea*	or Spoonbills.
		PHŒNICOPTERIDÆ.	*Phœnicopterus*	or Flamingo.
	GRALLÆ.	CHARADRIDÆ or PLOVER FAMILY.	*Charadrius*	or Golden Plover.
			Ægialitis	" Kill-deer, &c.
			Squatarola	" Black-bellied Plover.
		HÆMATOPODIDÆ.	*Hæmatopus*	or Oyster-catchers.
			Strepsilas	" Turnstones.
		RECURVIROSTRIDÆ.	*Recurvirostra*	or Avosets.
			Himantopus	" Stilts.
		PHALAROPIDÆ.	*Phalaropus*	or Phalaropes.
		SCOLOPACIDÆ or SNIPE FAMILY.	*Philohela*	or Woodcocks.
			Gallinago	" Snipes.
			Macrorhamphus	" "
			Tringa	" Sandpipers.
			Calidris	" Sanderling.
			Ereuntes	" Semi-palmated Sandpipers.
			Micropalma	" Stilt Sandpipers.
			Symphemia	" Willets.
			Glottis	" Greenshanks.
			Gambetta	" Yellow-legs.
			Rhyacophilus	" Solitary Sandpiper.
			Heteroscelus	" Wandering Tatler.
			Tringoides	" Spotted Sandpiper.
			Philomachus	" Ruff.
			Actiturus	" Field Plover.
			Tryngites	" Buff-breasted Sandpiper.
			Limosa	" Godwits.
			Numenius	" Curlews.
		RALLIDÆ or RAIL FAMILY.	*Rallus*	or Rails.
			Porzana	" Sora Rails.
			Crex	" Corn Crakes.
			Fulica	" Coots.
			Gallinula	" Gallinules.

[VERTEBRATES: BIRDS— *Continued.*]

ORDERS.		FAMILIES.	Genera.	
NATATORES or SWIMMERS.	ANSERES.	ANATIDÆ or DUCK FAMILY.	*Cygnus*	or Swans.
			Anser	" White-fronted Goose.
			Bernicla, &c.	" Wild Goose, &c.
			Anas	" Mallard, &c.
			Dafila	" Pintail.
			Nettion	" Green-winged Teals,
			Querquedula	" Blue-winged Teals, &c.
			Spatula	" Shoveller.
			Chaulelasmus	" Gadwall.
			Mareca	" Widgeons.
			Aix	" Wood Duck.
			Fulix	" Scaup Ducks.
			Aythya	" Canvas-Back and Red-Head.
			Bucephala	" Golden-eye and Dipper.
			Histrionicus	" Harlequin Duck.
			Harelda	" Longtail.
			Camptolæmus	" Labrador Ducks.
			Melanetta	" White-winged Coot.
			Pelionetta	" Sea Coot.
			Oidemia	" Scoter.
			Somateria, &c.	" Eider Ducks.
			Erismatura	" Ruddy Ducks.
			Mergus	" Sheldrakes.
			Lophodytes	" Hooded Merganser.
			Mergellus	" Smew.
	GAVIÆ.	PELECANIDÆ.	*Pelecanus*	or Pelicans.
		SULIDÆ.	*Sula*	or Gannets.
		TACHYPETIDÆ.	*Tachypetes*	or Man of-War-Bird.
		PHALACROCORACIDÆ.	*Graculus*	or Cormorants.
		PLOTIDÆ.	*Plotus*	or Snake-Bird.
		PHÆTONIDÆ.	*Phæton*	or Tropic Bird.
		PROCELLARIDÆ or PETREL FAMILY.	*Diomedea*	or Albatrosses.
			Procellaria	" Fulmar Petrels.
			Thallasidroma	" Stormy Petrels.
			Puffinus	" Shearwaters.
		LARIDÆ or GULL FAMILY.	*Stercorarius*	or Jagers or Skua Gulls.
			Larus, &c.	" Gulls.
			Sterna, &c.	" Terns.
			Rhynchops	" Black Skimmers.
		COLYMBIDÆ.	*Colymbus*	or Divers.
			Podiceps	" Grebes.
			Podylimbus	" Pied Grebe.
		ALCIDÆ or AUK FAMILY.	*Alca*	or Auks.
			Aptenodytes	" Penguins.
			Mormon	" Puffins.
			Uria	" Guillemots.
			Mergulus	" Sea-Dove or Dove-Kie.

THE CLASS OF REPTILIA OR REPTILES.

ORDERS.		FAMILIES.	Genera.	
TESTUDINATA or TURTLES.	AMYDÆ or FRESH-WATER AND LAND TURTLES.	TESTUDININA or LAND TORTOISE FAMILY.	*Xerobates*	or Gophers.
			Testudo	" Galapago Tortoise, and European Tortoise.
		EMYDOIDÆ or TERRAPIN FAMILY.	*Ptychemys*	or Red-bellied Terrapin.
			Trachemys	
			Graptemys	" Map Turtles.
			Malacoclemmys	" Salt-water Terrapins.
			Chrysemys	" Painted Turtles.
			Deirochelys	" Reticulated Turtles.
			Emys	" Blanding's Tortoise.
			Nanemys	" Speckled "
			Calemys	" Mühlenberg's Tortoise.
			Glyptemys	" Wood "
			Actinemys	
			Cistudo	" Box Turtles.
		CINOSTERNOIDÆ.	*Aromochelys* or *Ozotheca*	or Musk Tortoise.
			Thyrosternum	" Cinosternum or Mud Tortoise.
		CHELYDROIDÆ.	*Macroclemmys*	or Alligator Turtle.
			Chelydra	" Snapping "
		HYDRASPIDÆ.	*Hydraspis*	of South America.
		CHELYOIDÆ.	*Chelys*	or Matamata, of S. America.
		TRIONYCHIDÆ.	*Platypeltis*	or Soft-shelled Turtles.
			Amyda and *Aspidonectes*.	
	CHELONII.	CHELONIOIDÆ or LOGGERHEAD FAMILY.	*Chelonia*	or Green Turtles.
			Eretmochelys	" Hawk-bill Turtles.
			Thallassochelys	" Loggerheads.
		SPHARGIDIDÆ.	*Sphargis*	or Leather-backed Turtles.
SAURIA or SAURIANS.	DINOSAURS.		*Iguanodon*	(Fossil.)
			Hylæosaurus	"
			Megalosaurus	"
	CROCODILIANS.		*Alligator*	or Alligators.
			Crocodilus	" Crocodiles.
			Gavialis	" Gavials.
			Cetiosaurus	(Fossil.)
			Teleosaurus	"
	LACERTIANS.		*Thecodontosaurus*, *Palæosaurus*, *Proterosaurus*, and *Mososaurus* (all Fossil).	
			Ameiva	or Striped Lizards.
			Iguana, *&c.*	" Iguanas, Monitors, Green Lizards, Horned Toads, Geckos, Chameleons, Skinks, &c.
	ENALIOSAURS.		*Ichthyosaurus*	(Fossil.)
			Plesiosaurus	"
	PTEROSAURS.		*Pterodactyl*	(Fossil.)

[VERTEBRATES: REPTILES — *Continued.*]

ORDERS.	FAMILIES.	*Genera.*	
OPHIDIA or SERPENTS.	BOIDÆ.	*Boa*	or Boas and Anacondas.
		Python	" Pythons.
	COLUBRIDÆ or COLUBER FAMILY.	*Eutænia*	or Striped Snakes.
		Nerodia	" Water "
		Regina	
		Heterodon	" Hog-nose "
		Pituophis	" Pine "
		Scotophis	
		Ophiobolus	" Chain Snakes and Chicken Snakes.
		Georgia	" Indigo Snakes.
		Bascanion	" Black "
		Masticophis	" Coach-whip Snakes.
		Leptophis	" Southern Green "
		Chlorosoma	" Northern Green "
		Diadophis	" Ring-necked "
		Rhinostoma	" Scarlet "
		Rhinocheilus	
		Haldea	" Brown "
		Farancia, &c.	" Horn " &c.
	CROTALIDÆ or RATTLESNAKE FAMILY.	*Crotalus*	or Rattlesnakes.
		Crotalophorus	" Prairie Rattlesnakes.
		Ancistrodon	" Copperheads.
		Toxicophis	" Water Moccasins.
	VIPERIDÆ.	*Vipera*	or Vipers of the Old World.
	ELAPIDÆ.	*Elaps*	or Harlequin Snakes.
		Naia	" Cobra, Haje, &c.
	HYDROPHIDÆ.	*Hydrophis*	or Venomous Water Snakes of the Old World.

THE CLASS OF BATRACHIA OR BATRACHIANS.

ORDERS.	FAMILIES.	*Genera.*	
ANOURA or TAILLESS BATRACHIANS.	RANIDÆ or FROG FAMILY.	*Rana*	or Frogs proper.
	HYLOIDÆ.	*Hyla*	or Tree-Frogs or Tree-Toads.
		Hylodes	" Cricket-Frogs.
	BUFONIDÆ or TOAD FAMILY.	*Bufo*	or Toads.
		Scaphiophus	" Toad-Frog.
URODOLA or TAILED BATRACHIANS.	SALAMANDRIDÆ or SALAMANDER FAMILY.	*Salamandra* and *many other genera*	or Salamanders & Tritons.
	AMPHUMIDÆ.	*Amphiuma*	or Congo Snake.
		Menopoma	" Hell-bender.
	SIRENIDÆ or SIREN FAMILY.	*Siren*	or Sirens.
		Menobranchus	" Mud-Puppies.
		Siredon	" Axolotl.
		Proteus	
APODA or FOOTLESS BATRACHIANS.	CÆCILIIDÆ.	*Cæcilia*	or Cæcilians, Blind-worms.

THE CLASS OF FISHES.

ORDERS.		FAMILIES.	Genera.	
SELACHI or SELACHIANS or PLAGIOSTOMI.	RAIÆ or RAYS or SKATES.	CEPHALOPTERIDÆ.	*Cephaloptera, &c.*	or Vampires, &c.
		MYLIOBATIDÆ.	*Myliobates, &c.*	or Sea-Eagles, &c.
		TRYGONIDÆ.	*Trygon, &c.*	or Sting Rays, &c.
		RAIIDÆ.	*Raia, &c.*	or Rays or Skates proper.
		TORPEDINIDÆ.	*Torpedo, &c.*	or Electric Rays.
		RHINOBATIDÆ.	*Rhinobatus, &c.*	or Rhinobats.
		PRISTIDÆ.	*Pristis*	or Saw-Fish.
	SQUALI or SHARKS.	ZYGÆNIDÆ.	*Zygæna*	or Hammerhead Shark.
		SQUATINIDÆ.	*Squatina*	or Monk Fish.
		SCYMNIDÆ.	*Scymnus, &c.*	or Nurse or Sleeper, &c.
		SPINACIDÆ.	*Acanthias, &c.*	or Spined Dog-Fish, &c.
		NOTIDANIDÆ.	*Hexanchus, &c.*	
		RHINODONTIDÆ.	*Rhinodon*	
		CESTRACIONTIDÆ.	*Cestracion*	or Cestracionts.
		ALOPECIIDÆ.	*Alopecias*	or Thresher Shark.
		LAMNIDÆ.	*Lamna* *Selachus*	or Mackerel Shark. " Basking or Elephant Shark.
		GALEIDÆ.	*Mustelus, &c.*	or Topes and Hounds.
		CHARCARIDÆ.	*Carcharias*	or White Shark, Gray Shark, &c.
		SCYLLIDÆ.	*Scyllium*	or Dog-Fishes.
		†		
GANOIDEI or GANOIDS.		STURIONIDÆ.	*Accipenser, &c.*	or Sturgeons.
		AMIIDÆ.	*Amia*	or Mud-Fishes.
		POLYPTERIDÆ.	*Polypterus*	or Polypterus of the Nile.
		LEPIDOSTEIDÆ.	*Lepidosteus*	or Gar-Pikes.
TELEOSTEI or BONY FISHES.	LOPHOBRANCHII.	SYGNATHIDÆ.	*Sygnathus, &c.* *Hippocampus, &c.*	or Pipe-Fishes, &c. " Sea-Horses, &c.
		PEGASIDÆ.	*Pegasus*	or Flying-Horses.
	PLECTOGNATHI.	DIODONTIDÆ.	*Diodon* *Tetraodon* *Orthagoriscus*	or Puffers. " " " Sun-Fish.
		OSTRACIONIDÆ.	*Ostracion, &c.*	or Trunk-Fish, &c.
		BALISTIDÆ.	*Balistes, &c.*	or File-Fishes.

* The classification here adopted is essentially that of Müller, as modified by Owen, and is taken from the Encyclopædia Britannica.

† Owen places here Chimæridæ and Sirenidæ, the latter represented by the Lepidosiren.

[VERTEBRATES: FISHES — *Continued.*]

ORDERS.	FAMILIES.	*Genera.*	
TELEOSTEI or BONY FISHES — *Continued.* ACANTHOPTERI or SPINE-FINNED FISHES.	LOPHIIDÆ.	*Lophius*	or Angler.
		Cheironectes	" Hand Fishes.
		Batrachus	" Toad "
	BLENNIIDÆ.	*Blennius*	or Blennies.
		Zoarces	" Eel-Pout.
		Anarrichas, &c.	" Wolf-Fish, &c.
	GOBIIDÆ.	*Gobius, &c.*	or Gobies, &c.
		Cyclopterus	" Lump-Fish.
	AULOSTOMIDÆ.	*Fistularia, &c.*	or Flute-mouths.
	TEUTHYDIDÆ.	*Acanthurus, &c.*	or Lancet-Fish, &c.
	TÆNIIDÆ.	*Trachypterus, &c.*	or Ribbon-Fish.
	CHÆTODONTIDÆ.	*Chætodon, &c.*	or Chætodonts.
	ZEIDÆ.	*Zeus, &c.*	or Dories, &c.
	SCOMBRIDÆ or MACKEREL FAMILY.	*Scomber*	or Mackerels proper.
		Thynnus	" Tunnies.
		Xiphias	" Sword-Fish.
		Pelamys	" Skip-Jack.
		Naucrates	" Pilot-Fish.
		Temnodon	" Blue-Fish.
		Coryphæna, &c.	" Dolphins, &c.
	ATHERINIDÆ.	*Atherina*	or Silversides.
	MUGILIDÆ.	*Mugil, &c.*	or Mullets.
	LABYRYNTHIBRANCHIDÆ.	*Anabas, &c.*	or Climbing Perch, &c.
	SPARIDÆ.	*Sargus*	or Sheepshead.
		Pagrus, &c.	" Scupaug or Porgee, &c.
	SCIÆNIDÆ.	*Otolithus, &c.*	or Weak-Fish.
		Corvina	" Lake Sheepshead.
		Umbrina	" King-Fish.
		Pogonias, &c.	" Drums, &c.
	TRIGLIDÆ or SCLEROGENIDÆ, &c. or SCULPIN FAMILY, &c.	*Trigla*	or Gurnards.
		Prionotus	" Sea-Robins.
		Dactylopterus	" Sea-Swallows.
		Cottus	" Sculpins.
		Hemitripterus	" Sea-Raven.
		Scorpæna	" Sea-Scorpion.
		Gasterosteus, &c.	" Sticklebacks, &c.
	MULLIDÆ.	*Mullus, &c.*	or Surmullets.
	POLYNEMIDÆ.	*Polynemus, &c.*	or Paradise Fish, &c.
	THERAPONIDÆ.	*Pomotis*	or Breams.
		Boleosoma, &c.	" Darters, &c.
	PERCIDÆ.	*Perca*	or Perch proper.
		Labrax, &c.	" Bass, &c.
	URANOSCOPIDÆ.	*Uranoscopus, &c.*	or Star-Gazers, &c.

[VERTEBRATES: FISHES — *Continued.*]

ORDERS.		FAMILIES.	*Genera.*	
TELEOSTEI or BONY FISHES — *Continued.*	ANACANTHINI.	PLEURONECTIDÆ.	*Platessa*	or Flounders.
			Hippoglossus	" Halibuts.
			Rhombus & *Solea*, &c.	" Turbots and Soles, &c.
		ECHENEIDIDÆ.	*Echeneis*	or Remora.
		GADIDÆ or COD FAMILY.	*Morrhua*	or Cods and Haddock.
			Merlangus	" Pollack.
			Merluccius	" Whiting.
			Lota	" Burbot.
			Brosmius	" Cusk.
			Phycis, &c.	" Hake, &c.
		OPHIDIDÆ.	*Ophidia*	or Serpent-form Fishes.
	PHARYNGO-GNATHI.	AMBIOTOCIDÆ.	*Ambiotoca*	
		LABRIDÆ.	*Ctenolabrus*	or Conners.
			Tautoga, &c.	or Tautog, &c.
		SCOMBERESOCIDÆ.	*Belone*	or Gar-Fishes.
			Scomberesox, &c.	" Bill-Fishes, &c.
			Exocœtus	" Flying-Fishes.
	MALACOPTERI.	GONIODONTIDÆ.	*Goniodontes*	or Goniodonts.
		SILURIDÆ.	*Silurus*, &c.	or Silurus, &c.
			Pimelodus, &c.	" Cat-Fishes.
		CYPRINIDÆ.	*Cyprinus*, &c.	or Carps proper, &c.
			Leuciscus, &c.	" Dace, Shiners, &c.
		CATOSTOMIDÆ.	*Catostomus*	or Suckers.
		ESOCIDÆ.	*Esox*	or Pike and Pickerel.
		CYPRINODONTIDÆ.	*Fundulus* & *Hydrargyra*, &c.	or Mummachogs, &c.
		MORMYRIDÆ.	*Mormyrus.*	
		ELOPIDÆ.	*Elops*	or Silver-Fish.
		CLUPESOCIDÆ.	*Notopterus*, &c.	or Herring-Pikes.
		SCOPELIDÆ.	*Scopelus*, &c.	
		CHARACINIDÆ.	*Salminus*, &c.	or Salmon-like Fishes.
		SALMONIDÆ.	*Salmo*, &c.	or Salmon, Trout, Smelts, &c.
		CLUPEIDÆ.	*Clupea*	or Herring, Pilchards, &c.
			Alausa, &c.	" Shad, Alewives, &c.
		APHRODEIRIDÆ.		or L. Ponchartrain Fishes.
		HETEROPYGII.	*Amblyopsis*	or Blind-Fish.
		GYMNOTIDÆ.	*Gymnotus*, &c.	or Electric Eels.
		OPHISURIDÆ.	*Ophisurus*, &c	or Snake-tailed Eels.
		CONGERIDÆ.	*Conger*	or Conger Eels.
		ANGUILLIDÆ.	*Anguilla*	or Common Eels.
		MURÆNIDÆ.	*Murœna*, &c.	or Roman Murœna.
		SYNBRANCHIDÆ.		or Eel-like Fishes.

[VERTEBRATES: FISHES — *Continued.*]

ORDERS.	FAMILIES.	*Genera.*	
DERMOPTERI.	LEPTOCEPHALIDÆ.	*Leptocephalus*	or Slender-heads.
	PETROMYZONTIDÆ.	*Petromyzon*	or Lampreys.
	MYXINIDÆ.	*Myxine*	or Hag.
	AMMOCÆTIDÆ.	*Ammocœtes*	or Mud-Lampreys.
	AMPHIOXIDÆ.	*Branchiostoma*	or Amphioxus.

THE BRANCH OF ARTICULATA, OR ARTICULATES.

THE CLASS OF INSECTA, OR INSECTS.

INSECTS PROPER. HYMENOPTERA or BEES, WASPS, &c.	APIARIE or APIDÆ.	*Apis* *Bombus* *Xylocopha, &c.*	or Hive Bee. " Humble-Bees. " Carpenter Bees, &c.
	VESPARIÆ or VESPIDÆ.	*Vespa* *Polistes, &c.*	or Wasps and Hornets. " Wasps.
	CRABRONIDÆ.	*Crabro, &c.*	or Wood-Wasps, &c.
	BEMBECIDÆ.	*Bembex, &c.*	
	SPHEGIDÆ.	*Sphex, &c.*	or Mud Wasps.
	SCOLIETÆ.	*Scolia, &c.*	
	FORMICARIÆ.	*Formica, &c.*	or Ants.
	CHRYSIDIDÆ.	*Chrysis, &c.*	or Golden Wasps.
	PROCTOTRUPIDÆ.	*Platygaster, &c.*	or Egg-Parasites.
	CHALCIDIDÆ.	*Chalcis, &c.*	
	ICHNEUMONIDÆ.	*Ichneumon*	or Ichneumons.
	EVANIALES.	*Evania, &c.*	
	CYNIPSERA.	*Cynips, &c.*	or Gall Flies.
	UROCERATA.	*Tremex, &c.*	or Boring-Saw Flies.
	TENTHREDINETÆ.	*Selandria,* *Cimbex, &c.*	or Rose & Elm Saw-Flies, &c.
LEPIDOPTERA or BUTTERFLIES and MOTHS.	PAPILIONIDÆ.	*Papilio, &c.*	or Papilio Butterflies, &c.
	PIERIDÆ.	*Pieris* *Colias*	or White Butterfly. " Yellow "
	NYMPHALIDÆ.	*Limenitis, Danais, Argynnis, &c.*	
	SATYRIDÆ.	*Satyrus, &c.*	or Hipparchians, &c.
	LYCÆNIDÆ.	*Chrysophanus, &c.*	or Copper Butterflies, &c.
	HESPERIDÆ, &c.	*Hesperia, &c.*	or Skipper Butterflies.
	SPHINGIDÆ.	*Sphinx, &c.*	or Hawk-Moths, &c.
	ÆGERIDÆ.	*Trochilium, &c.*	or Peach-tree Borers, &c.
	ZYGÆNIDÆ.	*Eudryas, &c.*	or Wood Nymphs, &c.
	BOMBYCIDÆ.	*Bombyx, Telea, &c.*	or Silk-Worm Moths.
	NOCTUELITÆ.	*Agrotis, &c.*	or Dart-Moths, &c.
	PHALÆNIDÆ.	*Geometra, &c.*	or Geometers, Canker-worms, &c.
	PYRALIDÆ.	*Pyralis, &c.*	or Meal-Moth, &c.
	TORTRICIDÆ.	*Penthina, &c.*	or Apple-worm Moth, &c.
	TINEIDÆ.	*Tinea, &c.*	or Clothes Moths, &c.
	PTEROPHORII.	*Pterophorus*	or Feather-winged Moths.

[ARTICULATES: INSECTS — *Continued.*]

ORDERS.		FAMILIES.	Genera.	
INSECTS PROPER — *Continued.*	DIPTERA or FLIES, &c.	CULICIDÆ	*Culex, &c.*	or Gnats, Mosquitoes, &c.
		TIPULARIÆ.	*Ceidomyia, Tipula, &c.*	or Hessian Fly, &c.
		TABANIDÆ.	*Tabanus*	or Horse Flies.
		ASILICI.	*Asilus*	or Asilus Flies.
		BOMBYLIARII.	*Bombylius*	or Bee Flies.
		SYRPHIDÆ.	*Syrphus, &c.*	or Syrphus Flies.
		DOLICHOPIDÆ.	*Dolichopus*	or Long-legged Flies.
		ŒSTRIDÆ.	*Gasterophilus, Œstrus, &c.*	or Bot Flies.
		MUSCIDÆ.	*Musca, &c.*	or House Flies, &c.
		HIPPOBOSCIDÆ.	*Hippobosca, &c.*	or Spider Flies.
		PULICIDÆ.	*Pulex*	or Fleas.
	COLEOPTERA or BEETLES.	CICINDELIDÆ.	*Cicindela*	or Tiger Beetles.
		CARABIDÆ.	*Calasoma, &c.*	or Caterpillar Hunters.
		DYTICIDÆ.	*Dyticus*	or Water Beetles.
		GYRINIDÆ.	*Gyrinus*	or Whirligig Beetles.
		HYDROPHILIDÆ.	*Hydrophilus*	or Water-loving Beetles.
		SILPHIDÆ.	*Silpha*	or Carrion Beetles.
		STAPHYLINIDÆ.	*Staphylinus*	or Rove Beetles.
		HISTERIDÆ.	*Hister*	or Mimic Beetles.
		DERMESTIDÆ.	*Dermestes*	or Skin Beetles.
		BYRRHIDÆ.	*Byrrhus*	
		LUCANIDÆ.	*Lucanus*	or Horn-Bugs.
		SCARABÆIDÆ.	*Copris* *Geotrupes* *Macrodactylus* *Lachnosterna, &c.*	or Tumble Beetles. " Earth-borers. " Rose-chafers. " May Beetles, &c.
		BUPRESTIDÆ.	*Buprestis*	or Buprestians.
		ELATERIDÆ.	*Elater*	or Snap or Spring Beetles.
		LAMPYRIDÆ.	*Lampyris*	or Fire-Flies or Glow-worms.
		MALACHIDÆ.		
		CLERIDÆ.	*Clerus*	or Bee-destroyers.
		LYMEXILLIDÆ.	*Lymexylon*	or Wood-destroyers.
		PTINIDÆ.	*Anobius, &c.*	or Death-Watches, &c.
		TENEBRIONIDÆ.	*Tenebrio*	or Meal-worms.
		MORDELLIDÆ.	*Mordella*	
		MELOIDÆ.	*Cantharis, &c.*	or Cantharides.
		STYLOPIDÆ.	*Stylops, &c.*	or Bee-Parasites.
		CURCULIONIDÆ.	*Curculio, &c.*	or Curculios or Weevils.
		CERAMBYCIDÆ.	*Prionus, &c.*	or Capricorn Beetles.
		CHRYSOMELIDÆ.	*Chrysomela, &c.*	or Chrysomelans, &c.
		COCCINELLIDÆ.	*Coccinella*	or Lady-Birds.

[ARTICULATES: INSECTS—*Continued.*]

ORDERS.		FAMILIES.	*Genera.*	
INSECTS PROPER—*Continued.*	HEMIPTERA or CICADAS, BUGS, &c. 6	CICADARIÆ.	*Cicada*	or Cicadas or Harvest Flies.
		FULGORIDÆ.	*Fulgora*	or Lantern-Flies.
		CERCOPIDÆ.	*Membracis*	or Tree-Hoppers.
		APHIDÆ.	*Aphis*	or Plant-Lice.
		COCCIDÆ.	*Coccus*	or Bark-Lice, Cochineal, &c.
		NOTONECTIDÆ.	*Notonecta*	or Boat Flies.
		NEPIDÆ.	*Nepa*	or Scorpion-Bugs.
		HYDROMETRIDÆ.	*Gerris*	or Water-Measurers.
		COREIDÆ.	*Coreus, &c.*	or Squash-Bug, &c.
		THRIPSIDÆ.	*Thrips*	
		CIMICIDÆ.	*Cimex*	or Bed-Bug.
		PEDICULIDÆ.	*Pediculus, &c.*	or Lice.
	ORTHOPTERA or GRASSHOPPERS, &c. 2	FORFICULIDÆ.	*Forficula*	or Earwigs.
		BLATTARIÆ.	*Blatta*	or Cockroaches.
		PHASMIDA.	*Diaphomera, &c.*	or Walking-stick, &c.
		MANTIDÆ.	*Mantis*	or Mantis.
		GRYLLIDES.	*Gryllus, &c.*	or Field Crickets, &c.
		LOCUSTARIÆ.	*Cyrtophyllus, &c.*	or Katydid, &c.
		ACRYDII.	*Caloptenus, &c.*	or Red-legged Locust, &c.
		THYSANOURA.	*Lepisma, &c.*	or Spring-tails.
	NEUROPTERA or DRAGON-FLIES, &c. 3	TERMITIDÆ.	*Termites*	or Termites.
		PSOCIDÆ.	*Atropos*	or Book-Louse, &c.
		PERLARIÆ.	*Perla, &c.*	or Stone-Flies.
		EPHEMERIDÆ.	*Ephemera*	or May-Flies.
		ODONATA.	*Agrion, Æschna, &c.*	or Dragon-Flies.
		SIALIDÆ.	*Corydalis, &c.*	or Corydalis, &c.
		HEMEROBINI.	*Hemerobius, &c.*	or Lace-wings.
		PHRYGANIDÆ.	*Neuronia, &c.*	or Caddice Flies.
ARACHNIDA or SPIDERS, &c.		ARANEIDÆ.	*Lycosa, &c.*	or True Spiders.
		PEDIPALPI.	*Buthus, &c.*	or Scorpions.
		PSEUDO-SCORPIONES.	*Chelifer, &c.*	or Book Spiders, &c.
		PHALANGITA.		or Daddy-long-legs.
		ACARINA.	*Trombidium, &c.*	or Velvet-red Mites, &c.
MYRIAPODA or CENTIPEDES.		GLOMERIDÆ.	*Glomeris*	
		JULIDÆ.	*Julus*	or Galley-Worm.
		POLYDESMIDÆ.	*Polydesmus.*	
		LITHOBIIDÆ.	*Lithobius*	or Earwigs.
		SCOLOPENDRIDÆ.	*Scolopendra*	or Centepedes.
		GEOPHILIDÆ.	*Geophilus*	

THE CLASS OF CRUSTACEA OR CRUSTACEANS.

ORDERS.		FAMILIES.	*Genera.*	
DECAPODA or TEN-FOOTED CRUSTACEANS.	BRACHYURANS (including ANOMURANS) or CRABS.	MAHDÆ or SEA-SPIDER FAMILY.*	*Maia, &c.*	or Sea-Spiders, &c.
		CANCRIDÆ or EDIBLE CRAB FAMILY.	*Cancer, &c.*	or Edible Crab of Europe, &c.
		PORTUNIDÆ or EDIBLE CRAB FAMILY.	*Lupa, &c.*	or Edible Crab of U. S., &c.
		GECARCINIDÆ or LAND CRAB FAMILY.	*Gecarcinus, &c.*	or Land Crabs.
		GELASMIDÆ or FIDDLER CRAB FAMILY.	*Gelasmus, &c.*	or Fiddler Crabs.
		PAGURIDÆ or HERMIT CRAB FAMILY.	*Pagurus, &c.*	or Hermit Crabs.
	MACRURANS or LOBSTERS, &c.	PALINURIDÆ or SPINY LOBSTER FAMILY.	*Palinurus, &c.*	or Spiny Lobsters.
		ASTICIDÆ or COMMON LOBSTER FAMILY.	*Homarus* *Astacus, &c.*	or Common Lobsters. " Cray-Fishes.
	GASTRURANS or STOMAPODS or SHRIMPS, &c.	CRANGONIDÆ or SHRIMP FAMILY.	*Crangon, &c.*	or Shrimps.
		PALEMONIDÆ or PRAWN FAMILY.	*Palemon, &c.*	or Prawns.
		SQUILLIDÆ or SEA MANTIS FAMILY.	*Squilla, &c.*	or Sea Mantes.
		MYSIDÆ or OPOSSUM SHRIMP FAMILY.	*Mysis, &c.*	or Opossum Shrimps.

* Only the more common Families and Genera of Crustaceans are here given. The same is true in regard to Worms, on page XXI.

[ARTICULATES: CRUSTACEANS — *Continued.*]

	Orders.	Families.	Genera.	
TETRADECAPODA or FOURTEEN-FOOTED CRUSTACEANS.	ISOPODS (including ANISOPODS).	ONISCIDÆ or WOOD LOUSE FAMILY.	*Oniscus*	or Wood Louse.
		ARMADILLIDÆ or PILL-BUG FAMILY.	*Armadillo*	or Pill Bugs.
		CYMOTHOIDÆ.	*Cymathoa*	or Parasites.
		BOPYRIDÆ.	*Bopyrus*	or Parasites.
	AMPHIPODS (including LÆMIDOPODS).	ORCHESTIDÆ or SAND-HOPPER FAMILY.	*Orchestia*	or Sand-Hoppers or Beach-Fleas.
		GAMMARIDÆ or FRESH-WATER SHRIMP FAMILY.	*Gemmarus*	or Fresh-water Shrimps.
		CAPRELLIDÆ or MEASURER FAMILY.	*Caprella*	or Measurers.
		CYAMIDÆ or WHALE-LOUSE FAMILY.	*Cyamus*	or Whale-Lice.
	TRILOBITES. (Fossil.)			
ENTOMOSTRACA or ENTOMOSTRACANS.	CARCNOIDS (including SIPHONOSTOMES.)	CYCLOPIDÆ or CYCLOPS FAMILY.	*Cyclops*	
		ARGULIDÆ.	*Argulus*	
		CALIGIDÆ.	*Caligus*	
	OSTRACOIDS (including CIRRIPEDS).	CYPRIDÆ or CYPRIS FAMILY.	*Cypris*	
		DAPHNIADÆ or MONOCULUS FAMILY.	*Daphnia*	or Monoculus.
		LIMNADIADÆ.	*Limnadia*	
		LEPADIDÆ.	*Anatifa*	or Geese Barnacles.
		BALANIDÆ.	*Balanus*	or Acorn Barnacles.
	LIMULOIDS.	LIMULIDÆ.	*Limulus*	or Horse-shoe Crab.
	ROTIFERA.			

THE CLASS OF WORMS.

ORDERS.		FAMILIES.	*Genera.*	
ANNELIDES.	DORSI-BRANCH-IATÆ.	ARENICOLIDÆ or SAND-WORM FAMILY.	*Arenicola*	or Sand-Worms or Lob-Worms.
	TUBI-COLÆ.	SERPULIDÆ or SERPULA FAMILY.	*Serpula*	or Serpulæ.
	TERRI-COLÆ.	LUMBRICIDÆ or EARTH-WORM FAMILY.	*Lumbricus*	or Earth-Worms.
TREMATODS or SUCTORIA.		HIRUNDINIDÆ or LEECH FAMILY.		
	ENTO-ZOA.	CESTOIDÆ or TAPE-WORM FAMILY.	*Tænia*	or Tape-Worms.
NEMA-TOIDS.	HELMIN-THES or	GORDIIDÆ or HAIR-WORM FAMILY.	*Gordius*	or Hair-Worms.

THE BRANCH OF MOLLUSCA OR MOLLUSKS.

THE CLASS OF CEPHALOPODA OR CEPHALOPODS.

ORDERS.	FAMILIES.	*Genera.*	
DIBRANCHIATA or TWO-GILLED CEPHALOPODS.	ARGONAUTIDÆ or PAPER SAILOR FAMILY.	*Argonauta*	or Argonaut or Paper Sailor.
	OCTOPODIDÆ or POULPE FAMILY.	*Octopus*	or Poulpes.
	TEUTHIDÆ or SQUID FAMILY.	*Loligo*	or Squids.
	BELEMNITIDÆ or BELEMNITE FAMILY. (Fossil.)		
	SEPIADÆ or CUTTLE-FISH FAMILY.	*Sepia*	or Cuttle-Fishes.
	SPIRULIDÆ.	*Spirula*	or Spirulas.
TETRABRANCHIATA or FOUR-GILLED CEPHALOPODS.	NAUTILIDÆ or NAUTILUS FAMILY.	*Nautilus*	or Pearly Nautilus.
	ORTHOCERATIDÆ or ORTHOCERAS FAMILY. (Fossil.)		
	AMMONITIDÆ or AMMONITE FAMILY. (Fossil.)		

THE CLASS OF GASTEROPODA OR GASTEROPODS.

ORDERS.	FAMILIES.	Genera.	
GASTEROPODA or GASTEROPODS PROPER.	STROMBIDÆ or STROMB FAMILY.	*Strombus* *Pteroceras, &c.*	or Shombs. " Scorpion Shells, &c.
	MURICIDÆ or MUREX FAMILY.	*Murex* *Pyrula, &c.*	or Thorny Woodcock. " Pyrulas, &c.
	BUCCINIDÆ or WHELK FAMILY.	*Buccinum* *Harpa, &c.*	or Whelks. " Harp-Shells, &c.
	CONIDÆ or CONE FAMILY.	*Conus, &c.*	or Cones, &c.
	VOLUTIDÆ or VOLUTE FAMILY.	*Voluta* *Mitra, &c.*	or Volutes. " Mitre-Shells, &c.
	CYPRÆIDÆ or COWRY FAMILY.	*Cypræa* *Ovulum, &c.*	or Cowries. " Egg-Cowries, &c.
	NATICIDÆ or NATICA FAMILY.	*Natica* *Sigaretus, &c.*	or Naticas.
	PYRAMIDELLIDÆ or PYRAMID SHELL FAMILY.	*Pyramidella, &c.*	or Pyramid Shells, &c.
	CERITHIADÆ or CERITHIUM FAMILY.	*Cerithium, &c.*	or Cerithiums.
	MELANIADÆ or MELANIA FAMILY.	*Melania, &c.*	or Melanias.
	TURRITELLIDÆ or WENTLE-TRAP FAMILY.	*Turritella* *Scalaria, &c.*	or Tower-Shells. " Wentle-traps, &c.
	LITORINIDÆ or PERIWINKLE FAMILY.	*Litorina, &c.*	or Periwinkles, &c.
	PALUDINIDÆ or RIVER-SNAIL FAMILY.	*Paludina, &c.*	or River Snails, &c.
	NERITIDÆ or NERITE FAMILY.	*Nerita, &c.*	or Nerites, &c.
	TURBINIDÆ or TOP-SHELL FAMILY.	*Trochus, &c.*	or Top-Shells, &c.
	HALIOTIDÆ or EAR-SHELL FAMILY.	*Haliotis*	or Ear-Shells.

[MOLLUSKS: GASTEROPODS — *Continued.*]

ORDERS.	FAMILIES.	*Genera.*	
GASTEROPODS PROPER — *Continued.*	JANTHINIDÆ or VIOLET-SNAIL FAMILY.	*Janthina*	or Violet-Snails.
	FISSURELLIDÆ or KEY-HOLE LIMPET FAMILY.	*Fissurella, &c.*	or Key-hole Limpets.
	CAYLYPTRÆIDÆ or BONNET LIMPET FAMILY.	*Calyptræa, &c.*	or Bonnet Limpets.
	PATELLIDÆ or LIMPET FAMILY.	*Patella, &c.*	or Limpets.
	DENTALIDÆ or TOOTH-SHELL FAMILY.	*Dentalium*	or Tooth-Shells.
	CHITONIDÆ or CHITON FAMILY.	*Chiton*	or Chitons.
	HELICIDÆ or LAND-SNAIL FAMILY.	*Helix, &c.*	or Land-Snails.
	LIMACIDÆ or SLUG FAMILY.	*Limax, &c.*	or Slugs.
	LIMNÆIDÆ or POND-SNAIL FAMILY.	*Limnæa* *Physa* *Planorbis, &c.*	
	AURICULIDÆ.	*Auricula, &c.*	or Little-Ear Shells.
	CYCLOSTOMIDÆ.	*Cyclostoma, &c.*	or Round-Mouths.
	ACICULIDÆ.	*Acicula, &c.*	or Needle-Shells.
	TORNATELLIDÆ.	*Tornatella, &c.*	
	BULLIDÆ.	*Bulla, &c.*	or Bubble-Shells.
	APLYSIADÆ.	*Aplysia, &c.*	or Sea-Hares.
	DORIDÆ.	*Doris, &c.*	or Sea-Lemons.
	TRITONIADÆ.	*Tritonia, &c.*	
	ÆOLIDÆ.	*Æolis, &c.*	
	ELYSIADÆ.	*Elysia, &c.*	
HETEROPODA or HETEROPODS.	FIROLIDÆ.	*Firola, &c.*	
	ATLANTIDÆ.	*Atlanta, &c.*	
PTEROPODA or PTEROPODS.	HYALEIDÆ.	*Hyalea, &c.*	
	LIMACINIDÆ.	*Limacina, &c.*	
	CLIIDÆ.	*Clio, &c.*	

THE CLASS OF ACEPHALA OR ACEPHALS.

ORDERS.	FAMILIES.	Genera.	
LAMELLIBRANCHIATA or LAMELLIBRANCHIATES.	OSTREIDÆ or OYSTER FAMILY.	*Ostrea, &c.* *Pecten, &c.*	or Oysters. " Pectens or Scallops.
	AVICULIDÆ or PEARL-OYSTER FAMILY.	*Avicula, &c.*	or Pearl-Oysters, &c.
	MYTILIDÆ or SEA-MUSSEL FAMILY.	*Mytilus, &c.*	or Sea-Mussels.
	ARCADÆ.	*Arca* *Leda, &c.*	
	TRIGONIDÆ or TRIGONIA FAMILY.	*Trigonia, &c.*	
	UNIONIDÆ or POND & RIVER MUSSEL FAMILY.	*Unio* *Anodon, &c.*	
	CHAMIDÆ.	*Chama, &c.*	
	TRIDACNIDÆ or TRIDACNA FAMILY.	*Tridacna* *Hippopus*	
	CARDIADÆ or COCKLE FAMILY.	*Cardia, &c.*	
	LUCINIDÆ.	*Lucina, &c.*	
	CYCLADIDÆ.	*Cyclas, &c.*	
	CYPRINIDÆ.	*Cyprina* *Astarte, &c.*	
	VENERIDÆ or VENUS-SHELL FAMILY.	*Venus* *Cytherea, &c.*	

[MOLLUSKS: ACEPHALS — *Continued.*]

ORDERS.	FAMILIES.	*Genera.*	
LAMELLIBRANCHIATES—*Continued.*	MACTRIDÆ.	*Mactara, &c.*	
	TELLINIDÆ.	*Tellina, &c.*	
	SOLENIDÆ or RAZOR-SHELL FAMILY.	*Solen, &c.*	or Razor-Shells, &c.
	MYACIDÆ or CLAM FAMILY.	*Mya, &c.*	or Clams, &c.
	ANATINIDÆ or LANTERN-SHELL FAMILY.	*Pandora, &c.*	
	GASTROCHÆNIDÆ or WATERING-POT-SHELL FAMILY.	*Aspergillum, &c.*	or Watering-pot Shells, &c.
	PHOLADIDÆ or SHIP-WORM FAMILY.	*Pholas* *Teredo*	or Pholads. " Ship-Worms.

THE CLASS OF BRACHIOPODA OR BRACHIOPODS.

	TERREBRATULIDÆ.	*Terebratula, &c.*	
	RHYNCHONELLIDÆ.	*Rhynchonella.*	
	CRANIADÆ.	*Crania.*	
	DISCINIDÆ.	*Discina.*	
	LINGULIDÆ.	*Lingula.*	

THE CLASS OF TUNICATA OR TUNICATES.

	ASCIDIADÆ.	*Ascidium*	or Simple Ascidians.
	CLAVELLINIDÆ.	*Clavellina, &c.*	or Social Ascidians.
	BOTRYLLIDÆ.	*Botryllus, &c.*	or Compound Ascidians.
	PYROSOMIDÆ.	*Pyrosoma*	or Fire-bodies.
	SALPIDÆ.	*Salpa*	or Salps.

THE CLASS OF POLYZOA OR POLYZOANS.

THE BRANCH OF RADIATA OR RADIATES.

THE CLASS OF ECHINODERMATA OR ECHINODERMS.

ORDERS.	FAMILIES.	*Genera.*
HOLOTHURIOIDS or SEA-CUCUMBERS.		
ECHINOIDS or SEA-URCHINS.		
ASTERIOIDS or STAR-FISHES.		
OPHIURIOIDS or SERPENT STARS.		
CRINOIDS or LILY-LIKE ECHINO-DERMS.		

THE CLASS OF ACALEPHA OR JELLY-FISHES.*

Orders.	Families.	Genera.
CTENOPHORÆ or COMB-BEARERS.	BOLINIDÆ, &c.	*Bolina, &c.*
	OCYROEÆ.	*Ocyroe, &c.*
	MERTENSIDÆ.	*Mertensia, &c.*
	CYDIPPIDÆ.	*Pleurobrachia, &c.*
	BEROIDÆ.	*Beröe, &c.*
DISCOPHORÆ or DISK JELLY-FISHES.	RHIZOSTOMIDÆ.	*Rhizostoma, &c.*
	POLYCLONIDÆ.	*Polyclonia, &c.*
	AURELIADÆ.	*Aurelia, &c.*
	STHENONIÆ.	*Sthenio, &c.*
	CYANEIDÆ.	*Cyanea, &c.*
	PELAGIDÆ.	*Pelagia, &c.*
	THALLASANTHEÆ.	*Foveolia, &c.*
	TRACHYNEMIDÆ.	*Trachynema, &c.*
	LEUCKARTIDÆ.	*Liriope, &c.*
	CLEISTOCARPIDÆ.	*Manania, &c.*
	ELEUTHEROCARPIDÆ.	*Lucernaria, &c.*
HYDROIDÆ or HYDROIDS.	OCEANIDÆ.	*Oceania, &c.*
	EUCOPIDÆ.	*Eucope, &c.*
	ÆQUORIDÆ.	*Rhegmatodes, &c.*
	GERYONOPSIDÆ.	*Tima, &c.*
	POLYORCHIDÆ.	*Polyorchis, &c.*
	LAODICEIDÆ.	*Lafœa, &c.*
	MELICERTIDÆ.	*Melicertum, &c.*
	PLUMULARIDÆ.	*Plumularia, &c.*
	SERTULARIDÆ.	*Sertularia, &c.*
	NEMOPSIDÆ.	*Nemopsis, &c.*
	BOUGAINVILLEÆ.	*Bougainvillia, &c.*
	NUCLEIFERÆ.	*Turris, &c.*
	WILLIADÆ.	*Willia, &c.*
	SARSIADÆ.	*Coryne, &c.*
	ORTHOCORYNIDÆ	*Corynitis, &c.*
	PENNARIDÆ.	*Pennaria, &c.*
	TUBULARIDÆ.	*Tubularia, &c.*
	HYDRAIDÆ.	*Hydra, &c.*
	HYDRACTINIDÆ.	*Hydractinia, &c.*
	DIPHYIDÆ.	*Eudoxia, &c.*
	AGALMIDÆ.	*Nanomia, &c.*
	PHYSALIDÆ.	*Physalia, &c.*
	VELELLIDÆ.	*Velella, &c.*
	PORPITIDÆ.	*Porpita, &c.*
	MILLEPORIDÆ.	*Millepora, &c.*

* According to Agassiz.

THE CLASS OF POLYPI OR POLYPS.*

ORDERS.	FAMILIES.	*Genera.*
ALCYONARIA or ALCYONARIANS.	PENNATULIDÆ or SEA-PEN FAMILY.	*Pennatula, &c.*
	PAVONARIDÆ.	*Pavonaria, &c.*
	VERETILLIDÆ.	*Veretillum, &c.*
	RENILLIDÆ.	*Renilla, &c.*
	GORGONIDÆ or SEA-FAN FAMILY.	*Gorgonia, &c.*
	PLEXAURIDÆ.	*Muricea, &c.*
	PRIMNOIDÆ.	*Primnoa, &c.*
	GORGONELLIDÆ.	*Verrucella, &c.*
	ISIDÆ.	*Isis, &c.*
	CORALLIDÆ or RED CORAL FAMILY.	*Corallium, &c.*
	BRIARIDÆ.	
	ALCYONIDÆ.	*Alcyonium, &c.*
	XENIDÆ.	
	CORNULARIDÆ.	
	TUBIPORIDÆ or ORGAN-PIPE CORAL FAMILY.	*Tubipora, &c.*
ACTINARIA or ACTINARIANS.	ACTINIDÆ or SEA-ANEMONE FAMILY.	*Metridium, &c.*
	THALLASSIANTHIDÆ.	
	MINYIDÆ.	
	ILLYANTHIDÆ.	
	CERIANTHIDÆ.	
	ANTIPATHIDÆ.	
	GERARDIDÆ.	
	ZOANTHIDÆ.	
	BERGIDÆ.	

* According to Verrill.

[RADIATES: POLYPS — *Continued.*]

ORDERS.	FAMILIES.	*Genera.*
MADREPORARIA or MADREPORARIANS.	EUPSAMMIDÆ.	*Astroides, &c.*
	GEMMIPORIDÆ.	
	PORITIDÆ or PORITES FAMILY.	*Porites, &c.*
	MADREPORIDÆ or MADREPORE FAMILY.	*Madrepore, &c.*
	LITHOPHYLLIDÆ.	
	MÆANDRINADÆ or MEANDRINA FAMILY.	*Mœandrina, &c.*
	EUSMILLIDÆ.	
	CARYOPHYLL-IDÆ.	*Caryophyllia, &c.*
	STYLINIDÆ.	
	ASTRÆIDÆ or STAR CORAL FAMILY.	*Astræa, &c.*
	OCULINIDÆ or OCULINA FAMILY.	*Oculina, &c.*
	STYLOPHORIDÆ.	
	CYCLOCLITIDÆ.	
	LOPHOSERIDÆ.	
	FUNGIDÆ or FUNGUS CORAL FAMILY.	*Fungia, &c.*
	MERULINIDÆ.	*Merulina, &c.*

THE BRANCH OF PROTOZOA OR PROTOZOANS.

THE END.

CHARLES SCRIBNER. A. C. ARMSTRONG. A. J. PEABODY.

Condensed Catalogue

OF THE PUBLICATIONS OF

CHARLES SCRIBNER & Co.,

654 BROADWAY,

NEW YORK.

JANUARY, 1868.

*** *The figures in the last column of this Catalogue refer to* CHARLES SCRIBNER & CO.'S *Descriptive Catalogue, copies of which will be sent to any address upon application.*

*** *Blanks in the column of prices, except in the case of School Text-Books, the prices of which may be learned from* CHARLES SCRIBNER & CO.'S *Educational Catalogue, indicate that the Works are either out of print or in press.*

*** *In this list the names of Books just published are given in* SMALL CAPITALS; *those issued during the year* 1867, *as well as new editions of Works previously produced, are indicated by italics.*

*** *The prices here given are for the regular style of binding in cloth. Books furnished in other styles, and their respective prices, may be learned from the Descriptive Catalogue.*

*** *Any of these Books will be sent post-paid to any address upon receipt of the price.*

	Volumes.	Size.	Price.	Page
ADAMS, W., D.D.				
THANKSGIVING	1	12mo	$2 00	2
Three Gardens	1	12mo	2 00	1
AGASSIZ, PROF. LOUIS.				
Structure of Animal Life (The)	1	8vo	2 50	2
ALEXANDER, A., D.D.				
Moral Science	1	12mo	1 50	3
ALEXANDER, J. W., D.D.				
Alexander, Archibald, Life of, (Portrait) . .	1	12mo	2 00	3
Christian Faith and Practice (Discourses) .	1	12mo	2 00	6
Consolation (Discourses)	1	12mo	2 00	5
Faith (Discourses)	1	12mo	2 00	6
Forty Years' Correspondence with a Friend	2	12mo	4 00	4
Preaching, Thoughts on	1	12mo	2 00	5

	Volumes.	Size.	Price.	Page.
ALEXANDER, J. A., D.D.				
Acts (Commentary)	2	12mo	$4 00	8
Isaiah " (complete)	2	8vo	6 50	7
" " (abridged)	2	12mo	4 00	8
Mark "	1	12mo	2 00	8
Matthew "	1	12mo	2 00	8
Psalms "	3	12mo	6 00	7
New Test. Literature and Ecc. Hist. . .	1	12mo	2 00	9
Sermons (with Portrait)	2	12mo	——	9
ALLSTON, WASHINGTON.				
Lectures and Poems	1	12mo	2 50	9
ANDREWS, REV. S. J.				
Life of Our Lord	1	post 8vo	3 00	10
ARMSTRONG, G. D., D.D.				
Works of	3	12mo ea.	1 25	11
BAUTAIN, PROF. A.				
Extempore Speaking	1	12mo	1 50	12
BEECHER, REV. H. W.				
PRAYERS FROM PLYMOUTH PULPIT . . .	1	12mo	1 75	13
BOTTA, PROF.				
Dante as Poet, Patriot, Philosopher . .	1	crown 8vo	2 50	14
BRACE, CHARLES L.				
Hungary, with Experience of Austrian Police,	1	12mo	——	15
Home Life in Germany	1	12mo	——	14
Norse Folk	1	12mo	2 00	15
Races of Old World	1	post 8vo	2 50	15
Short Sermons for Newsboys	1	16mo	1 50	16
BUSHNELL, HORACE, D.D.				
Character of Jesus	1	18mo	1 00	19
Christ and his Salvation	1	12mo	2 00	19
Christian Nurture	1	12mo	2 00	18
Nature and the Supernatural	1	12mo	2 25	17
New Life, Sermons for	1	12mo	2 00	19
Vicarious Sacrifice, The	1	8vo	3 00	18
Work and Play	1	12mo	2 00	19
CHEEVER, G. B., D.D.				
Works of	3	12mo	——	20
CLARK, PROF. N. G.				
English Language, Elements of	1	12mo	1 25	21
COLLIER, J. PAYNE, F.S.A.				
Rarest Books in English Language . . .	4	small 8vo	12 00	21
CONYBEARE, REV. W. J., and HOWSON, REV. J. S.				
St. Paul, Life and Epistles of (illustrated)	2	8vo	7 50	22

	Volumes.	Size.	Price.	Page
FIELD, HENRY M., D.D.				
Atlantic Telegraph, History of	1	12mo	$1 75	37
FISHER, REV. PROF. GEO. P.				
Supernatural Origin of Christianity	1	8vo	2 50	38
Silliman, Benj., LL.D., Life of (Portrait)	2	crown 8vo	5 00	37
FORSYTH, WM.				
Cicero, New Life of (Illustrations)	2	crown 8vo	5 00	39
FOWLER W. C., LL.D.				
Sectional Controversy (The)	1	8vo	1 75	39
FROUDE, J. A.				
HISTORY OF ENGLAND	10	crown 8vo	3 00 ea.	40
SHORT STUDIES ON GREAT SUBJECTS	1	crown 8vo	3 00	41
GAGE, REV. W. L.				
Carl Ritter, Life of	1	12mo	2 00	42
GASPARIN, COUNT.				
America before Europe	1	12mo	2 00	42
GIBBONS, J. S.				
PUBLIC DEBT OF THE UNITED STATES	1	crown 8vo	2 00	43
GUIZOT, M.				
Meditations (First and Second Series)	2	12mo ea.	1 75	43
GUYOT, PROF. ARNOLD.				
Geographies, Wall Maps, Key, etc.	–	——	——	44–47
HALL, EDWIN, D.D.				
Puritans and their Principles	1	8vo	2 50	48
HALSEY, REV. LEROY J.				
Bible, Literary Attractions of	1	12mo	1 75	48
HEADLEY, J. T.				
Complete Works of	15			49 53
HEIDELBERG,				
Catechism	1	small 4to	3 50	54
HERBERT, H. W.				
Works of	3	12mo ea.	1 50	54
HOLLAND, DR. J. G. (Timothy Titcomb.)				
Bay-Path. A Novel	1	12mo	2 00	58
Bitter-Sweet. A Poem	1	12mo	1 50	56
Gold Foil	1	12mo	1 75	56
KATHRINA. A Poem	1	12mo	1 50	59
Lessons in Life	1	12mo	1 75	57
Letters to the Joneses	1	12mo	1 75	57
Letters to Young People	1	12mo	1 50	55
Miss Gilbert's Career	1	12mo	2 00	59
Plain Talks on Familiar Subjects	1	12mo	1 75	58

	Volumes.	Size.	Price.	Page
HOURS AT HOME.				
A FAMILY MAGAZINE	Per	annum	$3 00	6c
HURST, REV. J. F.				
Rationalism, History of	1	8vo	3 50	61
JAMESON, JUDGE J. A.				
Constitutional Convention	1	8vo	4 50	62
JOHNSON, ANNA C. (MINNIE MYRTLE).				
Cottages of the Alps	1	12mo	——	63
Germany, Peasant Life in	1	12mo	——	63
Myrtle Wreath	1	12mo	1 25	63
JONES, CHARLES COLCOCK, D.D.				
HISTORY OF THE CHURCH OF GOD . . .	2	8vo	ea. 3 50	63
KIRKLAND, MRS. C. M.				
Evening Book	1	12mo	1 25	65
Holidays Abroad	2	12mo	2 00	65
Patriotic Eloquence	1	12mo	1 75	64
School Garland	1	12mo	——	64
KNOX, REV. J. B.				
St. Thomas, W. I., History of	1	12mo	$1 25	65
LABOULAYE EDOUARD.				
Paris in America	1	12mo	1 25	69
LANGE, PROF. J. P., D.D.				
Commentary on Scriptures, Theolog. and Homiletical: Matthew (1); MARK AND LUKE (1); ACTS (1); JAMES, PETER, JOHN, JUDE (1); CORINTHIANS (1); THESSALONIANS, TIMOTHY, TITUS, PHILEMON, HEBREWS (1); GENESIS, (1)		8vo	ea. 5 00	66–68
LAWRENCE, EUGENE.				
British Historians, Lives of	2	12mo	3 50	69
LIBER LIBRORUM	1	16mo	1 50	70
LORD, JOHN, LL.D.				
OLD ROMAN WORLD	1	crown 8vo	3 00	7c
LYNCH, LIEUT. W. F.				
Naval Life	1	12mo	1 50	71
MCLEOD, ALEX., D.D.				
Life of, by Rev. S. B. Wylie	1	8vo	2 00	114
MACDONALD, J. M., D.D.				
My Father's House	1	12mo	1 50	72
MACLEOD, DONALD,				
Works of	3	12mo	ea. 1 25	71
MAGOON, REV. E. L.				
Living Orators in America	1	12mo	1 75	72
Orators of the American Revolution . . .	1	12mo		72

	Volumes.	Size.	Price.	Page
MAINE, HENRY S.				
Ancient Law	1	crown 8vo	$3 00	73
MARSH, HON. GEO. P.				
English Language, Lectures on	1	crown 8vo	3 00	74
" " Origin and History of	1	crown 8vo	3 00	74
Man and Nature	1	crown 8vo	3 00	75
MARSH, J., D.D.				
Temperance Recollections	1	12mo	2 25	76
MASON, J. M., D.D.				
Complete Works	4	post 8vo	——	76
MEARS, REV. J. W.				
Bible and the Workshop	1	12mo	1 00	76
MITCHELL, DONALD G. (IK MARVEL).				
Dream Life	1	12mo	$1 75	77
Dr. Johns: A Novel	2	12mo	3 50	79
Fresh Gleanings	1	12mo	——	78
Lorgnette, The	1	12mo	——	78
My Farm of Edgewood	1	12mo	1 75	79
Reveries of a Bachelor	1	12mo	1 75	77
RURAL STUDIES	1	12mo	1 75	80
Seven Stories	1	12mo	1 75	78
Wet Days at Edgewood	1	12mo	1 75	80
MITCHEL, PROF. O. M.				
Astronomy of the Bible	1	12mo	1 75	81
Planetary and Stellar Worlds	1	12mo	1 75	81
Popular Astronomy	1	12mo	1 75	81
MORRIS, GEO. P.				
Poetical Works (blue and gold)	1	16mo	1 50	83
MULLER MAX.				
Science of Language, Lectures on	1	crown 8vo	2 50	82
" " " (Second Series)			3 50	82
NEWCOMB, REV. H.				
Missions, Cyclopædia of (32 Maps)	1	8vo	——	83
ORLEANS, DUCHESS OF.				
Memoirs of (Portrait)	1	12mo		86
OWEN, J. J., D.D.				
Commentaries on the New Testament: *Matthew* (1); MARK AND LUKE (1); JOHN (1); ACTS (—), in press; GENESIS (—), in press	1	12mo ea.	1 75	84–85
PAEZ, DON RAMON.				
South America, Wild Scenes in	1	crown 8vo	2 50	86
PAULDING, JAMES K.				
Literary Life of (1); BULLS AND JONATHANS (1); TALES OF THE GOOD WOMAN (1); A BOOK OF VAGARIES (1); DUTCHMAN'S FIRESIDE . . . each	1	crown 8vo	2 50	87

	Volumes.	Size.	Price.	Page.
PERCE, ELBERT.				
Magnetic Globes			——	88
PERRY, PROF. A. L.				
Political Economy, Elements of	1	crown 8vo	2 50	89
PRENTISS, S. S.				
Life of, by Geo. L. Prentiss, D.D.	2	12mo	3 50	86
PRIME, S. I., D.D.				
Power of Prayer	1	12mo	1 50	90
PUBLIC PRAYER.				
Book of	1	12mo	2 00	13
RANDOLPH, A. D. F.				
Hopefully Waiting, and other Verses	1	16mo	$1 50	90
SCHAFF, P., D.D.				
Apostolic Church, History of	1	8vo	3 75	91
CHRISTIAN " "	3	8vo ea.	3 75	92–93
PERSON OF CHRIST	1	16mo	1 50	94
SEAMAN, E. C.				
Progress of Nations	1	12mo		94
SHEDD, W. G. T., D.D.				
Christian Doctrine, History of	2	8vo	6 50	95
HOMILETICS AND PASTORAL THEOLOGY	1	8vo	3 50	96
SHELDON, E. A.				
Model Lessons on Objects, etc.				97
SILLIMAN, BEN. M.D.				
Life of, by Professor Geo. P. Fisher	2	crown 8vo	5 00	37
SIMMS, W. G.				
Huguenots in Florida	1	12mo	1 50	—
SMITH, PROF. H. B., D.D.				
Christian Church, History of, in Tabular Form	1	Folio	6 75	98–99
SMITH, J. H.				
Gilead; or, The Vision of All Souls' Hospital	1	12mo	1 50	100
SMITH, MRS. MARY HOWE.				
Lessons on the Globe	1	12mo	0 50	88
SPRING, GARDINER, D.D.				
Personal Reminiscences, etc. (Portrait)	2	12mo	4 00	100
STANLEY, DEAN A. P.				
Eastern Church	1	8vo	4 00	101
Jewish Church, Lectures on (First Series)	1	8vo	4 00	101
" " " (Second Series)	1	8vo	5 00	101 / 102
Sermons	1	12mo	1 50	103
STRICKLAND, AGNES.				
QUEENS OF ENGLAND, LIVES OF	1	12mo	2 00	103

	Volumes.	Size.	Price.	Page.
TAYLOR, GEORGE.				
Indications of the Creator.	1	12mo	1 75	105
TENNEY, S., A.M.				
Natural History	1	crown 8vo	3 00	104
" " Library Edition	1	large 8vo	4 00	104
" " for Schools	1	12mo	2 00	104
TIMOTHY TITCOMB.				
Complete Works of. See HOLLAND.	—	—	—	55-59
TRENCH, RT. REV. R. C.				
Epistles to Seven Churches, Commentary on	1	12mo	1 50	106
Studies in the Gospels	1	8vo	3 00	106
Glossary of English Words	1	12mo	1 50	107
Synonyms of New Testament (1st and 2d Series)	2	12mo ea.	1 25	107
TUCKERMAN, H. T.				
America and her Commentators	1	crown 8vo	2 50	108
TULLIDGE, REV. H.				
Triumphs of the Bible	1	12mo	2 00	108
TUTHILL, MRS. L. C.				
Joy and Care. Book for Young Mothers	1	16mo	1 00	109
VAN SANTVOORD, GEO.				
Lives of Chief Justices of U. S.	1	8vo		109
WHITNEY, PROF. W. D.				
LANGUAGE, AND THE STUDY OF LANGUAGE	1	crown 8vo	2 50	110
WILLIS, N. P.				
Complete Works of				111 112
WISE, CAPT. H. A.				
Works of	2	12mo ea.	1 75	112
WOOLSEY, PREST. T. D.				
International Law, Introduction to	1	8vo	2 50	113
WYLIE, S. B., D.D.				
McLeod, Alexander, Life of	1	8vo	2 00	114
ZINCKE, F. BARHAM.				
Extemporary Preaching	1	12mo	1 50	114

ILLUSTRATED GIFT-BOOKS.

*** *In this list the prices of the respective works as bound in cloth, full gilt, are given. All of them, however, are furnished in Turkey morocco, and the prices of these editions may be learned by reference to the catalogue.*

	Price.	Page.
ÆSOP'S FABLES	$18 00	117
Bitter-Sweet	9 00	117
Book of Rubies	7 00	119
Christian Armor	15 00	120
Cotter's Saturday Night	5 00	119
FLORAL BELLES	25 00	116
FRED AND MARIA AND ME	$1 50	121
Folk Songs	15 00	115
MY FARM OF EDGEWOOD	10 00	121
Pilgrim's Progress	5 00	120
QUEENS OF AMERICAN SOCIETY	6 00	118

www.ingramcontent.com/pod-product-compliance
Lightning Source LLC
LaVergne TN
LVHW021224110826
845150LV00002B/237

* 9 7 8 1 4 2 5 5 6 4 0 2 5 *